An Int
of Psychology

An Intellectual History of Psychology

Third Edition

Daniel N. Robinson

The University of Wisconsin Press

The University of Wisconsin Press
2537 Daniels Street
Madison, Wisconsin 53718

10   9   8   7   6   5   4   3   2

Printed in the United States of America

Library of Congress Cataloging-in-Publication Data
Robinson, Daniel N., 1937–
    An intellectual history of psychology / Daniel N. Robinson
    —3rd ed.
       390 p.   cm.
    Includes bibliographical references and indexes.
    ISBN 0-299-14840-8.   ISBN 0-299-14844-0 (pbk.)
    1. Psychology—History. I. Title
BF81.R65   1995
150'.9—dc20   95-5697

Contents

Preface to the Third Edition

This is probably the final edition of a work begun some twenty years ago and intended then, as now, to fill a gap in the range of texts devoted to the history of psychology. One cannot undertake such a project without experiencing an odd mixture of respect for and disappointment with others who have written books of this kind, and I suspect that some of those others have had similar reactions to earlier editions of this work. There is ample room here for differences in emphasis and interpretation, and for differences in authors' estimations of the importance or relevance or staying-power of specific perspectives and methods.

In my own case, it seemed fairly clear to me years ago, and seems still clearer now, that the general outlines for a systematic psychology were drawn in Hellenic and Hellenistic Greece. If there is a defensible sense in which all philosophy, as Whitehead said, is a footnote to Plato, much in the history of psychology is a footnote to Aristotle. If this is not entirely obvious at the present time it is only because of the reluctance of contemporaries to attempt to absorb the moral and political dimensions of life into current theories and approaches. The reluctance itself is based on the well-formed habit of treating psychology as a natural science (this, too, another debt to Aristotle) but then conceiving of science as an enterprise far more constricted in its aims and methods than was Aristotle's own science.

If the general outlines of the discipline were drawn by ancient Greek philosophers and scientists, the more specific and detailed features were bequeathed by the century of Bacon, Newton, Galileo, and Descartes. It was to these figures that Locke would turn, as would the eighteenth-century *philosophes* and other luminaries of the Enlightenment so convinced that a science of the mind was within reach. By the time the methodology caught up with the metaphysics in the nineteenth century, the essential character of what we take to be "modern psychology" was in place. Thus do we continue to prosecute the agenda of nineteenth-century psychology. If behaviorism or the "cognitive revolution" or the psychobiological perspective seems new, it is only because we have forgotten or never learned the history of the subject. Failing in this respect, we run the risk of entering the same blind alleys that halted progress in an earlier time. Political and social history are not prophetic, for political and social events are unique and essentially nonrecurring. But intellectual history *is* prophetic. A poorly constructed argument leading to doubtful or distracting conclusions will retain these properties in any and every incarnation. I undertook the first and later editions of this work to put before the reader the main arguments and conclusions that guided psychological thought during the major epochs of Western intellectual history. It is the reader's task to judge which of them is sound and summoning, and which may be no more than distracting or even fallacious.

It is with deep appreciation that I acknowledge the uncommon support provided by the University of Wisconsin. I can only hope that Allen Fitchen's confidence in this and earlier editions has been redeemed. The editorial assistance so generously given by Elizabeth Steinberg and the close and thoughtful editing by Sylvan Esh have improved the book in scores of ways. Its remaining faults are all mine.

Twenty years ago I was indebted to my wife and dearest friend, Francine, for the atmosphere of love and warmth she created for us both. This book, like its author, is dedicated to her.

1 Philosophical Psychology

1   Defining the Subject

Not too many years ago it was part of the received wisdom that the history of scientific or philosophical subjects was entirely distinct from the subject itself; that there is the *doing* of psychology, sociology, economics, physics, etc., and then there is the *history* of such fields. We know better now. Although there are still important differences to be noted, we are more prepared to acknowledge that all intellectual pursuits arise within, and to some extent must bear the marks of, their historical and cultural contexts. This is so not only for the obvious reason that major issues are generally bequeathed rather than suddenly born, but also because the entire enterprise of creating knowledge and solving problems of a given kind is itself a continuing tradition, a culture in its own right. This does not prevent the development of new fields and modes of inquiry, nor does it deny the occasional genius a chance to fashion the unexpected. Rather, it is this culture of knowledge, this culture of thought, that comes to establish the very standards of novelty and usefulness and proceeds to confer meaningfulness and significance on just such efforts.

To refer to a "culture of knowledge" is to introduce a set of traditions that transcend ethnic, national, religious, and racial distinctions, traditions that may be said to pertain to something close to our basic humanity. Indeed, free and imaginative inquiry, seemingly idle speculation, argument and spirited discourse, the play of the mind—these are so central to what we take to be part of a flourishing life that we pity or condemn societies and governments unwilling to permit or nurture such activities.

Psychology as a discipline and as a profession is one of the creations of the culture of thought. Just how it was brought about, how the psychological life of individual persons was objectified and externalized and thus made a subject of study, forms an important chapter in the cluttered book of human initiative. But in its current form, psychology is so various, so partitioned into separate provinces, that the nonspecialist might pardonably conclude that there is no unified subject at all. This state of affairs, too, is the result of historical developments.

Inevitably, it is a subject's history that defines it, and this is perhaps especially the case with psychology, which William James playfully referred to as "that nasty little subject." It has been at the center of speculation and inquiry since ancient times. Contrary to an all too common claim, it is not in any sense a "young subject," even if its extensive experimental programs are relatively recent. On the latter point, too, one must remain mindful that experimental science within the context of university life is fairly recent in all subject areas. It was not until the first half of the nineteenth century that separate university facilities were provided for research in chemistry, the pioneer here being Liebig in Germany. Of course, academic philosophers and scientists had engaged in research long before this, but only in the nineteenth century do we find the beginning of a tradition that includes experimental science as a recognized and official part of the university's functions. Thus, the founding of psychology's first laboratory by Wundt at Leipzig in 1879 takes place only a little later than the establishment of such laboratories in any discipline.

Might psychology, however, be regarded as young at least in one sense? For was not research undertaken in medicine and in the physical sciences since ancient times, even if not within the context of university studies? This is a more difficult question, for it pertains to activities not readily dated, and to activities not easily defined. Just what, after all, qualifies as scientific research? And what must the subject of the research be to qualify as psychological? There is much in the surviving works of Aristotle to establish that he undertook systematic studies of perception, emotion, and behavior in the animal kingdom. His writings on psychological matters are often detailed and technical, and refer to processes now routinely featured in psychological experiments and courses. Accordingly, it would be entirely proper to date the origins of psychological research and of scientific theories about psychological processes with Aristotle, thus pushing back the birth of this part of the subject to the fourth century B.C. Comparable claims could then be made for most of the succeeding centuries, for in nearly all of them one can find a record of scientific or medical inquiry pertaining to psychological states, processes, or attributes.

The tendency, common in books on the history of psychology, to regard the subject as young or immature arises from uncritically accepted positions on just what the essence of psychology is. If on some distant reckoning the (unlikely) conclusion were reached that the issues of central concern to psychology are utterly unsuited to experimental modes of inquiry, there would surely be no reason to examine early experimental projects in order to date its birth. Only after taking a settled position on just what sort of subject psychology is does it make sense to focus on Wundt and the Leipzig laboratory or, for that matter, on Plato's dialogues.

But the tendency to consider psychology young arises also from an uncritical belief that similar actions taking place in radically different epochs can be understood as arising from similar aims and perspectives. The ancient astrologers constructed star-charts and models of celestial behavior, activities not unlike what is now found in the fields of cosmology and astrophysics. The ancient astrologers, however, based their efforts on a perspective and a set of assumptions now judged to be nothing more than superstitions. Any definition that takes psychology to be just what the psychologists of a given era might be doing is bound to be misleading and ill conceived. First, absent a more fundamental definition, there would be no way of identifying the participants as psychologists in the first place. Secondly, unless the activities and inquiries are merely reflections of personal interests, including quirky ones, what psychologists undertake must be addressed to some set of issues or problems bequeathed to them by earlier students of the subject. Clearly, then, important fields of study cannot be fully understood solely in terms of current practices, or by recording apparently similar inquiries taking place over long stretches of time. To comprehend the nature of psychology and find a definition that establishes both its uniqueness and its relatedness to other subjects, the study of its history becomes essential.

The Idea of History

It is generally thought to be of importance to a man that he should know himself: where knowing himself means knowing not his merely personal peculiarities, the things that distinguish him from other men, but his nature as a man. . . . Knowing yourself means knowing what you can do; and since nobody knows what he can do until he tries, the only clue to what man can do is what man has done.

R. G. Collingwood's words appear in the introduction to his *The Idea of History*.[1] Following his precepts one might say that the best sense of what psychology is and the most tutored guesses as to what it might become, are developed under the light of its history, the record of what has been *done* in its name (or in all but its name). But as Collingwood goes on to explain, there is more to the study of history than marking down names and dates. History, for Collingwood, is a species of science; it answers questions by identifying the causes and conditions by which historical events came about:

Science in general . . . does not consist in collecting what we already know and arranging it in this or that kind of pattern. It consists in fastening upon something we do not know, and trying to discover it.[2]

To know, then, that Wundt secured funds to establish a laboratory at Leipzig in 1879, or that Freud wrote lengthily on the interpretation of dreams at the end of that same century, is to be in possession of facts, but not of the explanations that give meaning to the facts. It is to know nothing that points to the rationale, to the impulses behind such undertakings; nothing that discloses the basis upon which either of these important figures in the history of psychology chose his subject or undertook such significant projects. None of this, however, should be taken as a plea for memoirs (as important as these sometimes are). Only if memory were perfect and impartial, and only if commentators were accurate and in full agreement, would the historian's mission be abbreviated or unnecessary, usefully replaced by memoirs. Memories are imperfect and incomplete, sometimes selective, often rhetorical in their deployment. Even the letter-perfect memory of the camera yields a record seen from one direction only. The camera sometimes faces away from the real action and, in any case, never has a view so wide as to capture that full context that gives meaning to the particulars.

To put the matter briefly, there is no dispensing with *interpretation* when seeking to understand complex, unfolding, incomplete events and ideas. Where intellectual history is at issue, the relevant contexts go beyond the compass of any observer's memory. Such a history is comprised of words, and these, unlike swords and tanks and the exchange-rates of currency, are used differently at different times, often intending more than they convey. Intellectual histories are thwarted by all the obstacles facing other forms of historical inquiry, and then some. The thin lines of evidence diverge while the actual witnesses, out of ignorance or fear or sloth or confusion, fail to reflect accurately the ideas of their period. In this way the words themselves obstruct a later reckoning; in addition, modern analysts then enter with their own biases and confusions, further darkening the record.

These biases were insightfully exposed by Giovanni Battista Vico (1668–1744),[3] one of the first great historians of the modern era. He warned of the constant temptation to revere antiquity beyond all sense of proportion. This prejudice not only colors our evaluations but may even lead to quantitative errors, errors in estimating wealth or population or longevity. He also recognized the subtle chauvinist who writes about his own people in large case.

Most common of all blunders, Vico noted, is the tendency to identify the character of an age by studying the thoughts of its scholars. Vico recognized that scholars often are not at all representative of the time in which they live. They differ from their contemporaries in interests, in schooling, even in basic values. The importance of understanding Aristotle's theory of the soul or Socrates' conception of virtue cannot be to learn "the Athenian" understanding of such notions. The average Athenian, more likely than not, seldom considered such matters and, had he been privileged to discuss them with Aristotle or

Socrates, might well have found the philosophers' treatments bizarre. If we would unearth the popular views on subjects of this sort, we would have to search the myths and rites of antiquity, the poems and correspondence and plays of the age. Once the commoner understandings are thus located, they are nearly invariably different from the sage's.

Vico also warned against imputing too much credibility to those who have provided eyewitness accounts. The fact that Tacitus wrote about events in his own time does not mean that he was conceptually closer to them. He wore the distorting lenses of his epoch. His travels were limited. He spoke and read in but two languages. His resources were saddled to a primitive technology. Lacking the methods of modern scholarship (methods only recently developed by linguists, archeologists, physicists, etc.), a Tacitus or a Herodotus or a Livy could examine earlier eras only through the eyes and fables of those who came before.

Finally, Vico presents the pervasive error of *scholastica successionis civitatium,* the "scholastic succession (or bequest) of states." This error moves us to believe that when an intellectual development occurs sequentially in two different countries, it occurs in the second through the direct influence of the first. Since, for example, Adelardus spoke in the twelfth century of the earth's motion and since Copernicus was generous in his reading habits, Copernicus must have got the idea from Adelardus (or Cicero or Aristarchus or Nicole Oresme, etc.). According to this view, an idea is the pebble of a uniquely enlightened time dropped by philosophers on the calm lake of slumbering minds. The idea disturbs first one culture, then the next, seriatim. The historian's challenge is to work back from the ripples to the source and, thereby, to discover the first source of all our wisdom. Vico, as with Aristotle earlier, knew that such orderly succession was seldom the case. Ours is a questioning and a theorizing species which does not rest at length. There is hardly a notion prized by any age that was not entertained by many others at many different times and quite independently. Accordingly, a mere chronology of ideas would not only fail to be a history of ideas but would fail to be true. We can say, with measurable confidence, that an event of a certain kind occurred for the first time or, at least, that no earlier record has been found. But we cannot date the first appearance of a thought or a point of view with similar precision.

The Greeks invented philosophy as a discipline, as a form of inquiry regulated by rules and organized around a set of questions. They were, in this narrow sense, the first philosophers. But it would be ridiculous to argue that Thales was the first human being to recognize that "it is hard to be good." We explore the history of ideas to learn how others have dealt with problems facing us, what others have considered important, and the factors that seemed to have caused events of consequence. One inescapable product of our study is a chronology, but this is surely not the goal. We expect the historian to tell us when things happened much as we expect the modern astronomer to tell us the position of the planets on the day Caesar was killed. But we expect much more. We believe the astronomer because his calculations are based on laws—laws fashioned in the light of criticism, objective measurement, and multiple observations—and we expect as much from the historian. Not that the explanations will unfold inexorably from the fixed "laws" of history, but that they will be framed according to the very best evidence available and will, therefore, yield the truest picture of what happened and why it happened.

In the epilogue to *The Idea of History,* Collingwood ventures, with characteristic decisiveness, to place psychology in perspective. He argues that psychology's chief mistake was to identify with the physical sciences and by that strategy to adopt methods wholly unsuited and even futile to its enterprise. Psychology, Collingwood insists, can answer its questions only by transforming its methods into those of history:

> To regard [psychology] as rising above the sphere of history, and establishing the permanent and unchanging laws of human nature, is therefore possible only to a person who mistakes the transient conditions of a certain historical age for the permanent conditions of human life.[4]

It is not necessary to accept these prescriptions for psychology in order to appreciate the subtler point he raises. A science concerned with the determinants of human conduct, with the character of human perception and emotion, will find in the annals of civilization a laboratory more varied and rich than any we could hope to re-create. This part of Collingwood's instruction is unimpeachable, and it is this part that urges us to study history if we are to study psychology.

In attempting what has been described as an intellectual history, therefore, one is committed to something different from chronology. And, in attempting an intellectual history of psychology, one must be concerned with far more than academic psychology. This said, the following pages may seem to belie the premise, for, in fact, a chronological succession does follow. This, however, is a concession to custom and convenience and is not to be viewed as the fallacy of *scholastica successionis civitatium.* One age does not impart momentum to the next as would a pair of colliding billiard balls, though the culture of thought does have its founders and followers. Ideas are kept alive, they do not merely assert themselves. And where important ideas are concerned, those developed within a productive age, it is misleading even to think that the age itself has died.

There is no question but that psychological scholarship, at any time since Socrates, has been influenced by Greek thought. Yet a study of the Socratics is not enough. Succeeding generations did not passively assimilate Plato's dialogues but interpreted them in altered contexts and for diverse purposes. Often enough the "Greek view" is advanced by a modern who discovers only later that someone else had said very much the same thing. Thus the following chapters could easily have been arranged not as a succession of epochs but as a collection of problems; for example, the problem of knowledge, the problem of conduct, the problem of values. Indeed, if there is anything that one age may be said to inherit directly from its predecessor, it is these problems, and a chronological assessment honors this even as it makes clear just where there has been no real progress at all.

There are, however, at least two disadvantages courted by the "great issues" model. First, it leads to wearying redundancy, since nearly every age has come to advance the same range of answers to the same great questions. To choose one example, informed opinions on the question of human goodness or wickedness are neither broader nor livelier in contemporary scholarship than in the days of Aristotle. Still, although the richness of opinion may not have deepened over the ages, the chains of fact and the rules of evidence have advanced, even if sluggishly. Knowledge, after all, in some compartments is cumulative, and of these compartments a chronological account is appropriate. Knowledge is most cumulative in the sciences and, to the extent that modern psychology seeks membership in the family of science, we are advised to explore the growth of those sciences on which psychology has depended.

Perhaps the more compelling reason for adopting the chronological model (and here an author's limitations are stripped of all disguise) is simply that it is easier to compile evidence and interpret it within one era at a time. We can say in one sentence that questions of importance and daring answers to them have appeared repeatedly and often independently in every civilized epoch. We can say his, but we cannot show it until books are scored as symphonies and until readers sight-read the simultaneous performances of strings, brass, woodwind, and tympany.

Having adopted, no matter how grudgingly, a sequentialist approach, and knowing that no caveat will be heeded by everyone, there is still the problem of identifying more or less discrete periods of intellectual history. Did the Hellenic Age, for example, ever really end? The world of thought and scholarship still speaks of justice, searches for truth, houses democracies of one stripe or another. Only the more literal-minded would require "Hellenism" to remain in Athens and express itself in the old Greek. And, since all forms of Roman arts and letters were utterly suffused by this same Hellenism, on what conceptual basis do we defend such expressions as *Roman* civilization or the Augustan age? Is the American Constitution an *American* constitution, or is it John Locke's vision of government? Is it, then, an English constitution or a modified form of Roman Republicanism?

Such questions can be raised to challenge any suggestion of historical discontinuity, and they are valid. The general point can be further illustrated with a popular example. Suppose we wish to identify a block of time as "the Italian Renaissance" and adopt the conventional dates, 1350 to 1600. In doing so, we are implicitly adopting at least four propositions: first, that there was, at some earlier time, a classical outlook later to be recovered for reborn; second, that this older age vanished or entered a period of prolonged hibernation; third, that at a definable time, the hibernation ended and a renascence of perspective and achievement took place; and fourth, that this rebirth itself ran a given course only to be replaced by something discriminably different.

These propositions can be sustained or rejected depending on what we choose to make use of when we decide to define an "age." Religiously, the Renaissance mind was probably closer to the neo-Platonism of the Dark Ages than it was to the Thomistic Aristotelianism of the later Middle Ages. If by "renaissance" we mean to imply a rebirth of what made classical Athens famous, we must account for the troubling fact that the Italian Renaissance did not produce a single philosopher of enduring consequence.[5] As for the vaunted humanism of the period, few Greek philosophers would have found much to recommend in it. But here we anticipate ourselves. The point to be made is that (and with due respect to Vico, Hegel, and Spengler) history is not cyclical, and that what sometimes makes it seem so is the mere recurrence of certain political and philosophical dispositions. The conditions responsible for these are always different the second (third, fourth, etc.) time around. So, in no historically defensible way can we understand the Renaissance as a re-creation of the Athenian zeitgeist, or the eighteenth-century Virginia plantation as "medieval." Such descriptions can be no more than metaphorical and can yield ludicrous conclusions.

If we are clear in our recognition of the artificiality of abrupt historical transitions, we can gain the advantage that labels give to memory without suffering the burden they impose on judgment. Since history is neither constant nor cyclical, we must be prepared to discover change without expecting duplication, orderliness but not historic necessity. Where the second billiard ball must move (if our laws in physics are to endure), the King may not. He may, for example, be sleeping. And, as we place people and events under sadly general categories, we must not forget that they are probably more similar to ourselves than different. We have no reason to assume that, because they thought about different matters, they thought in different ways, or that because they perceived their condition differently they either had no perception at all or lacked one that could be defended. "They," after all, are "we"; otherwise our inquiry would be reduced to a kind of comparative anatomy.

If the metaphor of the family is not too homespun, we might profitably view the several epochs as relatives, each with a set of distinguishing features but each nonetheless related to every other: related, but not in a mechanical fashion. Some relatives leave diaries, portraits, photographs, and even handsome bequests. Others leave only a certain record of good will and tolerance. Still others remain ever a mystery. They said little, traveled less, and wrote nothing. Occasionally, distant cousins show features more in common than brothers, and certain children seem to bear no resemblance at all to either parent.

When we treat historical inquiry as a kind of genealogy, we not only whet our intellectual appetite with a mixture of fancy and vanity but we also set certain traps. We invite the error of *scholastica successionis*. There is no gene for iambic pentameter, for twelve-tone harmony, or for painting madonnas. We do not understand the wealth of Uncle John and the poverty of his brother by exploring their pedigrees. To carry the metaphor to its limit, we do not explain the difference between ages by measuring their breeding practices. Therefore, the sense in which the following chapters are devoted to the various epochs of Western civilization is geographic only. Indeed, had it not been for medieval scholars in the Arab world, the chapter featuring Aristotle's psychological theories might be less than a page.

In light of all this, it should be clear that the idea of history is a kind of theory of explanation. It bears the same sort of relation to the facts of history that, for example, a concept or scientific law bears to the facts of experience. If we can imagine a being equipped with human senses but lacking human concepts, we might expect it to be flooded by sights and sounds that are never brought into a meaningful or even a coherent relationship. Think, for example, of how the events in the world would appear to a being utterly lacking the concept of causation. It is by virtue of our conceptual abilities that otherwise disjointed occurrences finally become explainable. We are able to distinguish between merely correlated and causally related events; between accidents and determined occurrences; between coincidence and lawful conformities. We refer to history as a similar collection of concepts and laws, or, at least, as a conviction that significant historical events can be explained only through a similar sort of rational-cognitive analysis.

At the root of the idea of history lies a judgment on what makes an event historical in the first place. We do not think of every falling leaf or every cry in the night historical, even though such events can be dated and their causes identified. We refuse to classify falling leaves as historical because the concept of history excludes purely physical transactions taking place among items of inanimate matter and embraces those occurrences that seem to require human participation, human striving. Indeed, the ordinary sense of "history" excludes everything that is merely physical. Even cries in the middle of the night take on historical standing only to the extent that they have a significant bearing on the actions of persons, institutions, or whole nations. As a theory of explanation, the idea of history arises from the judgment that events of social, political, moral, or intellectual consequence can only be understood by consulting the perceptions, reasons, intentions, motivations, and comprehensions of those who lived through the events in question. Even more, the judgment perceives these persons as constituting a sufficient reason for the occurrence of these events. That is, had persons of a different perceptual, intentional, motivational or cognitive makeup been substituted, the same event would not have occurred.

This, however, is not some useless truism. It is, of course, trivially true that had everything been different, things would have been different. But the judgment out of which the idea of history arises does not take stock of " "everything," only of some things—specifically, of uniquely human things. It takes for granted that history is something caused by human agency, something that receives its essential character from the specific moral, psychological, and conceptual attributes of actual persons. Note, however, that this is not a silent endorsement of the "great man" theory of history, nor is it an implicit defense of "psycho-history."

To say that events are judged as historical only when they require the assumption of human agency does not say that such agency proceeds from "great" individuals or even nameable persons. Even less does it say that the historical event must be understood in terms of someone's "psychological background," such as early childhood experiences, inferiority complexes, fear of water. Instead, it takes for granted (at least as a working hypothesis) that events in human history proceed from human actions, that the latter proceed from human judgments, and that the events become intelligible to posterity only when connections can be established between the understandings and the actions of the historical participants themselves. History as we know it is the disciplining of empathy by scholarship.

The subject matter of psychology is as old as reflection. Its broad practical aims are as dated as human societies. Human beings, in any period, have not been indifferent to the validity of their knowledge; they have not been unconcerned with the causes of their behavior or that of their prey and predators. Our distant ancestors, no less than we, wrestled with the problems of social organization, child rearing, competition, authority, individual differences, and personal safety. Solving these problems required insights, no matter how untutored, into the psychological dimensions of life.

If we are to follow the convention of treating psychology as a young discipline, we must have in mind something other than its subject matter. We must mean that it is young in the sense that physics was young at the time of Archimedes or in the sense that geometry was "founded" by Euclid and "fathered" by Thales. Sailing vessels were launched long before Archimedes discovered the laws of buoyancy. Pillars of identical circumference were constructed before anyone knew that $C = IID$. We do not consider the shipbuilders and stonecutters of antiquity physicists and geometers. Nor were the ancient cave dwellers psychologists merely because they rewarded the good conduct of their children. The archives of folk wisdom contain a remarkable collection of achievements, but craft, no matter how perfected, is not science, nor is a litany of successful accidents a discipline. If psychology is young, it is young as a scientific discipline, but it is far from clear that it has attained or should have sought to attain this status. In any case, the foundations of the subject predate its scientific aspirations. The foundations are philosophical, and it is to these that we turn in the search for origins.

Notes

1. R. G. Collingwood, *The Idea of History,* Oxford University Press, New York, 1972. This quotation is taken from page 10 of the Introduction. We owe the availability of this work to Professor T. M. Knox, a close associate of Collingwood's, who edited these essays after Collingwood's death in 1943 at the age of 52. His Preface to the edition is important to an understanding of Collingwood's program, of which *The Idea of History* is but illustrative.

2. Collingwood, *Idea of History,* p. 9.

3. The most accessible source for the works of Giovanni Battista (or Giambattista) Vico is the translated edition, *The New Science of Giambattista Vico,* by T. G. Bergen and M. H. Fisch, Cornell University, New York, 1970. This is an abridgement of the 1948 translation by the same translators. Vico's historiographic method is well illustrated in Book Three, the *Discovery of the True Homer.* On his complaint about earlier historians, a delineation of their flaws, and his own approach, see *Elements,* Book One, Sec. II. On the error of *scholastica successionis,* see XIII, 146 of the *Elements.*

4. Collingwood, *Idea of History,* p. 224.

5. This observation will be amplified and explained in Chapter 6, with apologies to Erasmus, Luther, Pico, et al.

2 Psychology in the Hellenic Age: From the Pre-Socratics to the Dialogues

Within an area approximating the size of Montana, and over a period of less than three hundred years, the Greek-speaking people of the ancient world would produce the philosophers Thales, Anaxagoras, Empedocles, Parmenides, Zeno, Pythagoras, Protagoras, Socrates, Plato, and Aristotle; the dramatists Aeschylus, Sophocles, Euripides, and Aristophanes; the historians Herodotus, Thucydides, and Xenophon; Hippocrates the physician; Pericles the political leader; Pheidias the sculptor; and the various schools and traditions that would become associated with these names. This span of time, a period covering roughly the appearance of the school or sect of the Pythagoreans in 530 B.C. to the flowering of Aristotle's scholarship two centuries later, comprises the Hellenic epoch of classical Greece, an epoch not anticipated by any prior age.

Having now taken the fateful step toward resurrecting this famous coterie of "dead, white, European males," as some in today's world would have it, a few words are in order concerning the nature of the debt to Hellenism in general. To begin, although the ancient Greeks were suspicious of strangers and more or less convinced of the superiority of their own culture, the leaders of thought in that age did not regard their achievements in racial or ethnic terms. It would take the unifying vision and energy of Alexander the Great to spread Hellenism throughout the conquered territories, but many of these same territories needed little by way of coercion. Moreover, the more enlightened Greeks understood Hellenism itself to be a set of values and attitudes about culture and civilization; about beauty and learning and education; about life within an ordered family and an ordered state. As Isocrates (436–338 B.C.) would express it in his *Panegyrics,* all who shared the Hellenic conception of culture (*paideia*) were "Greek" in the most relevant sense of the term. The extent to which ancient Romans came to share this conception was so considerable that Pliny would declare Rome to be a Greek city! Over the many centuries since that time, something of a unifying and common culture has been adopted and developed by the peoples of England, Europe, and their major settlements. This culture, too, has been so utterly indebted to Hellenism that Anglo-European civilization might also be termed "Greek," at least at its core.

The attention routinely directed to the ancient Greek world by students of nearly every academic subject is to be understood, therefore, not as a species of Western jingoism, but as the necessary means of identifying the sources of a profoundly influential intellectual tradition. Only through such inquiries does it become possible to fathom how a set of culturally transmitted patterns of thought can be developed and rearranged to yield the well-known achievements and limitations of civilizations in the post-classical period. Roman civilization itself, the very template for so much that would unfold in European history, was self-consciously "Hellenic" in many of its most defining features. To this day, the preoccupations of scientists, scholars, and intellectuals of every stripe are bequests of Hellenic civilization, including the very preoccupations that would find such fault with that civilization.

Nonetheless, an inquiry into Hellenism should not be understood as provincially "Western." Hellenic and Hellenistic influences reached beyond Rome and across great expanses of time and space, extending to and beyond the Indian subcontinent. The influences were fully discernible in art, in philosophy, and in the general culture long after the last vestiges of Alexandrian or Roman military power had receded from the near-eastern world. Hellenistic decorative arts, for example, remained the choice of affluent near-east Asian well into the fourth century A.D. In a word, there is nothing regional about Hellenic civilization except its birth, and, of that part of it that has endured into modern times, there is surprisingly little that is derivative. For the historian, therefore, and especially the historian of ideas, an understanding of that particular civilization must be a central project, and one without any alternatives of equal consequence. At the very core of this central project is the appearance of philosophy and the immense range of issues and subjects it would spawn. We are repaid by devoting attention to the factors that nurtured this incomparably productive civilization.

The Nurturing Conditions

Preliterate societies perpetuate themselves by developing non-literate aids to memory, such as paintings and monuments, songs and signs, poems and dances. As with so many such societies, the peoples who would emerge as the Greek-speakers of the ancient world must have used just such devices to record a settled history and to establish a stabilizing method by which to transmit customs and belief to later generations. In such a world the dance might evolve into the poem, the poem into a chorus, the chorus into drama, and the drama into the dialogue. In the Greek world, the dramatic and poetic depictions of life's confusions and possibilities were transformed into nothing less than philosophy itself. The foundations can be seen in those eighth-century epics of Homer that would shape Greek thought and imagination for centuries.

In the *Iliad* and the *Odyssey,* Homer presents what is already a developed conception of human nature and the determinants of the human condition. We find heroic undertakings explained in terms of specific emotions arising within and taking control of regions of the body (heart, chest, abdomen), and still other behavior ascribed to the gods who can send demons to disturb sleep or shape dreams or impersonate mortals. What is so vividly developed in these epic poems is the need to understand oneself, to externalize our thoughts and images and passions in such a way as to examine them in nearly clinical fashion. Also emblematic in these works is at once the tentativeness and the sheer multiplicity of explanations. There is no pretention to certainty or infallibility, no last word. Consider only Homer's treatment of the causes of the war with Troy. Was it the result of Helen's indiscretion? Was it caused by Paris, who chose Aphrodite over her competitors for the prize reserved to the "most beautiful," or was the cause the offense given Discordia by not inviting her to the feast in the first place? Was it the intemperate anger of the dishonored Hellenes seeking revenge at all costs? Was it the personal character of such men as Agamemnon and his brother, Menelaus, husband of Helen—a character seen in their father, Atreus, whose own rapacious actions had been at the expense of innocent others? In verse after verse, the audience discovers the complex chain of events that might unfold from the simplest and most innocent of beginnings; the manner in which our most thoroughly considered courses of action can be upset by chance or divine intervention.

In the end, the epics point to the importance of an ordered state, to the need for reasonable methods for settling disputes, to the necessity of character and fortitude in all we do. And though due consideration is paid to the gods and to the fates, Homer's explanations of human conduct are typically naturalistic, even biological. One acts owing to impulses arising in the body. One dies when vital organs are damaged. After death there is some shadowy but (except in rarest instances) mindless residual, briefly revived by the fresh blood of animals, but soon to revert to a wraith-like existence utterly removed from the human condition. True immortality is possessed only by the gods, and this owing to a special substance, *ichor,* which runs in their veins. They are otherwise distinguishable from mortals by their extraordinary but not total power over natural events, and by their strength, their cunning, and their craftsmanship.

The gods of Homer have their favorites among mortals and even occasionally breed with them, but in general the Olympians are preoccupied with their own affairs, often indifferent and even contemptuous of human lives and limitations. Thus they must be propitiated, never aroused to anger or envy. But they are also not looked to for answers to the abiding questions or for solutions to the problems of life and mind. In the matter of fundamental truths and their implications, we are left to our own devices, for, in these matters, the gods themselves are limited. Even mighty Zeus must consult the fates if he would know the end result of his designs. In these and related respects, Homer's immense contribution to the emerging culture of Hellenism was pre-philosophical but utterly compatible with an essentially philosophical outlook.

There is no question but that the intellectual tone and background of Greek life from the seventh to the fourth centuries derived inspiration from these works. But Homer's was not the only literary influence. Supplementing it were works by Hesiod slightly later in the same century. His *Works and Days* records the life of the modest farmer working with nature, within the cycles of the seasons, developing an ethics based upon the orderly and productive works of a properly tended and respected earth. The point of *Works and Days* is not merely a summary of the agrarian life, but a praise of its social and ethical character. Together, the Homeric and Hesiodic perspectives would impose on Greek thought a conservative and measured quality, a worldview that would locate human life within the larger framework occupied by all living things, by the elements, by the gods. When Achilles finally overcomes his indignation and prepares to avenge the death of his dear friend Patroclus, he urges his horses to their duty, and he hears from one of them that they will not let him down. If speech is a gift of the gods, who is to say on whom or what they will confer it? Human life and our special powers, on this view, are not above nature but inextricably bound up with it. Centuries later Aristotle will insist that the political state, the *polis,* is an entirely natural entity proper for a kind of animal that is social by nature. In this he reaches a philosophical position entirely at one with Homer's.

The Greeks honored the Olympians and believed in the prophetic power of oracles, but they did not have the equivalent of a "prophet" as this type was known in the eastern world, nor did they claim to possess a body of divine and unchallengeable truths. Epic poetry was rich enough in metaphor to serve as a daily guide to a life. Unlike their Jewish contemporaries or their Christian successors, the Greeks of the ancient world did not possess revealed truths or a code of undisputed moral precepts. The codes by which they lived had for them the authority of history and the additional authority of nature: To live in accordance with the model of nature was to be true to one's own nature, for the latter was but a part of the former.

The effects of this were at once positive and negative. On the positive side, the absence of such revelation permitted a freedom of interpretation and a creative approach to the spiritual dimensions of life. This surely had a subtle but pervasive part in the evolution of philosophy, nurturing an inquiring and even skeptical mind.

On the negative side, the absence of a codified body of religious principles not only compounded the ever greater problem of political unity of the Greek people but also encouraged an assortment of superstitious beliefs within various communities. Even as the classical achievement was becoming part of the historical record, the masses of Greek people remained quite committed to notions of witchcraft, prognostication, omens, and curses. That the better minds of the period had only contempt for all this is evident in Plato's rebukes found in the *Republic* (Book II, 364) and in *Laws* (Book X, 909b; XI, 933a). Nevertheless, in this (as in so many other ways) the philosophers were outside the common culture.

There is another side to this story that warrants brief mention. It has often been observed that, if the Greek world did not have an official and codified religion, it also never was a totally secular state. What the period we call "classical" displays, in such a great variety of ways, is integration: the integration of the religious, the political, the moral, and the aesthetic. Where strict religious orthodoxy does come into play, it is often aroused by philosophical challenges. In the main, however, the ancient Greeks grounded their convictions quite broadly, with religion and ethics, and family life and the good of the polis being interchangeable conceptions, interdependent features of life. Accordingly, the religious tone of sixth-century Greece was neither speculative nor personal. Rather, it was civic.[1] Philosophy, when it did appear, was less a threat to than an expression of a reverential and naturalistic theology already in place for centuries.

Philosophy: The Ionian Invention

The situation of the seventh-century Ionians was unique in the Greek world. Their settlement of the Aegean islands and coastal cities of Asia Minor was prompted initially by scarcity on the mainland and by attractive trading prospects. Their native homeland was for the most part Attica, meaning that Athens was their mother city. If the history of immigrant experiences in modern times is applicable, we can conjecture that they displayed that same "pride of race" witnessed among displaced citizens the world over. Fear of assimilation seems to be an abiding condition of wandering people. That the earliest Homeric scholars should appear in Ionia is, therefore, in keeping with this tradition. The further one removes oneself from the culture of one's childhood, the more strenuously one asserts its virtues and protects its uniqueness. The Ionians of the seventh century were quite possibly more "Greek" than their Athenian cousins.

Given the general cultural background, the Ionian Greeks were further inclined toward philosophy by daily contact with radically different cultures based on radically different concepts and traditions. Commerce introduced them to foreign goods and foreign beliefs. Their relations with Egypt were surely the source of their later development of monumental architecture and may well have conveyed certain religious doctrines that the philosophically minded Greeks would transform in a speculative science. It is not surprising, then, that the Ionian colonies and those established in Italy would yield such inventive philosophical pioneers: Thales, Anaximander, and Anaximenes were from Miletus; Heraclitus from Ephesus; Anaxagoras from Clazomenae; Pythagoras from Samos; and Xenophanes from Colophon. In the Italian reaches of the Ionian influence, we record Philolaus from Tarentum, Empedocles from Acragas, and Zeno and Parmenides from Elea. Then north of the Ionian cities, in Thrace and the centers along the Black Sea, the names of Protagoras, Democritus, and Leucippus are added. These are the figures who outlined the issues with which Plato and Aristotle would contend and which have caught the imagination and frustrated the minds of scholars for two millennia. The movement would culminate in Athens, which we learn on the authority of Athena herself is the place the gods would choose to live above all others save for Olympus.

Pre-Socratic Psychology

It is especially important when assaying the outlook and contribution of the pre-Socratics to keep two considerations in mind. First, we have very few of their writings (of those who actually did have their thoughts recorded), and, second, many of the later commentators on whom we must rely are far from unimpeachable. If Thales or Pythagoras ever set thought to papyrus, all evidence is lacking. What survives of Anaximander's scholarship is some five sentences; and only one remains by Anaximenes. We have nearly 140 fragments of Heraclitus, of which perhaps several dozen require serious attention. We are more fortunate where Empedocles and Democritus are concerned, less fortunate in the case of Protagoras. All told, the major figures of the pre-Socratic achievement can be investigated directly through fewer than two hundred pages of modern text. Indirect evidence, however, suggests that the group was responsible for perhaps as many entire books. Liberally, then, we may lay claim perhaps to one percent of their contributions.[2] The balance of what we know of them must be reconstructed from what later philosophers said of them, often in the context of severe criticism developed in conjunction with attempts to fashion opposing systems. It will be sufficient in the present context to consider the pre-Socratics only to the extent that they set part of the agenda for the psychological theories and studies of the Socratic and later schools.

Cosmocentric-Anthropocentric

It is often said (and was said by Socrates and other ancient commentators themselves) that Socrates, unlike his predecessors, directed philosophical inquiry away from the heavens and toward human concerns: that he shifted attention from cosmos to anthropos. Although it is undeniable that the richly developed psychological theories passed down by Plato are without precedent, they are not utterly divorced from an earlier philosophical tradition. That earlier tradition, with debts to far older and Eastern mythologies, focused on the created cosmos, the source of all that exists, the elementary composites out of which ever more complex entities come into being. This earlier tradition is exemplified in the teachings of Pythagoras and of Parmenides. Brief consideration of their philosophies conveys a sense of the issues and methods bequeathed to the Socratics. But even before this, it is worth noting some of the pressures that might have inclined Socrates and his circle away from cosmological matters. The reasons are not purely intellectual.

Protagoras and Anaxagoras, among the pre-Socratic philosophers, had both suffered the wrath of orthodox Athenian society by investigating the secrets of the heavens. The ordinary Athenians were willing to tolerate philosophical speculation as long as things were going well, but once the political or financial or military fortunes of the city were in jeopardy, they were quick to find scapegoats. Thus, that the Fragments of the pre-Socratics are often ambiguous, that the Pythagoreans were committed to keeping their wisdom private, and that the Age of Pericles itself was rife with censorship offer proof enough that the Hellenic epoch was never really tolerant, or, if tolerant, was never completely nurturing of speculation.

A passage from Plutarch's *Lives* drawn from the biography of Nicias is suggestive here:

> People would not then tolerate natural philosophers, and theorists, as they called them, about [the laws of the heavens]; as lessening the divine powers by explaining away its agency. . . . Hence it was that Protagoras was banished and Anaxagoras cast in prison . . . and Socrates, though he had no concern whatever with this sort of learning, yet was put to death. . . . It was only afterwards that the reputation of Plato . . . because he subjected natural necessity to divine and more excellent principles . . . obtained for these studies currency among all people.[3]

The interests of the pre-Socratics arose from the most basic issues and problems. In general terms, the issues that attracted their attention were metaphysical, concerned with nothing less than questions of actual being. Just how is it that anything exists, and what is the source of such existence? It will be instructive to review even hastily how Pythagoras and Parmenides approached questions of this sort.

At the risk of putting words into the mouth of a philosopher whose entire sect was sworn to public silence, one might still speculate with some confidence that Pythagoras was inspired initially by the marked relationship between mathematical abstractions and actually existent and observable events and things. The most well-known of these relationships is the Pythagorean theorem which at once is an abstract axiom of plane geometry and, at the same time, a relationship obtaining among all actual rectilinear triangles.

An even more subtle form of the relationship is found in music, especially the relationship between number and music. Pythagoras is credited with the theory of musical harmony, but he was clearly interested less in this than in the larger implications arising from it. If, in fact, the full range of musical perceptions experienced as harmonious is generated by notes that stand in strict numerical relationships, it follows that other—even all—aspects of observable nature have a similar origin. Taking the first four integers (the tetractys) 1, 2, 3, and 4 as primal, Pythagorean teaching proposed that the first corresponded to the point, the second to the line, the third to the plane and the fourth to the solid. The four integers as abstractions are somehow generative of a world of solid objects. This is achieved by means that cannot be physical for the physical world is produced by such means. Thus, it must be in the realm of soul that abstract numbers are, as it were, absorbed and used to produce the material world. A similar line of reasoning then would confer similar constructive powers on the souls temporarily embodied in living persons; not that the mental or spiritual side of human nature actually creates the physical world, but that the changeable and imperfect objects of nature become understood in terms of abstract conceptions possessed by the soul. This tension between the abstract and the objective, the ideal and the "real," the spiritual and the material will be confronted time and again in the history of psychology and centrally in the philosophical psychology of Plato.

We next consider Parmenides of Elea (b. 515 B.C.?) as illustrative of the most original thinking of the period, though here again the available record is fragmentary and amenable to a variety of interpretations. Choosing to express his theories in poetic form, Parmenides set forth his philosophy in a work that may originally have been titled "On Nature," in which he critically examines various paths to truth.

The function of critic in the poem is taken by a goddess who leads Parmenides toward truth by way of a dialectical method. The most important conclusions Parmenides reaches through close analysis are that whatever has real being—whatever really is—must be eternal and unchanging, and that such real existence can never be discovered by the senses. The thesis seems to rest on the claims that it is contradictory to attribute existence to that which is never the same at different times, and that it is incoherent to argue that any existing thing came into being from nothingness. If, in fact, it were the case, as Heraclitus insisted, that one can never descend into the same river twice (i.e., that everything is everywhere in a state of constant flux), then we could not say that anything actually exists, for whatever we would put forth as a candidate would be changing even as we referred to it. It would, in a manner of speaking, be gone before it was noticed. But then we would have to conclude that all that does exist came from that very non-thing whose nonexistence is guaranteed by its continuous change. To escape the bind is first to reject the sensible realm as the realm of real being, for only the sensibles display such inconstancy. What is left, absent the removed sensibles, is a realm of abstractions, stable and eternal truths accessible if at all only to reason.

Plato's debts to this line of reasoning, too, will become obvious later in this chapter. It is sufficient here to note Parmenides, along with Pythagoras, Anaxagoras, Heraclitus, Anaximander, and others as the founders of a new intellectual movement appearing in the Greek world of the sixth and late fifth centuries B.C. Their teachings would come to be the subject of painstaking dissection and study by the Socratics and later by Aristotle. Throughout this development the most general issue would remain that of "real being." In its original Parmenidean (Eleatic) form it was an issue tied to an implicit theory of human psychology. The theory, which would be defended later by Plato, is that the human senses are confined to the realm of appearances, whereas what is true occupies a realm beyond the senses.

Although their teachings and even their methods of philosophizing differed, the pre-Socratic philosophers at large shared a set of common concerns that might be economically reduced to four, the same four that have guided speculation ever since. In the broadest sense of the terms these are theology, physics, ethics, and psychology. The philosopher's mission was understood as the discovery of those unifying principles that would illuminate and solve the many different problems associated with these four areas of concern. In the dialogues of Plato, there is a sustained attempt to integrate all four areas into the most general of theories regarding not only real being, but the very point of existence itself, including and especially human existence.

Plato's Psychology

The circle of philosophers forming around Socrates (469–399 B.C.) initiated discussions of the most fundamental nature and thus set in motion much of what would come to be the subsequent history of philosophy. The group was assembled in the wake of a devastating war which Athens lost to Sparta. The philosophers associated with Socrates were, it must be recalled, on the losing side of that war. It is not surprising that so much of their attention would be given over to questions of ethics, politics, and the virtuous life. Had there been no social pressures working against cosmological speculation, recent history alone would have inclined such active minds toward these humanistic subjects.

It is after the second and decisive Peloponnesian War (431–404) that we find these young and not so young Athenian intellectuals searching for the point at which things took a wrong turn; a turn away from the (glorified?) humanism of the Age of Pericles. Most of what is known of their concerns reaches us through dialogues composed by Plato who founded the famous Academy and composed his works years after Socrates' death. But, as with the Evening Star and Morning Star, the identities of Socrates and Plato dissolve into one another when either is observed long enough.

Socrates was the master, and by his own admission a "gadfly" flitting through the markets and assemblies of Athens and pricking the conscience of the complacent. It is likely that Plato not only knew him in childhood but that he was introduced to many members of the Socratic circle. Except for neighborly comments offered a generation later by Xenophon, our only record of Socrates' life and ideas is that recorded by Plato in the monumental dialogues. It is by no means certain, however, that Plato served as a loyal student in this regard, whatever he claims for himself in his Second Epistle:

> [T]here is not and never will be a work of Plato; the works which now
> go by that name belong to Socrates, embellished and rejuvenated.[4]

Much in the dialogues must be Plato's own invention, as much certainly establishes quite basic changes in perspective from the earliest to the latest dates. Accordingly, the inescapable generalizations we seek in the dialogues are, when found, unavoidably arguable. What is clear is that the failures of the state, the moral decline so evident in the wake of the Age of Pericles, help to account for the conservative tone of the teachings and for the reluctance at least on Plato's part to become absorbed into the nonphilosophical affairs of the world. A child of wealth, groomed for leadership, Plato relates his decision to forgo any and all political options:

> [O]ur city was no longer governed according to the customs and in-
> stitutions of our fathers. . . . [T]he letter of the laws, and our customs
> were giving way to an even greater corruption and disrespect. . . .
> [W]hen I considered these things, seeing everything being driven
> helter-skelter, my head was in a whirl. . . . I realized that all existing
> states without exception had irremediably bad constitutions. . . . So,
> in praise of true philosophy, I was obliged to say that through it alone
> can we recognize what is right for states as well as individuals.[5]

The intellectual range of the dialogues is universal. They analyze all the problems introduced by the pre-Socratics and in the process advance others that only the Socratics seem to have conceived. We shall investigate but a narrow set of these topics, the set that pertains to matters of consequence in the history of psychology and that still survives in contemporary psychological scholarship. Broadly defined, the set contains four core problems: (1) the problem of the knowable, (2) the problem of knowledge, (3) the problem of conduct, and (4) the problem of governance. The first of these contains the Platonic position on the truly existing. The second considers the role and limitations of sensation, perception, and memory in relation to the knowable. The third addresses psychological and psychosocial development, the determinants of behavior, and the tension between the rational and emotional dispositions. The fourth examines the foundations of society, the family, and the moral dimensions of interpersonal life.

The Problem of Knowledge and the Knowable

Few of the dialogues are confined narrowly to a single problem. Those that are—for example, *Charmides* (temperance), *Lysis* (friendship), and *Laches* (courage)—tend to be brief, even casual. The major works are intense and far-ranging. Very nearly all of them at least touch on the problem of truth and of what can be said truthfully about anything.

In its most innocent and most vexing form, the problem of knowledge is this: How can we specify what is knowable without implicitly claiming to know it? How can we sincerely claim to search for what we have not yet identified as knowable? And even if we do identify some domain of the potentially knowable, what sources of knowledge are the most valid and which ones are likely to deceive? How, in the end, can we ever be sure that we know anything? Maybe a complete skepticism is the most philosophically defensible position (assuming such a skepticism even leaves room for a position being defensible!).

For a philosophy such as that of the Socratics, concerned with virtue, with government, and with social organization, it is obvious that these ontological and epistemological problems must be engaged. It makes no sense to inquire into the question of justice or courage or love until we have established that inquiry itself is valid and that its results are true. Thus, the initial development of a workable epistemology is indispensable to the larger Socratic mission. It is this epistemology that will distinguish the knowable from the unknowable, truth from mere opinion, reality from illusory appearance.

The position taken by the Socratics on the question of knowledge was part invention and part reaction. The reaction was to the teachings of the major Sophists. We observe this most vividly in the opening scene of the dialogue Meno, named after the young aristocrat recently returned to Athens from Thessaly, where the sophist Gorgias is influential. He is accompanied by a servant who proves to be the pivotal figure in the dialogue.

In something of a teasing manner, Meno tells Socrates that he would like to be able to return to Thessaly equipped to instruct his Thessalian friends in the virtuous life and, accordingly, would hope that Socrates might share the truth with him. Socrates, ever on guard, explains that until he knows what Meno construes virtue to be, he surely could not prescribe a virtuous life. Meno, we might guess, has learned a trick or two from Gorgias and now poses a typical Sophist paradox:

> And how will you enquire, Socrates, into that which you do not know? What will you put forth as the subject of enquiry? And if you find what you want, how will you ever know that this is the thing which you did not know? (80)[6]

Meno's challenge is based on the contention that all inquiry is impossible, since (1) if we are ignorant, we have no starting point for an inquiry and (2) if we are informed, no inquiry is needed. Socrates' reply takes advantage of the fact that Meno's servant is a young, uneducated "barbarian." Socrates begins to question the servant about geometric forms which Socrates draws in the sand. The boy answers a number of questions with "yes" or "no" as Socrates leads him on to a version of the Pythagorean theorem. Little by little, and not before hosting several conceptual errors, Meno's servant approaches and finally attains the understanding of the relationship between the diagonal of a square and its area. On no occasion does Socrates actually give away the answer. He merely paces him through a series of logico-mathematical steps until the insight appears. The exercise is meant to prove the first principle of Socrates' theory of knowledge, that knowledge is a reminiscence. The knower has the truth, he doesn't learn it: rather, he is able to recall it through philosophical guidance.

The centrality of this position to the entire Socratic psychology (and epistemology) cannot be overdrawn, for if we believe that knowledge is memory we will accept the dialectical method of uncovering it. Moreover, we will give to experience a place of no special importance; to reflection, a place of unparalleled importance. The dialectical method, after all, is not simply a kind of conversation. Rather, it is a careful delineation and criticism of premises, an analysis of meanings, and an assessment of implications. Through it the student is expected to learn not only what true is but why he has failed to discern this truth previously. The "learning" is, of course, but recollection.

With respect to the place of experience in the quest for knowledge, the dialogues could not be more consistent. Under the indirect influence of Pythagoras and the direct influence of Parmenides and Zeno, Socrates rejects the senses as routes to truth. We discover in the *Theaetetus* (161), as well as in the *Timaeus* (43), and again in the "Cave Allegory" of the *Republic,* that Socrates defines the dominant mission of philosophy as the rejection of the world of appearance.

To appreciate more fully the theory developed within the dialogues it is helpful to consider one influential competing perspective. The famous Sophist, Protagoras, had contended that "Man is the measure of all things," a contention that was skeptical as to the very existence of certain knowledge. It was a thesis leading to a nearly wanton subjectivism. If human knowledge is limited to our actual and possible experiences (sensations and perceptions and memories of same), and if these are also private and unique to each individual, then not only is man generically the measure of all things, but so is each person.

It is in the *Theaetetus* that Socrates directly challenges Protagoras' maxim:

> I am charmed with his doctrine, that what appears is to each one, but I wonder that he did not begin his book on Truth with a declaration that a pig or a dog-faced baboon, or some yet stranger monster which has sensation, is the measure of all things; then he might have shown a magnificent contempt for our opinion of him by informing us at the outset that while we were reverencing him like a God for his wisdom he was no better than a tadpole. . . . For if truth is only sensation, and no man can discern another's feelings better than he, or has any superior right to determine whether his opinion is true or false, but each, as we have several times repeated, is to himself the sole judge, and everything that he judges is true and right, why, my friend, should Protagoras be preferred to the place of wisdom and instruction, and deserve to be well paid, and we poor ignoramuses have to go to him, if each one is the measure of his own wisdom? (161)[7]

As the dialogue proceeds, Socrates offers examples of the weaknesses of the sensationist position. Not only do brute animals have keen senses but so, as well, does the infant. Still, these beings cannot be said to know merely because they can see. Or, examined in another way, we continue to know that which is no longer visible. If knowledge were only and always perception, knowledge would cease when the objects of perception were removed. Not only this, but our knowledge would be in constant flux because of the ever-changing world of sense.

Thus the sick Socrates, the standing Socrates, and the reclining Socrates would all be different. But those who know Socrates are, in fact, not confused by these changing appearances. Rather, the soul, and not the senses, is able to read through these changing features and discover the real, unchanging, essential Socrates, that is, the true form of Socrates that survives all change: "Then knowledge does not consist in impressions of sense, but in reasoning about them." (186)[8]

To the extent that the method of dialectical analysis can succeed or fail, the *Theaetetus* fails. Indeed, at the end, Socrates is still frustrated by his inability to establish what knowledge is as opposed to what it is not. He succeeds in distinguishing between the specific factual knowledge acquired through the senses and the more significant general principles known to the mind (185). He also notes the difference between opinion, which may be and often is false, and knowledge which, by definition, cannot be false (188). Socrates also uses the *Theaetetus* to present his theory of memory according to which experiences are recorded as something like wax impressions. The durability of the impression depends on the frequency of the experience and the purity of the wax (191–195). The former claim is traditional associationism, while the latter is but another expression of the Socratic emphasis on the hereditary differences among people with respect to character and quality of mind. Notwithstanding the originality of these ideas, we still must judge the *Theaetetus* as incomplete in those parts of it attempting to develop a cognitive psychology. However, where it fails, the *Republic* succeeds in clarity and scope.

Plato's *Republic* is generally regarded as a foundational work in political science, which it is, but on still another account it should be read as a quite developed and systematic Psychology. Recall that in Book II, when several of those assembled (Glaucon, Thrasymachus, and Adeimantus) implore Socrates to analyze the principal features of justice and to establish how the just man is always happier than the unjust, Socrates warns them that such a task requires "very good eyes" (368). He explains his remark by invoking the metaphor of a nearsighted person who can read distant letters only when they are greatly enlarged. Thus, to examine the nature and functions of the individual, the philosopher, whose vision is also less than perfect, must enlarge the object of study. Accordingly, it is the State (the *polis*) as the enlargement of man's personal nature that must be studied if one is to have a clearer picture of human nature. By constructing the perfect State, the philosopher will necessarily comprehend those characteristics that would produce the perfected human being (Book II, 368–369).

There is a temptation, in attempting to summarize the psychological features of this dialogue, to cull a group of suggestive quotations and with them argue that Plato anticipated all modern schools of psychology, or, on the other hand, that he subscribed zealously to some narrow view of human life within an equally narrow society. We might, for example, offer those sections of Book III in which the citizens are categorized as being made of gold, silver, brass, or iron, framed so differently by the gods that some necessarily shall rule and others serve (415). Here is a passage (which, by the way, Socrates introduces as one of the convenient fictions a leader may have to foist on the people) that stands in close accord with Book V (459–460), in which prearranged marriages and controlled breeding are proposed in order to create the class of Guardians. In this Socrates is found following the practice of breeders of hunting dogs!

There is no question but that the psychological theories advanced in *Republic* are nativistic. They are based on the assumption of powerful hereditary influences in the formation of human character and intelligence. Socratic psychology, at least in the earlier dialogues, is nativistic also in locating the "true forms" within the soul prior to all experience; indeed, prior to our very birth (*Phaedo,* 73–76).[9] But quite as many passages can be cited in which the emphasis is on education, experience, and commerce with worthy and contemplative friends. Meno's servant, surely no man of gold by Socratic standards, still reveals (recalls) his knowledge of the "true form" of the right triangle.

As Socrates himself requires an enlargement of our humanity in order to see it more clearly, we too must step back from the lines of these dialogues in order to comprehend their meaning. The *Republic* can be understood at one level by considering again the conditions surrounding its authorship. The Athens of Pericles had been vanquished by the Spartans. In the dialogue *Laws,* it is claimed that Greece survived the assaults of Persia not because Athens was victorious at Salamis and Artemisium, but because the Spartans prevailed at Marathon and Plataea (IV, 707). In the same dialogue Plato worries that maritime wars (those on which Athens rested her security) foster a form of cowardice. The Spartans, self-denying, regimented, orderly, and traditional, have become the model by the time *Republic* is composed.

We need not search the annals of Orphism or the mystery cults of Sicily to find the roots of Platonic asceticism and puritanism. In his *Seventh Epistle* and in *Laws,* Plato is unequivocal in his respect for the political order and military achievements of fifth-century Sparta. He can only struggle to achieve a reconciliation of the basic cultural and moral tension between an Athens and a Sparta: the tension between law and freedom, pragmatism and moral absolutism, fact and value, passion and reason, and power and justice. As the various institutions and actions of the ideal republic are argued into being, we witness a coterie of valiant scholars striving to find a place in the great buzz of a declining age, to find a hopeful and reasonable order in their fallen house of cards. They are no longer sure of themselves, their loyalties, or their obligations. Unlike the confident Ionians of the previous century, they seem to have been abandoned by the gods of Homer. The proverbs of Thales, and the optimism of Anaxagoras, to which in his last hours Socrates looks with longing and disappointment (*Phaedo,* 97), are seen to be too simple. And so, read in this light, the *Republic* becomes a manifesto of the discontented and disaffected who will give up on this world,

> [u]ntil philosophers are kings, or the kings and princes of this world have the spirit and power of philosophy, and political greatness and wisdom meet in one, and those commoner natures who pursue either to the exclusion of the other are compelled to stand aside. (*Republic,* Book V, 473)

To describe the Socratics as discontented and disaffected is not to say they were sullen or even pessimistic. Good humor obtains even in *Phaedo,* the saddest of the dialogues. Rather, we are to recognize Socrates and his pupils as the enlightened and reflective critics of an age, and to realize that such philosophers, in any period, will perceive themselves as unheard by, even inaudible to, "those commoner natures." As critics they observed power falling into the hands of polemicists whose only talent was the ability to tell the masses what they wanted to hear. Seeing an entire population deluded by the trappings rather than the essence of greatness, they rejected perception as a means by which knowledge might be won. Watching a world tossed in seas of change, they searched for that which never changed and called it truth. Noting the sad fate of a people moved by passion, they devoted themselves to impersonal reason and argued well enough for what came to be known as rationalism to make it the dominant philosophy until well into the seventeenth century.

The *polis* described by Plato is to be led and served by those possessing the virtues of temperance, courage, and justice. Their character is guaranteed by heredity and instruction. Law is blind in the received sense, and justice is swift. Population is carefully regulated, each city limited to 5040 citizens (a number defended because it can be divided evenly by every integer from 1 to 10). Plato does not mention the plague of 430, but it is likely that the recommendations against crowding derived in part from such concerns. We also detect in this emphasis upon orderly division a retrospective appreciation of Pericles' land reforms, if not a lingering attachment to Pythagorean numerology. Education is most carefully orchestrated. Those epic poems disclosing a lack of virtue among the gods are forbidden, as are works of fiction and "panharmonic" music. No wonder that Lord Russell, reflecting on this State in the 1940s, would dismiss the entire enterprise as unblushing totalitarianism.[10] As with utopias ever since, Plato's republic would solve the abiding problems through the application of carefully constructed formulas and practices even if, in the end, all are threatened with death through boredom.

We are less concerned here with Plato's utopianism than with the theory of knowledge advanced in the *Republic*. This is illustrated most clearly in the "Cave Allegory" of Book VII, where Socrates speaks of prisoners chained in a deep cavern facing a wall on which, unbeknown to them, shadowy forms are projected. From the prisoners' perspective, the shadows are real, their movements self-controlled. By good fortune, however, one of the prisoners escapes and makes his way up to the light of day. Now, for the first time, he sees reality and recognizes that all his former understandings were but illusion. Returning to share his discovery, this newly born "philosopher" is chided by those still confined below for having been blinded on his journey and for being no longer able to enjoy life in the cave.

We can now sketch the essential features of the Socratic theory of knowledge by combining the cave allegory with both the attack on Protagoras in the *Theaetetus* and the servant's lessons in *Meno*. First, true knowledge is a knowledge of the permanent principles of reality and not of changing appearances; it is a knowledge conveyed not by the senses but by reason analyzing experience. For convenience, we refer to this doctrine as "rationalism," and construe it to mean that knowledge of the world is of a cognitive, rather than a perceptual nature.

All true knowledge, in this sense, is recollection. If it can be possessed, but is not received through perception, then it is not acquired but unearthed. In other words, and as demonstrated by Meno's slave, we possess the eternal truths prior to experience. They are locked within the soul and become available to consciousness only through philosophical (dialectical) training. The doctrine is a species of nativism, for it proposes that what is discovered through philosophy is what was there in the soul all along. Moreover, the theory assumes significant and innate differences among human beings such that certain minds will not be as quick as others, nor certain characters as virtuous, no matter how nurturing the environment. Rather, some will be naturally and ineradicably superior to others in intelligence and virtue. They will, properly and inevitably, lead and protect the less fortunate. This corollary of nativism is what is sometimes called elitism. Indeed, the entire program of teaching featured in Plato's Academy was designed not as a form of public instruction for the *hoi polloi*, but as accessible only to a relative handful of the best persons, the *kaloi kagathoi*. Finally, since the eternal truths do not depend on the material senses, since they are within the soul before birth and survive within the soul after death, they are truths of a nonphysical nature. They are abstract relations, ideas of a special form, and they constitute the ultimate reality of the universe. This doctrine is a version of philosophical idealism which has arrested the attention of nearly every major philosopher of the past twenty-three centuries.

What about these Platonic "ideas"? Of all the facets of Plato's writings, probably none has received closer scrutiny than the so-called theory of ideas, this despite the warning of the renowned translator of the dialogues, Benjamin Jowett, that Plato's "theory" is not clearly set forth, that it changed over the years, and that Plato himself treated it as a sort of guess.[11] On the authority of Jowett, then, we will not attempt to force a unified theory upon Plato, especially in light of the different philosophical positions he reaches during the time from the early to the later dialogues. But this much is clear in his dialogues and is nowhere contradicted in them: the Platonic "ideas" do not refer to facts, are not about things, and neither arise within the body proper nor die with it. In analyzing the sources of language Socrates insists that we could not invent and agree on names for things unless we shared a kind of intuitive idea of them (*Cratylus*, 389).[12] Because of their permanence, such ideas must originate in God's mind (*Timaeus*, 28),[13] which the cleansed soul shall join after abandoning the dead body (*Phaedo*, 81).[14] Unlike the attributes (e.g., goodness, justice) of things, these ideas behold the things themselves (e.g., the good, the just), which is simply another way of removing them from the arena of perception.

The Socrates we meet in the *Parmenides* shrinks back from treating all reality as nothing but idea because to do so would cause him "to fall into a bottomless pit of nonsense" (130). But in less nervous moments he "sometimes begin(s) to think that there is nothing without an idea" (130). The Socrates of the *Phaedo* has learned, as Parmenides had predicted years earlier, not to fear such bottomless pits, and he now seems quite prepared to reject all that is not idea.

From a psychological as opposed to a metaphysical point of view, we need not try to achieve coherence in the theories about so-called Platonic ideas. We need recognize only that with respect to origins, the ideas in question would have to be innate, for the truths they possess could not be conveyed by the shifting and often deceiving data of sensory experience. We might say that whereas the "commoner natures" know only what they see, philosophers see that which they know a priori.

Although the Platonic ideas are in this sense innate, the overall theory of knowledge is not static. We have already noted the attention given to education, and when we look into the specific program of education recommended for the Guardians we see that the Socratics explicitly subscribed to a stage-theory of cognitive development. The Athenian stranger who is the teacher in the *Laws* notes that virtue and vice in childhood are known to the young only as pleasure and pain (II, 653). Since children instinctively love what is pleasurable and hate what is painful, the principal task of the educator is to make sure that true virtue becomes the object of love, vice the object of hatred. Moreover, there are critical periods of development when the lessons of virtue are most effectively conveyed by music, since virtue fundamentally is a harmonious relationship between body and mind (*Laws,* II, 653–654; VII, 790–791). This aim is furthered by close contact between parent and infant, by the rhythmic rocking of the young:

> [N]ursing and moving about by day and night is good for them all and . . . the younger they are the more they will need it; infants should live, if that were possible, as if they always were rocking at sea.
> (*Laws,* VII, 790)

The same theme is sounded in *Republic* (III, 377, 441–442). The young are out of harmony, so to speak. Reason and passion have yet to establish that unique accord that constitutes virtue. Music, dance, and other gymnastics must be employed because the very young mind is not yet able to assimilate rational principles directly. Thus, early education uses metaphor, not literal lessons. The success of education depends, of course, on the "quality of the wax" and, only under rare circumstances will the children of a lower class qualify for the life of a Guardian (*Republic,* II, 375–376). Instruction notwithstanding, heritable differences prevail and foreordain one's receptivity to education and the life it affords.

This focus on heredity is entirely consistent with Plato's epistemology in general. Having established to his satisfaction that knowledge of principles cannot be acquired by direct perception, he must look beyond experience to find the sources of virtue. All that is left, once experience is dismissed, is heredity. Subsequent education avails itself of worthy genetic endowments absent which the education must fail. However, given the right constitution, the individual will pass through stages of receptivity culminating in that adult stage in which philosophy can make its appeals to the latent knowledge of the soul.

The ambiguity surrounding such notions might be reduced by considering the Socratic approach to a specific and essential element of knowledge, our knowledge of space. In *Timaeus* the spokesman of the same name has been given the task of lecturing on the origin of the universe and the creation of living things, including human beings. Timaeus raises the question of the reality of ideas (51) and proceeds to distinguish between the true opinions formed by perception and ideas beheld by the mind but neither received through nor confirmed by the senses. Having treated that which appeals to sense and that which is contemplated only by the intelligence, he moves to a third "nature," which is space,

> and is apprehended without the help of sense, by a kind of spurious reason, and is hardly real; which we, beholding as in a dream, say of all existence that it must of necessity be in some place and occupy a space. (*Timaeus*, 52)

We will confront a more extensive discussion of this theory of space perception in Kant's philosophy, where the non-sensory nature of space is a central thesis. But here in an ancient dialogue we find Timaeus insisting that space is not an object *out there*, nor is it a mere concept in the sense of an opinion or belief. It is certainly not learned, nor is there any feature of the objects we locate in it that permits inferences about it. Quite simply, it is not an *it* but a species of *intuitive knowledge* or, in Timaeus's terms, a kind of spurious reason.

The properly instructed and experienced child will surely become more accurate in space perceptions, becoming, for example, more expert as an archer. Nonetheless, we would not say that instruction and experience have in any way conveyed space perception itself. It existed a priori as a native endowment. Experience of a world of visible objects requires such an endowment, even as it arouses capacities otherwise latent. To say, then, that the theory under consideration is nativistic, but that it also gives education a major role, is to say that rationality is in this respect like space perception. The mind is rational by nature and is in possession of intuitive truths. Philosophical study, however, is the route to these.

Might so brief a summary as this do justice to the Platonic theory of knowledge and the psychological theory contained in it? Alas, no, for there is no single theory, or any one theory that rises above debate. Disagreement is still the rule among scholars considering the very nature of the "true forms" in Plato's theory and whether the teaching of the Academy regarded them as having real existence. Moreover, although the relevant passages in all the dialogues are skeptical about perception and the world of appearances, it is also clear that Socrates and his friends were practical and realistic men who took the evidence of perception as counting for something. Perception, they reasoned, yielded knowledge of specific facts. It failed to provide knowledge of general concepts. Thus it might serve as the foundation of belief, and a given belief might even be true. But philosophical knowledge (wisdom) is not a species of opinion or belief. A doctor, for example, may believe that a patient has diabetes on the basis of certain symptoms. A mystic may believe that this same patient has diabetes on the basis of something seen in a crystal ball. If, in fact, the patient is a diabetic, we can say that both the doctor and the mystic harbor "true beliefs," but clearly their knowledge differs. On the Socratic account, philosophical knowledge differs from each of these. It is unlike the physician's in that it concerns itself not with probabilities but with certainty. And it is unlike the mystic's because it proceeds from rational first principles, the gift of *logos*.

The Problem of Conduct

> [Y]ou fancy that the shepherd or neatherd fattens or tends the
> sheep or oxen with a view of their own good and not to the good
> of himself or his master . . . that the rulers of states are not
> studying their own advantages day and night . . . and so entirely
> astray are you in your ideas about the just and unjust as not even
> to know that . . . the just is always a loser in comparison with the
> unjust.
>
> *Republic,* I, 343

Here is the challenge Thrasymachus lays down against Socrates' theory of the good. In nearly modern terms, Thrasymachus points out that the rich avoid taxation through wile, whereas the poor and just person sinks further into debt. The tyrant who, were he a mere citizen, would be imprisoned for his conduct, is praised for the power, wealth, leisure, and loyalty he commands. The unjust partner too gains advantage over the just. In every sphere,

> injustice, when on a sufficient scale, has more strength and freedom
> and mastery than justice. (*Republic,* I, 344)

Later, in Book II, Glaucon resumes the attack with the legend of Gyges' ring. The story is borrowed from Lydian lore and involves Gyges (reputed to be an ancestor of Croesus), who has discovered a magic ring by which he can make himself invisible. He soon uses it to take command of the realm by killing the king and raping the queen. He does what he pleases once he is sure that he can never be found out, and this, argues Glaucon, is precisely what anyone will do when there is no fear of punishment.

Socrates' task is not an easy one. He must prove, first, that the way things are—the way most persons might be expected to behave—is not the way they have to be or should be; and second, that the breach between what is and what ought to be is created by a failure of reason. He begins his rebuttal by likening the soul to the state. As the state contains three classes (merchants, auxilliaries, and counsellors), the soul is occupied by three principles: the rational, the appetitive, and the passionate (Book IV, 441). The virtuous person is one who has harmonized these three principles such that reason controls appetite and, as an auxiliary to reason, passion strengthens resolve (Book IV, 443). The view of reason and appetite as opposing forces is as old as the Homeric epics and as current as psychoanalytic theory. Harmony is an abiding theme in Plato's philosophy, and the *Republic* makes frequent use of musical metaphors in stressing the need for the soul's powers to be harmonized. Ancient Greek medicine employed music therapeutically, and the Spartan ephors were very careful in regulating the kind of music heard by the citizens.

Debts to Pythagoras are fairly obvious here and elsewhere as Plato undertakes an examination of the right sort of life. The goal the Pythagoreans grandly conceived was the purification of the soul. The Socratic emphasis on harmony and music was not the overworking of mere metaphor or the sign of uncritical superstition, but an integral feature of a general cosmological theory. The world of appearances on this account is the metaphor, whereas the ultimate realities are found in abstract relationships. Just as the true form of the right-angle triangle is given not by some graphic display of appearances, so the true form of man is given by a kindred (harmonic) relationship among reason, appetite, and passion.

The soul whose elements are in discord is sick and dying. The question of "happiness" in such a condition is entirely beside the point. One will not drink a potion, no matter how cool and sweet, if one knows it is lethal. The pleasure of the moment will not serve as a rational excuse for suicide, and we would judge the alternative view as one that only a crazed person could find compelling. Thus, one who drinks this potion is either mad or is ignorant of the consequences. Those like Gyges, whose reason is a slave to appetite, are simply not living a life expressive of the nature of a rational being.

Acts of injustice and wrong-doing are perpetrated by the mad or the ignorant. As was stated in the *Timaeus* (86), madness and ignorance are both diseases of the soul. To inquire, then, whether the unjust are happier than the just is as ridiculous (*Republic,* IV, 445) as asking whether the diseased are happier than the healthy. If they are, they are mad or fools or ignorant or children. We see, then, that the problem of conduct is but another side of the problem of knowledge. The unjust are those whose appetites rule their reason. Just as those limited to perceptual knowledge fail to know what they should, those driven by sensuous pleasures fail to do what they should. Moral relativism could be true only if the epistemology of Protagoras were correct, for if it were the case that each person is somehow the measure of all things, it would follow that each person's conduct would be justified on the basis of personal likes and dislikes. Meno's slave would have as much of a right to judge of triangles before instruction as after, even though his judgments would contradict each other. If Socrates must reject the epistemological relativism of Protagoras, so too must he reject the moral relativism of Thrasymachus, Glaucon, and Adeimantus, and for the same reason. The Pythagorean theorem does not rest its truth on public opinion. The laws of harmony are not subject to human desire. Both become accessible through philosophical examination, which will lead also and inexorably to a life of temperance, justice, and spiritual health. As Meno's slave needed instruction, so do all the citizens of the State. This line of argument, then, establishes the framework for the Socratic theory of governance.

The Problem of Governance

In his *History of Western Philosophy,* Bertrand Russell seeks to demythologize Plato's heroic standing:

> It has always been correct to praise Plato but not to understand him. This is the common fate of great men. My object is the opposite. I wish to understand him, but to treat him with as little reverence as if he were a contemporary English or American advocate of totalitarianism.[15]

Russell is quite correct in following the traditional description of Plato's *Republic* as totalitarian, but he courts the same difficulties as those facing all unwavering opponents of paternalistic governments. Plato was not ignorant of the virtues of democracy. He grew up in one quite as enobling, we may submit, as Russell's own.[16] He was also well aware of how similarly the citizen's life proceeds under forms of government with merely different names. Plato's respect for Sparta and his high hopes for the tyrants of Syracuse are not plausibly reduced to a fascistic temperament. Rather, in vesting full authority in the State, Plato merely drew the obvious implications from his psychological and metaphysical theories. As the soul is driven by rational, appetitive, and passionate faculties, so too is the *polis.*

So how does Plato answer Lord Russell? To begin, we observe that the just man and the just State are those living under the rule of reason; those able to control the appetites and harmonize them with the wisdom of the intellect. While men and states are born with the capacity for such harmony, the capacity is actualized only under the leadership and guidance of the philosophically enlightened. Without this guidance the pleasures and pains of the flesh, which are the only sources of control for the child, continue to dominate the life of the adult. Mere sensuous experience in this world of appearances only strengthens the body's hold on the soul, the body being the soul's prison (*desmoterion*).

The conflict between body and soul is further illustrated in Plato's *Philebus,* which includes a discussion of three different types of pleasurable and painful experiences. There are feelings of an entirely *bodily* nature such as an itch that can be soothed by scratching; there are feelings in which both body and soul participate, such as a painful hunger happily relieved by the anticipation of food; then there are the various feelings developed within the soul alone, such as longing and love (*Philebus,* 46b). In cases in which body and soul are jointly engaged, there is a strong cognitive element in the emotion, but there can be feelings of the body without this element.

The soul will be governed either by a harmony of its faculties or by the tyranny of one of them over the others. The State is similar. The question, then, is how and by whom and to what end are political communities to be organized and directed. To answer "democratically," "by the people," and "for their happiness" is to miss the entire thrust of Plato's social and philosophical analyses. There is, first of all, no conceivable state in which the will of every single citizen can be honored. Infants cannot be consulted, nor can the dumb, the mad, the criminal. Even if all were queried, we would find few decisions recommended unanimously. The possibility, therefore, of tyranny by the majority is inescapable. What benefit is there for the minority to have a voice which, though heard, is never followed? This can sow only the seeds of rebellion. Moreover, history offers little to support any prospect of "truth by majority." The same majority on whom Pericles conferred power was quick to turn on him, to accuse him of vile and venal things. Plato would recall for Lord Russell's benefit that it was with equal dispatch that the majority sentenced to death Socrates,

> concerning whom I may truly say, that of all the men of his time whom I have known, he was the wisest and justest and best. (*Phaedo,* 118)

In his counsel to kings, Plato represented the interests of the multitude, and his advice, he insists, would have been the same no matter what the form of government. But he was convinced of the difference between the best interests of the people and their knowledge of it. People can be happy in their vices, their madness, their ignorance, even their servility. Contentment can be no criterion of virtue in the State or the citizen. If true happiness, which is the harmony of the soul leading to a life of wisdom, justice, and temperance, is to be enjoyed by the citizens, they must be led to it. It will not be found accidentally. This is the proper end of all just republics. As it can be achieved only under conditions of peace at home and with adequate defenses against quarrelsome neighbors, the rule of law and the might of armies must be unchallengeable. Only education can enlighten and only the law can compel the citizen to receive instruction. If the young are to reap the benefits of philosophical guidance, their homes and games, their adult models and leaders, must be exemplary. The child who is taught the idea of justice cannot then be turned loose in a town whose adults are driven by lust, whose poets demean virtue, whose leaders live by deception and guile. If harmony is the goal, the sources of dissonance must be removed. Some of these are genetic accidents of nature (i.e., those born with infirmities) and must be "exposed" (i.e., exposed to the elements). Regulated breeding can reduce these accidents to a minimum. By the standards of contemporary democracies, Plato's imaginary republic seems harsh indeed. Yet, even in America, there have been laws proscribing consanguinous marriages, marriages of the mentally defective, and even marriages by those with certain hereditary defects. Public education is compulsory, curricula are established by state boards, religious rites are disallowed, genetic counselling is widespread, and the abortion of "defectives" is encouraged. The list could be expanded considerably. These are the now traditional practices of a twentieth-century republican democracy more liberal than any in history. Then, too, there are laws against obscenity, libel, treason, indecent exposure, and discrimination on racial, sexual, or religious grounds. Each of these proscriptions entails the constriction of personal freedom, and some are based on explicitly paternalistic and moral foundations.

Plato would be among the first to recognize that the "totalitarianism" is not an approach restricted to any specific form of government. It is, instead, that approach which vests in any person or corporate entity the power to control the religious, aesthetic, intellectual, and moral expressions of the rest of the citizenry. Since, on this analysis, only anarchy can free the citizen of *all* such constraints, any form of government control is dictatorial at least within those regions in which that control is exercised coercively.

Under any form of government there is an explicit or implicit covenant formed between citizens and their officials, for "[a] State arises . . . out . . . of the needs of mankind; no one is self-sufficing, but all us have many wants" (*Republic,* II, 369). Plato, we see, advances a Social Contract theory of sorts and certainly recognizes the factors that initially bring people into communal affiliations. If this motivation is ignored or frustrated by any subsequent government, either the government will be replaced or the social organization will fall to pieces. Thus, his "totalitarian" regime does not ignore the implicit covenant; it seeks to expand it, to extend the range of contributions the State can make to the life of every citizen while still remaining a State. That Plato never envisaged a State controlling every detail of daily life is clear from his discussion of early education and the later life of the well-bred citizen:

> When they have made a good beginning in a play, and by the help we have given have gained the habit of good order, then this . . . will accompany them in all their actions and be a principle of growth to them. . . . Thus educated, they will invent for themselves any lesser rules which their predecessors have altogether neglected. (*Republic*, IV, 425)

That he did not recommend an unbending discriminatory code based on class is even clearer:

> we never meant when we construed the State that the opposition of natures should extend to every difference but only to those differences which affected the pursuit in which the individual was engaged." (*Republic*, V, 454)

It is the problem of governance that finally requires a theory of knowledge based on eternal memories. The confrontation in the *Protagoras* makes this clear, and it is only in the *Republic* that Socrates finally extricates his argument from a web of irksome contradictions. In the earliest stages of his debate with Protogoras, Socrates succeeds in chastening the great Sophist. Representing the interests of his young friend Hippocrates, Socrates reviews the bases on which one studies with a master. If, for example, Hippocrates wished to be a sculptor, he would want to study with Pheidias. Or, if medicine were his goal, Asclepius would be the proper tutor. What is it, then, that Hippocrates would want to be were he to submit his mind to the influence of Protagoras? When Protagoras replies that Hippocrates would wish to excel in virtue, Socrates has his opening: Yes, but can virtue be taught?

Socrates' unconvincing dialogue with Protagoras is the result of a skepticism that is absent from the later *Republic*, which was composed during his middle period. In the early work, he is hoisted by his own petard, as it were, the moment he questions a pedagogical approach to virtue for, if it is not teachable, the entire philosophic movement loses its raison d'etre. In the *Republic*, however, we discover a more developed approach to the problem. Virtue, defined as a kind of harmony within the soul, is not taught any more than one is taught to hear harmony. The capacity to recognize harmony and distinguish it from dissonance is an integral feature of the undiseased senses. However, experience with music is necessary if this native capacity is to be realized and is to become useful to the listener. Virtue is the same. Except for those "accidents of nature," every person enters the world with a soul capable of comprehending the good. Under the guidance of an enlightened community, the child's moral development must pass successively through stages of enlightenment culminating in a love for and recognition of virtue. This notion, only touched upon in connection with space perception in the *Timaeus* and with geometry in the *Meno,* finds its fullest expression in the *Republic* and *Laws.* The philosopher is not so much a tutor or instructor as an educator, one who leads the pupil to a confrontation with the otherwise camouflaged truths of the eternal universe. The pilgrimage can begin only after a renunciation of the world of sense, only after the dialectical method has revealed the contradictions and sophistries that hitherto had passed for wisdom. Thus, in the most basic respects, the solution to the problem of governance is education (*paideia*), understood as acculturation. The character of the State and all its citizens is completely determined by the nature of this education. It cannot be optional, nor can it be administered by just anyone who happens to come along.

Hippocrates

Socrates and his disciples loom so large historically that we often neglect to note many other creative enterprises of the Hellenic era. Of these, Greek medicine may rank as high as Greek drama, for, in significant respects, the Hippocratic method was as deliberately original as were the literary methods of Sophocles, Aeschylus, and Euripedes. In fact, when we come to examine the biological turn taken by Aristotle's natural philosophy, we must keep in mind the background influences of the Hippocratic school and Greek science in general.

Although the Hippocratics are discussed only fleetingly in the dialogues, the medical scholarship of ancient Greece suffuses Aristotle's naturalistic writings at many points. We observed in the discussion of the pre-Socratics that no rigid boundary divided philosophical and scientific inquiry. Philosophy and physics, morals and cosmology, were but altered aspects of the same underlying *logos.* The Pythagoreans accepted this notion, so much so that their ethics and theology were thoroughly intermingled with geometry. It appears as if the Greek physicians shared this integrated conception of nature. Until the dialectical and speculative elements of philosophy were made dominant by the members of Plato's Academy, and until the "true forms" replaced observable nature as the subject matter of philosophical importance, medicine and philosophy were handmaidens to each other. Platonic idealism, whose central message included skepticism toward the evidence of experience, was less than compatible with both the practice of medicine and those findings uncovered in the medical clinics every day by a host of practitioners.

Hippocrates himself is hard to date. Plato's dialogues make reference to Hippocratics, suggesting that Hippocrates was older than the dialogues. It is customary to assign his floruit to about 400. What we have of his medical treatises we owe principally to Galen (c. 130–200 A.D.), but the oldest manuscripts available which purport to be translations of Galen's works are by authors living eight centuries after Galen himself.[17] While this renders more difficult the task of authenticating the various treatises attributed to Hippocrates, there is sufficient presumptive evidence from many different sources, including ancient ones, for us to reconstruct the essential character of the Hippocratic theory and overall approach.

The approach is uncompromisingly empirical. As philosophy became more speculative, the Hellenic physicians may have become ever less interested in philosophers' notions about health and disease. What survived of philosophy in Greek medicine were remnants of Pythagoreanism, which treated disease of any sort as a lack of harmony. Hippocrates and the famous oath that bears his name may well have substantial debts to Pythagorean teaching. As practical biologists, the followers of Hippocrates reasoned that the body itself, and especially the humors of the body, required a delicate and harmonious balance. It was not uncommon for delirium and fits to be treated through the combination of specific foods and music. Diet was especially integral to therapy. In the Hippocratic works on *Epidemics* there is frequent reference to fever and cold, to "fluxes" or storms of the humours. Similarly, remedies focus on sleep and rest, quietude and temperance. In the next chapter, we will see that Aristotle's theories of memory and perception also take recourse to the notion of restlessness or quietude of the mind and the perceiving faculties. But more important than the specific observations made by the Hippocratics was their commitment to observation. Greek medicine was quite advanced. It served as a constant reminder that the world of sense and the empirical method of fact-gathering were suspended only at the peril of those who wished to learn and understand.

Hippocrates (or whoever is responsible for the teachings attributed to him) had recorded the fact that injuries to either side of the head resulted in spasms on the contralateral side.[18] There are also Hippocratic treatises on the venous supply to the brain and a nearly modern respect for the role of this organ in perception, movement, and any number of psychological processes and functions. The Hippocratic *Humoural* theory of psychological dispositions accorded an unchangeably phlegmatic nature to some (and therefore to their children!), a bilious one to another, a choleric to yet another. The justification for such attributions was not rigorously empirical but was surely based on family resemblances in the matter of temperament. Indirectly, such a genetic type-theory of personality lent support to the eugenic theories of Plato's Academy. Still, in the main, Hippocrates and his followers come close to the modern spirit of experimental science and clinical observation. They specifically rejected that Platonic version of "hypothesis" according to which all discourse must begin with self-evident truths, regarding such hypotheses as antithetical to the good care of patients and an understanding of their diseases. In the place of these the

Hippocratics compiled a veritable handbook of symptoms, therapies, and results. In the process, and through the translations and the influence of Galen, they had an effect on the practice and theory of medicine for over two thousand years. More subtly but just as surely, they required of any psychological philosophy that it address itself to the biological facts of human life and to the relationship between those facts and any theory that might be advanced to explain a psychological process.

Résumé

The Socratic philosophers, whose achievements were immortalized and "rejuvenated" by Plato, advanced many of the central problems with which all subsequent psychological scholarship would have to contend. They insisted on a distinction between the factual knowledge gleaned by error-prone perception and that knowledge of general principles made possible by reason alone. Combined with this was the assertion that knowledge, properly understood, is a recollection of that which is implanted within the soul (mind) natively. These theories have modern counterparts in the form of theories of cognition and genetic theories of intellect and emotion. The Socratics, to a far greater extent than one finds in contemporary psychology, emphasized the part played by the political and moral dimensions of the State in forging the character of the young. Their recognition of the interdependence of politics, morality, art, and psychology awaits recovery. Coeval with the development of idealistic philosophies was the development of Greek medicine along rigorously practical and observational lines.

Notes

1. On the Greek religious attitude, consult W. K. C. Guthrie's fine study, *The Greeks and Their Gods,* Beacon Press, Boston, 1950. My interpretation is not, however, identical to Professor Guthrie's.

2. On the influence of Miletus, consult Herodotus, Book I, 17–22; and Book V, 23–25. See also *The Cambridge Ancient History,* edited by J. S. Bury et al., Macmillan, New York, 1926, pp. 87–97; James Henry Breasted, *A History of Egypt,* Charles Scribner's Sons, New York, 1912; and Kathleen Freeman, *Ancilla to the Pre-Socratic Philosophers,* Basil Blackwell, Oxford, 1952.

3. Plutarch, *The Lives of the Noble Grecians and Romans,* translated by John Dryden, 1864 ed.; reprinted by Random House, New York, 1932. Xenophon's *Memorabilia* includes a sketch of Socrates that is important for corroborative purposes. It was written at least thirty years after the death of Socrates.

4. The Second Epistle, from which the quotation is taken, is of questionable authenticity. But the very same modesty is expressed in the Seventh Epistle (Sec. 341) whose authenticity is well established. In translating the Second Epistle, we may substitute "idealized" for "embellished." On this reading Plato may be admitting that he did more than merely adorn the teachings of Socrates.

5. For the collection of Plato's *Epistles*, consult Glen R. Morrow, *Plato's Epistles*, Bobbs-Merrill, Indianapolis, 1962.

6. The edition of the dialogues used in the present work is *The Dialogues of Plato*, 2 vols., translated by Benjamin Jowett. Random House, New York, 1937. Here and following, however, the parenthetical page numbers refer to the standard Greek manuscript by Stephens.

7. Ibid.

8. Ibid.

9. Ibid.

10. Bertrand Russell, *A History of Western Philosophy*, 14th ed., Simon and Schuster, New York (first printing, 1945). See especially pp. 10–105.

11. *The Dialogues of Plato*, p. 874. The reader will gain useful insights into Platonic philosophy by reading Professor Jowett's Index. The Index is not merely an alphabetical list of terms but is, in addition, an analysis of the meaning of the terms in context.

12. Ibid.

13. Ibid.

14. Ibid.

15. Bertrand Russell, *A History of Western Philosophy*, p. 105.

16. See, for example, *America: 1938–1944*, Vol. II of *The Autobiography of Bertrand Russell*, Atlantic Monthly Press, Little, Brown, Boston, 1968. Lord Russell confronted an intellectual intolerance quite as severe as we might have found in the Athens of Pericles.

17. The authoritative edition of the Hippocratic system is the translation by W. H. S. Jones, *Hippocrates*, 3 vols., Putnam, New York, 1923. Professor Jones's introduction is most useful.

3 The Hellenistic Age: Aristotle, the Epicureans, and the Stoics

Aristotle (385–322): His Philosophical Development

Vico's warning against the glorification of antiquity tends to fall on deaf ears when the full sweep of Aristotle's achievement presents itself. Here is a philosopher-scientist who has had a more direct and enduring effect on more departments of scholarship than any other single figure. More than one reverent historian has described him as the last person to have known everything that was known in his own time.

What we know of his life is spotty. He was a devoted student of Plato's for some twenty years, perhaps from the founding of the Academy (367 B.C.) until the master's death (347), although he participates in only one of the dialogues, the *Parmenides*. He was born in Stageirus, on the eastern edge of the Chalcidice, and was of Ionian heritage. His father was personal physician to the Macedonian king, Amyntas II, father of Philip II and grandfather of Alexander the Great. Thus, a measure of the prestige enjoyed by the Academy can be gleaned from the fact that a family of means and connections would send their seventeen-year-old son there for instruction.

Following Werner Jaeger, it is customary to separate Aristotle's philosophical development into three stages.[1] In the first, he is a Platonist to the core, the author of now lost dialogues, the student of the "ideas." The next phase of his life comes when the direction of the Academy falls to Plato's cousin, Speusippus, in 347. This was either the cause or a mere coincidence attending Aristotle's departure from Athens. During the following twelve years (347–335), his busy life included his marriage to Pythias, his teaching of Alexander (334–336), his direction of the group of Platonists at Assos, and what must have been a reassessment of his own position on basic philosophical and scientific questions. The third phase begins with the death of Philip II and the accession to power of Alexander (335). Now Aristotle returns to Athens, establishes his own school, the Lyceum, and devotes the remaining thirteen years of his life to that body of thought posterity would call "Aristotelianism" (although the latter would come to include any number of features otherwise not visible in Aristotle's own surviving works). On the news of Alexander's death in 323 Aristotle retired to Chalcis in the Euboea, recognizing that the Athenians would now be free to express their fervent nationalism by attacking any and all Macedonian sympathizers. We are told that he left Athens so the city "would not sin against Philosophy a second time."

Indirect evidence suggests that Aristotle arrived at the Academy after the Platonic circle had completed the *Theaetetus* and while it was engaged in the *Parmenides*. Werner Jaeger has argued convincingly that it is unlikely the challenges offered by the Aristotele featured in this dialogue would have come from a teenager from Stageirus.[2] Indeed, though we do not have Aristotle's own early dialogues, we have considerable circumstantial evidence that they were consonant with the prevailing theories of the Academy. Several of them were, we may believe, merely updatings of older Platonic works (e.g., the *Eudemus* vis-à-vis Plato's *Phaedo* and the *Gryllus* vis-à-vis Plato's *Gorgias*).[3] Among these earliest works, we do have fragments of the *Protrepticus* ("striving"),[4] which, although not a dialogue, is nearly orthodox Platonism. It is in this work that we find a defense of Anaxagoras' commitment to "contemplate the heavens" and recourse to the authority of Pythagoras. The world of sense is respected, but the world of idea is proclaimed supreme:

> Hence we should do all things for the sake of the goods that reside in man himself, and of these, that which is good in the body we should do for the sake of that which is good in the soul. . . . [W]isdom is the supreme end.[5]

> [W]e ought either to pursue philosophy or bid farewell to life . . . because all other things seem to be but utter nonsense and folly.[6]

Mingled with his devotion to the soul and its ideas, however, are the subtle intimations of a coming revolution:

> . . . if life is preferred and valued on account of sense-perception, and if sense-perception is a sort of knowledge. . . .[7]

It is clear that, on the whole, Aristotle's earliest period of intellectual commitment was as a Platonist. The revolution he initiated in his later years might be seen not so much as an attack on Platonism as it was a unique critical synthesis. It was quite common in the centuries following the death of Plato and Aristotle for philosophers to regard their two schools as complementary and compatible, not at all at odds at the most fundamental level.

Nonetheless, there are differences worth noting between the Aristotle of the missing dialogues and the Aristotle known through the surviving works of 335–322. First, the earliest Aristotle is a pupil in the Academy confronting the already traditional problems of philosophy, but through a medium and format not readily applicable to the Hellenistic Age. Aristotle, considering both his own interests and Alexander's commitment to spread and impose Greek culture throughout the empire, must have recognized that the dialectical method was less than useful. The materials prepared for instruction at his Lyceum would be lectures in scripted form, not the fruits of late-night colloquies.

Aristotle departed from Platonic tradition also in the very range and details of his inquiries. There can be no deeper study than what is found in Plato's dialogues, but they are often lacking in the substantive content now taken (thanks to Aristotle!) to be the essence of scientific and scholarly pursuit. This gap between general principles and factual information is one that Aristotle's teaching strived to fill. Where the Platonists of the *Republic* politely argued a constitution into being, Aristotle examined some 158 different and actual constitutions, analyzing their premises and the factual evidence standing in their support. Of this effort only *The Constitution of Athens* has turned up. The exacting study brought to bear on it by Aristotle and his students suggests how monumental an effort went into research on other constitutions. Again, where Timaeus in Plato's dialogue would content himself with an explanation of space perception based only on a "spurious kind of reason," Aristotle would devote entire books to the subjects of vision, touch, taste, sensory integration, and the senses of nonhuman animals.

Finally, the Platonists were too willing to dismiss all but abstractions as irrelevant. Neither Aristotle nor his contemporaries could share this luxury. Those rare passages in Aristotle that address Platonism in a contemptuous tone are invariably directed at the exclusionary elements of the Platonic program. This is hinted at in the two versions of Aristotle's *Metaphysics:* an earlier one in which the first-person plural is used in the discussion of ideas, and the later version in which the third-person plural is used.[8] This grammatical shift is of significance. In the later version those who subscribe to a separate sphere of reality, the sphere of ideas, and who regard the sphere of appearance to be but a distortion of the truth, have become "they," where earlier they had been "we."

In his *Metaphysics,* Aristotle praised Socrates for developing inductive reasoning and for advancing the theory and the problem of universals.[9] In the *Theaetetus* and the *Meno,* Socrates distinguished between perception of particular instances and general ideas, the latter of which cannot be received through perception. The word "cat" is illustrative. We can see a particular cat but we only know (cognitively) that it is a cat because it answers to the description given by our general idea of cat. In the simplest terms, the problem of universals is this: since no single cat can be so identified unless there is some supra-individual class (i.e., a universal) of which it is an instance, is the universal cat real?

In later Scholastic philosophy, those who accorded real existence to such universals were known as *realists,* and those who insisted that terms such as "cat" were simply names given to particular things were dubbed "nominalists." Now, Socrates did not suggest that somewhere in the cosmos a perfected, four-legged, mewing ideal cat sits sipping ideal milk. He was not that sort of realist, at any rate. What he did suggest, however, was that true knowledge of what a cat is depends on the true idea of the cat as a universal; the *true form* of cat is not something any particular cat can match. Since any particular cat can be no more than an approximation to the ideal, our perceptual knowledge of cats can only be an approximation to true knowledge of cat. But perception will only be of the attribute and, unaided, will not discover the ideal. The ideal, of course, is the idea: that in which each particular instance shares but only shares imperfectly and incompletely.

On Aristotle's account, Socrates was credited with retaining the connection between a universal class and its perceptible instances, but he rebuked the later Platonists for dissolving this connection:

> Socrates did not make the universals or the definitions exist apart; they, however, gave them separate existence, and this was the kind of thing they called Ideas.[10]

Aristotle would resist such idealistic notions in attempting to ground his own system of philosophy in the observable facts of animal and inanimate nature.

Aristotle on the Soul

Aristotle's most sustained discussion of human and nonhuman psychological processes is developed in the treatise *On the Soul* (*De Anima*), a treatise whose conjectures are intended to be fully "compatible with experience" (402b). What is to be avoided is a method of inquiry that is merely based on argument (*dialectikos*), another sign of his movement away from the traditions of the Academy.

In the very first chapter of the treatise he identifies the emotions (anger, courage, appetite) and the sensations as conditions of the soul and insists that they can only exist through the medium of a body. All such "affections" of the soul are to be conceived as "materialized formulable essences," which is sometimes more transparently translated as "properties derived from the body." The full sentence in which the thought is included is translated as,

"Clearly it follows from this that the soul's empathic feelings are to be understood as properties deriving from the material body" (*De Anima,* Book I, Ch. I, 403a; 24–25). Simply stated, Aristotle is insisting that an understanding of human and animal psychology depends upon and is informed by our knowledge of the material (biological) conditions of life.

The essentially physiological character of his psychology appears frequently throughout his works, including those that are not particularly concerned with psychological matters. In his *Physics,* for example, he argues that the affairs of the soul are brought about "by alterations of something in the body" (248a). In the same work he opposes the Platonic notion of intellectual processes as turbulent and offers, instead, the theory of quiescence: perception and intellect, if they are to be precise and organized, require that the soul approach a condition of rest. He attributes the inferiority of perception on the part of children and the aged to a "great amount of restlessness," and all such deficits are to be explained as some kind of alteration "of something in the body." [11] In his treatise on sleep and dreams (*De Somniis*) he again accounts for the phenomena by appealing to sensory-biological processes and offers the surprisingly modern hypothesis that dreams are the result of our conscious experiences and our emotions.[12] Even that most complex of psychological features, memory, is treated in purely biological terms. In *De Memoria et Reminiscentia* (*On Memory and Reminiscence*), which we will note again shortly, he discusses recollection as the "searching for an image in a corporeal substrate" (453a) and again explains the memory deficiencies of children, the aged, and the diseased in terms of biological anomalies.

When we study the philosophers before Aristotle, we can uncover, here and there, a psychological orientation that is naturalistic and biological in tone. This is certainly true of several of the pre-Socratics and it is also true of particular passages in the *Timaeus.* However, no predecessor could lay equal claim to the title of physiological psychologist. Aristotle was the first to specify the domain embracing the subject matter of psychology and, within that domain, to frame explanations in terms of the presumed biology of organisms. That the entire body of Aristotelian psychology, not to mention philosophy, does not fit into a radical materialistic mold is clear; the philosopher himself goes to some lengths to make it clear. But on the narrower issues of learning, memory, sleep and dreams, routine perceptions, animal behavior, emotion, and motivation, Aristotle's approach is naturalistic, physiological, and empirical.

At the same time, he rejected the extreme materialism of Democritus, who judged soul and mind (*nous*) to be the same and insisted that both were reducible to atoms (*On the Soul: De Anima,* 405a). At 407b he dismisses the Platonic theory of soul-as-harmony or as proportion, since such hypothesized harmonies are not evident in biological systems. And at 408b he specifically distinguishes soul from mind, the latter dwelling somehow within the soul but, unlike the soul, being imperishable. That mind itself is not merely certain mental faculties is clear from his treatment of the process of senility:

> [I]n old age the activity of the mind or intellectual apprehension declines only through the decay of some other inward part; mind itself is impassible. (408b)

Aristotle was not satisfied with any of the traditional theories of the soul, and he devoted Book I to a critique of them. He concluded by attributing their deficiencies to the error of believing that there is only one function of the soul. He, on the contrary, noted that biological systems exist in varying degrees of complexity and suggested that for each essential function there must be a corresponding psychic (i.e., motive) principle. The functions identified by Aristotle were the nutritive, the perceptive, the locomotor, and, in human beings alone, a "universalizing" faculty (*epistemonikon*) by which abstract principles can be comprehended. Any animal possessing an advanced function (e.g., intellect) will also have the less advanced functions (perception, locomotion, and nutrition). Each of these functions may express itself in one faculty or in a variety of faculties. Perception may, for example, be limited to touch, or all five senses may be well developed (Book II, Ch. 3). Even in the case of rationality, there may be only mere imagination or one may also possess the powers of calculation as well.

When Aristotle begins his *Metaphysics* with the observation that "all men by nature desire to know" and adds immediately that an example of this "is the delight they take in their senses," we begin to see the importance that will be attached to perceptual modes of knowing. We learn in the treatise *On the Soul* that sensation may be taken as part of the essential definition of "animal" in that every member of the animal kingdom is presumed to have sensory experiences. This is a major feature of Aristotle's general psychology. Nature does nothing without a purpose; animals of all kinds have special sensory organs; animals of all kinds must have the experiences such organs are clearly designed to mediate.

The perceptual theory developed by Aristotle emphasizes sensory *integration,* a process which is itself not sensory. Consider a cup of coffee. It is a hot, lightly sweetened, dark brown, liquid. The various specialized sensory organs respond respectively to each of these attributes: the temperature, taste, color and consistency. But the *experience* is not of separate attributes. Rather, there is a whole, fully integrated experience such that coffee will be distinguished from any number of other stimuli that also happen to be dark, hot, sweet liquids. So, in addition to the "five senses" delineated by Aristotle, there is also in his theory a *common* sense (*koine aisthesis, sensus communis*). This latter is not to be taken as itself a sense but as some process common to all the senses, by which experiences are forged out of the separate contributions from the different sense organs.

After reviewing the general and special character of the five senses in *On the Soul* (Book II), Aristotle concerns himself with the mind and with the problem of how informed thought ever occurs in the first place if, indeed, it is not native, as the theory of Ideas supposed. His solution was to distinguish the actual from the potential: mind has the potential for thought, but for this to be actualized, it must be acted upon by the world. In a sentence that might have come from John Locke in the seventeenth century, Aristotle describes the mind prior to the experience and learning thus:

> What it thinks must be in it just as characters may be said to be on a writing tablet on which as yet nothing actually stands written: this is exactly what happens with mind. (429b–430a)

On this account, the mind is taken to be a collection of complex processes supplied with information via the senses from the outside world. Without such mechanisms and processes, there would be nothing for the perceptual-cognitive processes to work on. Thus, the external world must cause physical responses in the sensory organs, these responses coming to depict or represent or stand as codes for the objects that cause them.

To this day scholars debate the extent of Aristotle's commitment to an essentially materialistic psychology. Was it in fact his position that all aspects of mind can finally be reduced to properties of the body such as those just cited? Much of *De Anima* certainly lends support to this conclusion, but there are telling exceptions, or at least suggestive ambiguities. The theory we assign to Aristotle will depend in part on how we understand his overall psychological system and also on how one translates several of the crucial terms he employed in his analysis. For example, when mind is discussed, such words as *nous, psyche,* and *epistemonikon* are used, but these are not synonyms. As is clear in *De Anima* and in his essays on natural history, Aristotle acknowledged the intellectual powers of the advanced species. His careful ethological observations led him to conclude, as Darwin would centuries later, that animals possess not only the "psychic" faculties of nutrition, reproduction, sensation, and motion, but also intelligence, feelings, and motives. What he reserved to human beings is that power or faculty that goes beyond the ability to learn and remember specific things; a faculty that rises to the ability to grasp universal concepts. This is evidence of reason proper (*nous*), the capacity for abstract rationality.

The term for this, *epistemonikon,* as it appears in *De Anima,* does not lend itself to easy translation. It is best rendered by the phrase, "that by which universals are grasped." Most interestingly, Aristotle asserts that the faculty in question does not move: "epistemonikon ou kineitai." Recalling that, on Aristotle's account, the very concept of matter is equated with motion, we are led to conclude that Aristotle is proposing a unique faculty of the soul; one possessed only by the human soul, and one that is incompatible with the very essence of matter. On this construction, it may be argued that Aristotle's theory of mind is not uniformly materialistic but *dualistic* at the highest level of rational activity. The issue remains controversial in Aristotle scholarship, as does the entire question of mind-body relationships, a question that will be confronted frequently in later chapters.

Aristotle on Learning and Memory

Except for the power of abstract rationality and the political and moral possibilities created by it, Aristotle's theories of human and nonhuman psychology are thoroughly naturalistic and biological. By "naturalistic" is meant that the assumed processes and principles governing psychological life are drawn from those operating in the biological world at large. Nothing "supernatural" or spiritual needs to be invoked to account for the phenomena. Aristotle's approach is also broadly ethological. What is to be explained is the manner in which the special physical design of creatures results in powers and functions that render them fit for the forms of life they characteristically live. Aristotle had a passion for observation and for classification. His overarching conception of nature was *teleological* in that he took the operations of the natural world to be aimed toward an end, or *telos*. Accordingly, his explanations of natural phenomena are generally framed in the language and logic of teleology. Animals and human beings do something *for the sake of* attaining something or *in order to* achieve a given goal. Similarly, the inanimate world is governed by laws and this very fact indicates a rational design and purpose. Various processes and functions are then understood according to the larger objectives or goals they serve.

The basic principles of learning are developed in one of Aristotle's already cited briefer works, *On Memory and Reminiscence*. In this essay Aristotle proposed repetition as the source of the strength of memories and proposed an associational mechanism by which items to be learned are held together. According to this theory, sensations set up certain motions within the soul. These subside in time, but, if they have been produced often enough, they can be re-created or, at least, a likeness of them can be re-created under comparable conditions in the future. Through repetition (by custom or habit), certain movements reliably follow or precede others. Our attempts to recall past events are only attempts to initiate the right *internal* events. This is why, when attempting to recall a sequence of events or objects, we must find the beginning of the appropriate series. When we do, an entire train of previously established associations is set in motion.

Aristotle advanced similar notions in his treatise on natural science (*Physics*), where he claims that all sentiments are finally traceable to sensory experience, going so far as to bring the moral sentiments themselves under the control of biological processes: "[A]ll moral excellence is concerned with bodily pleasures and pains" (247a). Thus, to the principle of associationism he adds a "pleasure principle" and thereby moves even his moral philosophy some distance from the traditional teachings of the Academy.

If all we possessed of Aristotle's works were these treatises on the soul and on memory and reminiscence we would conclude that Aristotle's theory of knowledge was empiricistic, his theory of learning associationistic, and his psychology essentially a form of behavioral biology. In a word, his psychology would be more or less contemporary. That this is not the case is established by his other works; his writings on logic, politics, ethics, and rhetoric. It is in these that we find a narrowing of the differences between his thought and Plato's. We understand this aspect of his psychological theory by examining his own approach to the problem of knowledge and the knowable.

Aristotle on the Problem of Knowledge

Aristotle's views in different treatises might make him seem inconsistent, for in some the emphasis is on experience and practice, in others on certain innate dispositions of the mind. The mission of the different treatises vary, however. In *De Anima,* for example, where it is claimed that, absent perception, there would be no knowledge at all (423a 3–10), Aristotle examines questions of an essentially psychological and psychobiological nature. Emphasis here is placed on perceptual knowledge. In his *Posterior Analytics,* however, the subject is scientific methodology. Whereas *De Anima* was devoted to natural processes, the latter is concerned with rules of evidence and argument. In *De Anima* attention is focused on the mechanisms of perception and learning, motivation and emotion. In *Posterior Analytics* we discover how a world of continuous changes can still be known with certainty.

Aristotle was well aware of the theory of fluxes advanced by Heraclitus. He knew also that the Academy had turned away from the senses in part because of this theory. Heraclitus had drawn attention to the ever-changing nature of the sensible world and argued for a relativistic attitude toward knowledge. The Socratics, however, found little merit in a life of study that produced no more than probabilities. So they looked behind the shifting world of Heraclitus, in which "no one enters the same river twice," and discovered an "ideal" world of true forms and timeless mathematical relations.

In the *Posterior Analytics* Aristotle, too, attempts to locate abiding truths, but without subscribing to the received theories of the Academy. The position he takes instead begins to unfold toward the end of Book I:

> Scientific knowledge is not possible through the act of perception. . . . . one must perceive [things] and at a definite present place

and time: but that which is commensurately universal and true in all cases one cannot perceive. (87b)[13]

Now, to know "that which is commensurately universal and true," we must have certain standards of truth; standards of an essentially cognitive nature which allow us to organize the teeming and shifting facts of the physical world. That is, to be knowable, any physical object or event must tap or trigger some disposition or capacity of the mind which will give to the object or event features of a cognitively meaningful sort. Aristotle does not suggest that objects and events lack such features and that the mind invents them. Rather, he reasons that, to be knowable and known, these objects or events must have features somehow compatible with our very methods of knowing.

The features of the external world that can be known fall into ten categories[14]: (1) substance (according to which a thing is), (2) quality (the thing is white), (3) quantity (the white thing is three feet tall), (4) relation (the white thing is taller than the red thing), (5) place (the thing is there), (6) time (it was there yesterday), (7) position (it is standing in the corner), (8) state (that horse is shod), (9) action (he lanced the blister), and (10) affection (he was lanced, i.e., affected by the lance). These categories are applied to objects and events in order that we might classify, judge, compare, and locate them in space and time. They are the exhaustive categories by which any object or event can be known. Our perceptions are drawn from these categories, but are not the categories themselves. Indeed, were the categories not available prior to the act of perceiving, the objects of perception could not be classified according, for example, to time, place, quality, and the rest.

Aristotle's *Categories* was intended as an essay on logic. It does not purport to be a cognitive or psychological treatise. But even in the nonpsychological context in which Aristotle introduced them, the categories appear as mental dispositions, as principles according to which the world of matter and the world of knowledge merge, or at least enjoy the possibility of becoming integrated. The categories, however, are just that: categories of knowledge and not the scientific knowledge of that "which is commensurately universal and true."

It is only through what Aristotle called demonstration that such scientific knowledge becomes possible. As the term "demonstration" is employed in modern times, one might be tempted to think that Aristotle was advancing a purely empiricist method for obtaining true knowledge. This is not the case. Scientific knowledge, according to Aristotle, is *demonstrative* knowledge, not in the modern sense of experimental demonstration, but in the sense of rational or logical demonstration. At first blush, this position seems naive and characteristically ancient. But Aristotle was a tireless observer. Recall that he invented embryology by breaking open chicken eggs at different stages of development[15] and that he established the shape of the earth by observing its shadow on the moon during eclipses.[16] This was not a scholar who thought logic was a substitute for the hard work of data-gathering. He was a gifted scientist, but he was a philosopher first, and as such he knew the difference between mere facts and demonstrative proofs.

The demonstrations Aristotle sought were grounded in the syllogistic modes of argumentation that he invented. The major premise in such an argument may be a law of nature, the minor premise, a fact of nature, and the conclusion, a necessary and demonstrated conclusion. Socrates is not said to be mortal merely because he died. He was known to be mortal while he lived. All persons are mortal and Socrates is a man. His death, on this account, is not a mere fact but a reasoned fact. Scientific knowledge, then, differs from perception by possessing the general (universal) principles which cover each and every particular instance.

The search for general laws is guided by another, central feature of Aristotle's theory of knowledge, its *teleological* character. To explain a phenomenon requires more than an identification of its antecedent cause. It requires the identification of its purpose, the very point of it, how it fits into the larger scheme of things. As he notes in his *Physics,*

> It is absurd to suppose that purpose is not present because we do not observe the agent deliberating. Art does not deliberate. If this ship-building art were in the wood, it would produce the same results by nature. (*Physics,* 199b)

A physical-causal law cannot supply the full explanation of an event, for the causal law will not reveal the goal or plan or purpose to be realized by such lawful relationships. The most developed forms of human knowledge, then, are not confined to the scientific discovery of general laws but must go beyond these to an essentially *metaphysical* inquiry into reasons and purposes. Nature is not wasteful or pointless. The perfect orderliness of the heavens, the cyclical repetition of the seasons, the sequence of birth, growth, death and decay—the entire spectacle of nature conveys evidence of intelligent design. To explain anything completely, therefore, is to demonstrate how it fits into the grand design, and this calls for rational powers and methods that go beyond mere observation and reach universal principles.

Aristotle on Universals

The problem of the universals is an abiding one, "solved for all time" in every period of heightened philosophical speculation. Aristotle recognized that at least one form of it is merely semantic. (Since the thirteenth century, there have been philosophers contending that it is *only* semantics). Still, there are perceptual and cognitive dimensions of this problem, and a good many psychologists have staked their research and theories on the premise that the problem is more than a quibble about words. We can catch a glimmer of these cognitive and perceptual elements by recalling the "universal cat" and determining just how one might go about teaching a three-year-old what a cat is.

Which animals should be called "cat," "dog," "wolf," "lion," etc.? The first and the simplest step is to bring in the family pet (a cat in this case), show it to the child and say "cat." If this is done often, we have it on the very best authority that most children will learn to say "cat" anytime Tabby is displayed. But suppose this pet, a large gray one, is replaced by a Siamese. Will the child still say "cat"? If so, what was learned in the first place? In just a few years the same child will correctly identify all species of things despite great intraspecies differences in size, weight, color, temperament, and condition of health. The temptation, to be sure, is to attribute this successful performance to "stimulus generalization" (or induction, as Aristotle would say). But to generalize, the child must be in mental possession of some class or genus over which particular instances can be generalized. In other words, we run the risk of accounting for the performance by assuming the very "universal" we set out to eliminate.

To examine Aristotle's approach to the problem we return to the *Posterior Analytics,* noting that the treatise begins with the declaration that "All instruction given or received by way of argument proceeds from preexistent knowledge" (99b–100a). It ends by addressing the sense in which knowledge is preexistent. It is in the last chapter that Aristotle is found defending a nativistic theory of knowledge while seeming to reject it. First he scorns the nativistic account and then notes its compelling element:

> Now it is strange if we possess them [states of knowledge] from birth for it means that we possess apprehensions more accurate than demonstration and fail to notice them. If on the other hand we acquire them and do not previously possess them, how could we apprehend and learn without a basis of pre-existent knowledge? (99b)

His only answer is that we do not have the knowledge at birth (thereby rejecting the nativistic theory of the Ideas), but we do have the capacity which is subsequently actualized through sense perception. And, while the senses themselves only detect the particular, their "content is universal" (100b), meaning that the mind is able to construct the universal from the data of experience. This is not orthodox Platonism by any means. Nonetheless, as a solution to the universal-particular aspect of the problem of knowledge, it is a compromise of the sort found in the *Theaetetus.*[17] The senses do not convey knowledge per se, but they do convey that from which the reasoning can extract knowledge.

Aristotle and Platonism

It is clear both from his *Nicomachean Ethics*[18] and from his *Rhetoric*[19] that Aristotle envisaged human psychology as an evolving process. The Platonists were two-stage theorists in their treatment of the mind. This followed naturally from the theory of Ideas. Since the Ideas are eternal and are present in the soul before birth, there are only two possible psychological states: one in which the individual is ignorant or mad, and one in which the person is enlightened. The senses contribute nothing to true knowledge, and so the mere experiences of a lifetime may find the old person as deeply in the cave as a child. Aristotle, forfeiting this binary conception and willing to accept the facts of everyday experience, advanced a dynamic theory of psychological development. As an associationist, he emphasized the role of practice and of rewards and punishments in learning and memory. As a social observer, he noted the differences among the young, the mature, and the aged as regards emotion, reason, courage, loyalty, and motivation. He knew that senility brought a decline of the faculties of perception and, consistent with his larger theory, that this necessarily produced changes in all spheres of intellectual endeavor. For Aristotle, growth and decay were the abiding correlates of the natural world, including human psychological prowess. While some rational principle may survive the grave, the individual soul and its more prosaic faculties would not.

The origins of reason were intermingled with our biological character, with the progression of our development, with the incessant commerce between mind and sense. Things that exist, exist for a purpose, and our purpose is to be, to become, to reason, and to die. That there is a "prime mover" responsible for setting the machinery in motion, a "cosmic reason" that has provided the overall plans and purposes of reality, was taken by Aristotle as established by the very order of the heavens. But with all this, Aristotle never conveyed the optimism of the Socrates of the *Phaedo*. For Aristotle, the affairs of man and of the prime mover are separated by an unrelenting breach. There is a plan, but one not likely to disclose itself fully. Reason evolves, experience teaches, and societies come and go. The best strategy is to live a rational form of life and realize those potentialities that are part of one's essential nature. Stoicism is on the horizon.

The Problem of Conduct

In Book VII of his *Politics,* Aristotle presents the principles of statecraft and those of individual excellence as virtually indistinguishable.[20] If Plato chose his Republic to serve as an enlarged model of the individual, Aristotle's attention to the individual is almost a preliminary stage leading to a discussion of the State. In his *Nicomachean Ethics,* man is first "a political creature whose nature is to live with others" (1169b). Every action of the individual (1094a) and the State (*Politics,* 1252a) aims at some good; otherwise, the action is involuntary and therefore to be judged either as accidental or performed in ignorance (*Nicomachean Ethics,* 1110b).

This position, that all voluntary actions have some good as their goal, is but an instance of Aristotle's general teleological theory of nature. Over the centuries this theory has fallen on hard times, often because of a failure to appreciate either Aristotle's statement of it or the facts of nature it was designed to explain. Since it is a central aspect of his social, political, and psychological speculations, it warrants further attention here.

In Book I of the *Posterior Analytics* (especially 73a–73b), Book II of the *Physics* (196a–197b), and Book I of the *Metaphysics* (all), Aristotle states his position on causation and on the theory of necessity. His position is maintained consistently, so for convenience we can examine the version of it offered in the *Physics*. We begin with the second chapter of Book II, where there appears an almost perfunctory acknowledgement that things can be said to be understood only when their causes are known. Aristotle offers the distinction between knowing *that,* and knowing *why* (194b), and then goes on to discuss the different senses in which we say we know the "why" of an event. There are four general senses of "why" in the explanations we offer, and these can be illustrated by explaining a statue, say, Bernini's "Fountain of the Rivers":

1. If we examine a statute, there is a sense in which we attribute the cause of it to the substance of which it is made, for example, stone. Here, we have what Aristotle referred to as the "material cause."

2. The difference between a lump of stone and a statue is that the latter has a certain form; not any single uniform one, but one that is not random. If we refer to this as the cause of the statue, we are invoking the Aristotelian notion of a formal cause. The same is true when we answer questions such as, "Why is the sum of the angles of a triangle 180 degrees?" The answer is that a triangle is defined as a figure containing 180 degrees; that is its essence, or its "formal cause."

3. Returning to the statue, one might attribute the cause of it to the changes produced by the hammer and chisel. Blow by blow, these changes lead to the finished work. A causal explanation based upon these actions expresses what Aristotle called the "efficient cause."

4. But if we stand in the Piazza Navona and ask someone the cause of the "Fountain of the Rivers," it is not likely that the reply will include references to stone, or definitions, or hammering. More likely a one-word answer will be given: Bernini. The cause of these statutes is the sculptor's vision or genius: that which he intended even before selecting the stone, the site, or the cutting tools. In this sense, the cause of the "Fountain of the Rivers" is the goal or end of the artist.

This last is what Aristotle referred to as the "final cause." The concept of final cause is the teleological element in Aristotle's theory of causation and the element that medieval theologians and philosophers exploited in giving Christian belief a scientific cast. Science, at least since the influence of David Hume (see Chapters 7 and 10 below), has been more or less confined to the study of efficient causes, appraising any form of teleological explanation as "metaphysical," if not superstitious. Nevertheless, we must contend with final causes here, as, indeed, we must in nearly every walk of life, even in experimental science, which also is understood in terms of its goals and purposes. Indeed, as long as we hold ourselves and others responsible, we are accepting an Aristotelian notion of final causes. The cause of a praiseworthy or blameworthy actions is taken to be an ulterior goal or purpose. The goal itself need not be good in some objective or unarguable way. It is sometimes alleged that Aristotle's final cause refers only to good intentions, but he makes it clear in a number of places that the final cause, the "that for the sake of which something is done," may be good or only apparently good (*Nicomachean Ethics,* Book III, Ch. 4).

Aristotle reasoned that since final causes are common in the affairs of the human world—itself part of nature—nature, too, then must possess final causes. On this account, birds build nests in order to care for their young, and elephants kneel in the presence of kings as a sign of respect. From a modern perspective, Aristotle may (again) appear naive. We are tempted to excuse him, in a patronizing way, for failing to realize that motives and intentions are themselves (efficiently) caused by immediately preceding agents. Bernini's vision, after all, can be traced back to his training, his religion, the price of stone, and so forth. But the Aristotle of *De Anima* and *Politics* was quite aware of "conditioning" and "behavior modification" and surely would have argued that these are irrelevant to the issue. The issue is not how a given goal or intention was established. Rather, the issue or proposition is that outcomes are never completely understood until the final cause is apprehended, no matter what "caused" the final cause. That we may be "conditioned" to prefer Bach to Berlioz does not eliminate the fact of the preference. Without laboring over the virtues and vices of teleological explanations, one need only examine a modern discussion of embryological development or evolutionary theory or cosmology to ascertain whether science has eliminated final causes from its conceptual apparatus.

Aristotle invokes teleology but resists necessitarian theories. The latter cannot account either for error or for genius. All events will seem necessary until exceptions are noted. And, by denying purpose, the necessitarian will have to find the ship in the wood: that is, without the shipbuilder. Eventually the necessitarian and the teleological propositions merge. Whether things are as they are by necessity or by design, the ultimate outcome is never in doubt. We may not know what it is, but we can be sure, on either account, that it is. Thus, even in his *Physics,* with a lusty attack on necessity, Aristotle laid the foundations for the coming age of Stoic resignation.

We are now able to return to the Aristotelian approach to conduct and to governance, an approach stripped of many of the Platonic ingredients. In abandoning the necessitarian theory, Aristotle was able to advance the nativistic theory of virtue: "None of the moral virtues rises in us by nature" (*Nicomachean Ethics,* II, Ch. 1, 19–20). In its place is offered the empirist alternative: "We are made perfect by habit" (19–20). The perfection in question was the good defined as "activity of the soul in accordance with virtue" (19–20, I, Ch. 7, 1098a). Virtue in the Aristotelian system consists of intellectual and moral dispositions. The former is divided into philosophical wisdom and intelligence. The moral dispositions are those that conduce to liberalism and tolerance (I, Ch. 7, 1098a, 1103a). Both classes of virtue result from "study and care" (1099b) and from teaching, growth, and habit.

Aristotle even refers to the roots of the ethics in the word *ethos,* which means habit or an exercised skill (1103a). Opposing the central role given to the passions by members of the Academy, Aristotle insisted that all virtue is based on intention and choice, and that since acts impelled by passion are involuntary the passions do not figure in an account of virtue (1106a). This was one of the bases upon which he denied to nonhuman animals such virtues as courage and temperance. Behavior controlled by the passions lacks the element of rational. Nor was the patient with idle speculation as a putative form of virtue. The virtuous man is one who acts virtuously because he intends to; that is, virtue is his goal and is, therefore, the final cause of his behavior (1105b).

Consistent with this behavioristic criterion was Aristotle's conviction that actions aim at pleasure and avoid pain (1104b). This pattern is established in infancy and is therefore very difficult to alter (1105a). Only rigorous education can succeed in transforming pleasure-seeking into virtue-seeking conduct (Ch. 1, 179b–180a). Traces of Platonism can be found not only in the definition of the good as an accord between the soul and virtue (i.e., a kind of harmony) but also in Aristotle's emphasis on moderation. Virtue, he asserts, is "a mean between two vices" (1107a): excess and deficiency. The virtuous life is one guided by Aristotle's golden mean. Vice in the person and the State is invariably an expression of excess or deficiency and an extreme expression of that which in moderate degree is an absolute good. The good monarchy, transformed by excess, becomes tyranny; aristocracy becomes oligarchy; timocracy, in which political power is based on property, becomes democracy (1160b).

The Aristotelian Legacy

No figure from antiquity until the seventeenth century would be as important to the history of psychology as Aristotle. His most general contribution was to locate the intellectual and motive features of mind in the natural sciences, while reserving the moral and political dimensions of human life to a much enlarged conception of nature itself. At the level of basic processes, his psychology was biological and ethological, grounded in considerations not unlike those which Darwin would develop centuries later. If his own version of empiricism did not go so far as to submit scientific truths merely to confirmation by the senses, it did establish the validity and importance of the world of sense. In the process, Aristotle presented the senses themselves as objects of study.

In this same empiricist vein, Aristotle proposed the first laws of learning, loosely drawn around the principle of association and fortified by principles of reinforcement. Except for his retreat toward a somewhat fatalistic hereditarianism in the *Politics,* he consistently emphasized the part played by early experience, education, practice, habit, and life within the *polis* itself in the formation of the psychological dispositions. In this way, he presented human psychology as a developmental subject whose parent-science was at once civics and moral philosophy.

Aristotle's influence on science in general has aroused much speculation, and it was to some extent mixed. Subsequent scholars invested his works with unchallengeable authority for which, of course, Aristotle is not to blame. He promoted a rationalistic attitude in those who would seek the first principles of things, and on this the history of science must vindicate him. But this orientation soon was extended to matters that Aristotle himself treated empirically and understood to be beyond the reach of logical certainties. His ethical teaching refers to conditions and propositions that obtain "in general," "more or less," and "for the most part." Though persuaded that the moral and political terms of life are to be governed by universal principles, he still appreciated the complexities of life and the resulting inability to impose certainties on it in advance. Although his doctrine of final causes became a central theme in religious controversy, his four-fold theory of causation was of indisputable value to generations of scientists. And, on a loose reading, even the final causes are not alien to theoretical science as long as they are theologically neutral.

Aristotle lived at the end of Greece's classical approach, and his works are a fitting tribute to the names that made this age what it was. And what it was in fact, was nothing short of the birth of Western civilization. Aristotle labored to assemble and dissect all that was known and knowable. Had his effect upon us been less, we would not be so proud to discover his errors. It may be true, as one critic has insisted, that "practically every advance in science, logic, or in philosophy has had to be made in the teeth of the opposition from Aristotle's disciples." [21] But neither science nor logic nor philosophy is easily conceivable without him.

We might join in the criticism of his political writings as departing not only from a healthy empiricist outlook but as sanctioning the cruelest forms of subjugation and elitism. Consider only his comments about "natural slaves" and women. What is often less scrupulously noted, however, is the place of his *Politics* in the development of that very constitutionalism by which liberties have been won and preserved. In both the *Politics* and the *Nicomachean Ethics* he struggled to balance the rights of the citizen and the needs of the State. The particular balance he achieved depended, of course, upon the view he held of essential human nature and the State's essential function. In considering persons to be innately rational, inquisitive, and social, he was given to believe that discipline, self-sacrifice, and principled authoritarianism would find universal endorsement. Only those lacking reason (e.g., the "natural slave," children, the insane) would rebel. Aristotle was undeviating in his support of a society ruled by law, but this law itself was ruled by an ethics whose first principle was the welfare of the citizens, understood in terms of their realizing their full human potentials. It is this interconnection of society, law, ethics, and the public good that comprises constitutionalism and that installs Aristotle's psychology as perhaps the most integrated and systematic ever developed.

Stoic and Epicurean Psychology

As a general introduction to the post-Aristotelian philosophies, it is worth noting that debts to Aristotle are to be found nearly everywhere in them, including those parts devoted to critiques of Aristotle. J. R. Rist, in his *Stoic Philosophy* (1969), has observed that the very phrase "post-Aristotelian" really should be taken as referring to philosophies *dominated* by Aristotle's works.

In the previous chapter we noted the pre-Socratic interest in the physical properties of the world and contrasted this interest with the value-oriented philosophy of Socrates and his disciples. The pre-Socratics of Ionia developed their philosophical systems in a climate of commerce and growing prosperity. Socrates and the Academy of Plato flourished in the aftermath of the humiliating Athenian defeat at the hands of Sparta. While we must resist the allure of mechanical laws of causation in history, we must also acknowledge the regularity with which materialistic philosophies prosper in periods of empire and spiritualistic-idealistic philosophies emerge in the wake of destruction. With notable exceptions, intellectuals are apologists as well as critics, and so-called systems of thought are very often little more than rationalizations of the prevailing facts of life. The influence operates in both directions: political and economic forces affect the character of philosophy, and the latter works to maintain or modify those outlooks and institutions that define a period.

The major spokesmen for Stoic philosophy were Zeno of Citium (Cyprus), who lived between 336 and 265 B.C., and, both from Asia Minor, Cleanthes (ca. 331–ca. 232 B.C.) and Chrysippus (ca. 280–ca. 206 B.C.). Later Stoics of the Christian era include Seneca (ca. 4 B.C.–A.D. 65), Epictetus (ca. 50–ca. 138 A.D.), and Emperor Marcus Aurelius (121–180 A.D.). Epicurus (341–270 B.C.) was born on Samos and, although his parents were Athenians, his early education was received in Asia Minor from teachers versed in the philosophies of Plato and Democritus. His most famous spiritual descendant was the poet Lucretius (99–55 B.C.), whose *De Rerum Natura* set to verse the tenets of Epicurean thought.

Although it is true that Stoicism and Epicureanism were rival philosophies, the two shared several characteristics of note, especially in their earliest development. Both were moved by the need to answer the increasingly influential schools of the Cynics and Skeptics. These were the schools that elevated Socrates' modest admission of ignorance to metaphysical heights and fashioned the cleverest of arguments to prove that nothing can be known about anything. Thus, the initial motivation for both Zeno and Epicurus was grounded in this need to restore philosophy to respectability. Then, too, although Stoicism throughout its history never took on the radical materialism of the Epicureans, both systems were based upon a natural-science perspective which, with reservations, may be called *physicalism*. Both were monistic philosophies in that they conceived of the universe as being ultimately reducible to a single agency, force, element, or "stuff." For the Epicurean it was atomic; for the early Stoics, something akin to fire or ether. The Stoics often referred to this "creative fire" as *logos* and thought of the divine in the same physicalistic terms. *Logos,* which might be translated as "word" (as in, "In the beginning was the word") or "plan," also lends itself to a meaning or "point," as in "the very *point* of something."

Both the Stoic and the Epicurean philosopher tended also to be committed to the development of an all-inclusive system of philosophy whose basic principles were as applicable to physical phenomena as to those arising out of political and moral concerns. Where the Socratics tended to ignore the material world and focus on abstract problems in epistemology and ethics, and where Aristotle had liberated his political and moral theories from the exact sciences, morals, and politics, Stoicism and Epicureanism treated such boundaries as artificial.

But this perspective needed a favorable climate to make progress. Democritus (c. 460–370 B.C.), an influential contemporary of Plato's, had tried to promote a radical, materialistic philosophy in the years following the Peloponnesian ordeal. He was one of the fathers of atomism, teaching that all of nature was no more than aggregations of particles of matter distributed in infinite space. At the time, Democritus was considered a skeptic. His failure to be taken seriously by the Socratics was not due to his being provably wrong, but to his being provably irrelevant. The fundamental challenge to the Athenian philosopher from 404 to 350 B.C. was to explain the demise of Athenian power and to lay the groundwork for Athens' rebirth. Atomism, in this setting, was useless. But by 300 B.C. the world had neither a use for nor ready exemplars of the Platonic "ideas," or true forms. Alexander was the first bona fide potentate of the West and was followed by men of even greater organizational talent. From

300 B.C. to the first century of the Christian era, the Western world was impe-
rialist. Its defining ethic was expansionism. Its language, accordingly, was the
language of law, administration, and finance. The voices of philosophy that
spoke for this period are those of the Stoics and the Epicureans.

Epicurean philosophy begins with a psychological claim: that all knowledge
originates in sensation. Aristotle would not reject this, though he would also
not have meant the same thing by it. For Epicurus, the material organization of
the body is such that experience becomes recorded in memory and can be revived in the form of concepts. By association, these concepts come to stand for
the items originally given in experience. Thus, the frequent pairing of an actual
apple with the word "apple" results in the expectation (*prolepsis*) of an apple
when the word is heard. Since our only mode of verification is experience itself,
questions regarding the truth or validity of experience are meaningless.
All experience is the outcome of interactions (collisions) between material entities, between the matter of the world and the matter of the sense organs. Random collisions can produce random outcomes, as disease can produce pathological outcomes. Nevertheless, the properly functioning sense organ responds
to what is there. The soul itself is a type of material organization made up of
common elements (fire, wind, and air) and a fourth unnamed element that
makes it possible for stimulation to be distributed over and within the body.
This last element is the most subtle, but is material nonetheless. As a result of
its activity, we have the capacity for pleasure and for pain. The business of
life—the very preservation of life—centers on these facts and functions.

Those who set out to keep pain to a minimum will necessarily forgo any
number of pleasures. They will abide by the laws to avoid punishment and will
do what is "right" to garner praise. However, it is prudent for them to remove
themselves from those political and social arenas in which the likelihood of
frustration, danger, and disappointment is great. Happiness, it turns out, is tranquility, the absence of turmoil (*apatheia*) secured by friendship, moderation,
and the abandonment of vain ambition. Human life is but a brief episode in the
eternal history of atomic collisions. As with all other forms of being, this life
reflects simply one of the infinite number of possible atomic configurations. In
infinite time, all possible configurations come into being at least for the moment. (Hence, the Epicurean conviction that there are other worlds not much
unlike our own). In the circumstance, the only prudent goal of life is to derive
whatever pleasures the interval affords, and to exercise that degree of moderation consistent with the minimizing of pain and suffering.

How different from this is the Stoic outlook? In details, the two systems are
sometimes radically opposed. The Stoic is committed to the view that a rational
principle (the *logos*) guides the universe, while the Epicurean holds that the
entire affair is statistical. In this same connection, the major Stoics subscribed
to a cyclical theory of universal history; a theory of endlessly repeated phases

of creation and dissolution proceeding from the life-producing principle (*logos spermatikos*) of the universe. The Epicureans, of course, found nothing in experience to confirm such a view. And, since the Stoics found this rational principle expressed in human life, they concluded that each person has a duty (*kathekon*) to promote the cause of reason in the affairs of society and State. This, we note, is a duty the Epicurean could well live without.

But at a broader level, the similarities between the two become rather more apparent, as does the source of their appeal to the ancient Roman world. Both finally conduce to a life of order, harmoniousness, civility, and modesty. Both lay the groundwork for this ethics in a cosmological theory based upon physical laws. In this way, both present nature as the model for life and for government. Even in the details of their respective psychologies, the Stoic and Epicurean display as much agreement as rivalry. Knowledge on the Epicurean account begins with sensation. On the Stoic account it also begins as the mental image (*phantasia*) of the sensory event. Both subscribe to a form of associationism, and both propose that elementary sensations give rise to general concepts (*katalepseis*). Indeed, in attempting to establish the criteria of validity for our perceptions, the Stoics often advocated a kind of statistical test based on clarity and utility. They did not go as far as to claim that nature itself was statistical in its operations, as the Epicureans did, but they were often content to accept that our knowledge of nature did not rise above the level of probability.

Central to Stoic thought, of course, is an ethics of rationality; an ethics pitting the emotional side of life against the powers of reflection and self-control. Even the word *pathos,* which might generally be taken to refer to a passion or emotion, was generally intended within Stoic contexts to be regarded as a kind of *disease* of the soul, or something on the way toward psychopathology. Plato and Aristotle had both emphasized the nature of the tension between rational and impulsive forces. Aristotle, however, recognized that both were part of the very nature of animal life and that what finally mattered for rational creatures was how they are disposed to express the passions. The Greek for "disposition" is *hexis* (pl. *hexeis*), and one may be said to have either a good or a bad *hexis* for, e.g., anger, love, etc. One's very character is expressed to a considerable extent by one's disposition (*hexis*) toward expressing a given emotion (*pathos*). It is with Stoicism that his line of argument concludes with a renunciation of emotion itself! Here again the Stoic and Epicurean movements of thought become ever more congruent, even if the foundational arguments differ.

What remains, in light of the foregoing, is the issue of materialism as one that would seem to make Stoic and Epicurean psychologies incompatible. We should note first that Zeno himself was credited with something of a materialistic psychology by ancient commentators and that the transcendental features of Stoicism come somewhat later in its development. But this is beside the point. Rather, the issue of materialism is far more important to a comparison of Stoic and Epicurean cosmology and theology than it is to their psychological theories. That the Stoics argued in favor of a rational first cause, which the Epicureans generally denied, has no bearing on the fact that both took human rational capacities for granted. Similarly, both based their ethical canons on the quest for happiness. Epicurus and Lucretius were more explicit than Zeno in tracing pleasure and pain to our material organization, but both schools accepted the "pleasure principle" as entirely natural, and as ordained by the very laws of our nature. Thus, at a global level of analysis, early (pre-Christian) Epicurean and Stoic psychologies both present human mental life as something that takes the data of experience as primary, and from these data proceeds to form general concepts in such a manner as to organize conduct around the goals of happiness and harmonious contentment. For both, but in different ways, human nature is the mirror of nature and is to be understood in precisely the same terms employed in the description of nature.

The philosophies taught by Zeno and Epicurus in Athens came to have their greatest impact not on the Greek but on the Roman mind. Indeed, although these two philosophies evolved in Athens, they were not Athenian in their principal philosophical features. To be sure, each carried traces of Platonic and Aristotelian influence, particularly the Plato of the *Laws* and the Aristotle of the ethical treatises. Zeno and Epicurus were ethicists even more than they were natural philosophers. But unlike Aristotle, who succeeded in keeping physics and ethics apart, Zeno, Epicurus, and their disciples proposed a philosophical unity that had not been seen since the pre-Socratics. Their philosophies drew inspiration from nature such that government, reason, perception, and the entirety of the human enterprise were to be comprehended in terms of natural law. And human laws were to be patterned on this.

Though not Athenian, the Stoic and Epicurean philosophies do reflect the evolution of Greek theological attitudes which, as we have noted, were only gently touched by the Academy and the Lyceum. As Greek economy and philosophy evolved, so, too, did the Greek religious outlook. In *From Religion to Philosophy*,[22] F. M. Cornford traces this evolution from the early Homeric period that found the gods obedient to destiny (*moira*) and at the mercy of chance (*lachesis*) to the Age of Pericles, by which time reason had replaced destiny and Zeus had arrogated to himself absolute power over the affairs of men and gods. In this same development, the earliest concept of law (*nomos*) as a dispenser (*nomeus*) of one's fair share or just proportion became transformed into the belief that the will of Zeus was itself the law of nature. Even from this all too superficial sketch of the changes that took place in the popular religious outlook from the sixth to the fourth centuries, we can discern the fundamentalist or revivalist tone of Stoicism. The Stoic philosophers saw their contemporaries as having wandered too far from the older and purer respect for *moira*, *lachesis,* and *nomos.* While retaining the current emphasis upon the will, they insisted that its freedom was of a limited sort: we are free only as long as our will is reconciled to destiny and harmonized with the immutable *nomos.* As we see in the Vatican Fragments, Epicurus, too, urges that nature not be violated but obeyed (Fragment XXI); that we must not complain about what we lack but realize that all we have is the gift of fortune (Fragment XXXV); and that death is our common bond and shared future (Fragment XXX).[23] In this same tradition but four hundred years later, Epictetus (60–120 A.D.) summarized the central idea with that moving simplicity that identifies the Stoic mind:

> Never say about anything, "I have lost it," but only, "I have given it back."[24]

From its origin in the fourth century, Stoic philosophy struggled to foster a moral commitment while adhering to naturalism. Lucretius proposed conservationist laws of matter and momentum, and the Epicurean tradition rested its argument on the veracity of common sense and tutored perception. This same Lucretius, whose book attempts nothing less than an explanation of all that exists and occurs, can offer not a sentence of advice to Caesar and Pompey. He cannot tell us how to distinguish the just act from the unjust, nor can he discover any final cause of life other than the grave.[25] Even Marcus Aurelius, whose Stoic *Meditations*[26] were to teach us those principles of conduct and good taste that might bring some order to our lives, finally offers little more than a paean to quiet resignation.

It is often averred that Stoicism was in some way natural to the Roman mind, to the spiritual and political character of the empire. But Stoicism was a varied system whose major architects focused on different problems and wrote in different periods. Moreover, there are points of conflict not only among the Stoics but within the writings of a given spokesman. Still, each one does reveal a side of Roman psychology, and we will benefit from a review of the connection between Stoic thought and the facts of Roman life.

The popularity of the early Stoics in Rome is explicable in terms of the dominant feature of the Roman epoch, law. Law, if it is to rise above the mere threat of reprisal, must be principled. It must make daily and intelligible contact with what the citizen believes to be right and wrong. As a form of constraint, it must justify itself in terms of a higher good. The dilemma facing the Roman of the third and second centuries B.C. was no less than the conflict between historic moral restraint and sudden opportunities for colossal wealth. Rome, in its religion and society, had asserted the virtue of a quiet and conservative republican life. As a limitless empire, however, Rome made the goods and resources of the entire world accessible to the crafty and acquisitive citizen.

In focusing on the reflection of the cosmic *logos* in human rationality, the Stoics came to regard the human community as a breed apart. It is with Stoic philosophy that we begin to hear about "human rights"—moral or political options inherent in human nature by virtue of its being *human*—and about the world as the common dwelling-place (*oikos*) of the community of rational beings. The universal brotherhood of Christianity is anticipated in this, as is the ever deeper separation between humankind and the natural world with its other precious inhabitants.

Epicurean materialism, on the other hand, could not accept a transcendent soul. Rather, if the soul can affect the body, it must be because it is a kind of body itself and is thus destructible. Personal immortality, therefore, is out of the question, as is fear of punishment in the afterlife. Accordingly, our only concern must be with life as it is lived from day to day. The goal is happiness which, in the last analysis, is freedom from pain. Since the quest for fame and power and riches usually leads to frustration and grief, the recommendation is for moderation, for a rejection of passion itself and the recognition that the passionate state is a kind of madness. As with Socrates, Plato, and Aristotle, the Stoic and Epicurean both emphasized contemplation as the means by which true happiness is secured. Unlike Plato, the Stoics endorsed contemplation less for the good of an immortal soul than for comfort in a troubled life. Thus did the Roman aristocracy proceed to study philosophy with all the enthusiasm one brings to other pastimes. In their thirst for culture, they imported hundreds of Greek scholars and thereby furthered the cause of Hellenization. Cicero stamped Roman law with Stoic reasonableness (even as he criticized Stoic philosophy), and Nero traduced the lessons of Epicurus into a grotesque caricature. Situationism, ethical relativism, and materialism lent righteous complacency to periods of prosperity but left the Roman morally defenseless when more powerful forces began to dismantle the Empire. As Rome approached the peak of her achievement, Lucretius could observe: "[N]early everything men need for life lies right within their grasp and . . . existence is safe and sound." [27] But two centuries later, when Augustan greatness had receded and the barbarians' steps grew louder, Marcus Aurelius would look back over the entire epoch and explain:

> Thou hast endured infinite troubles through not being contented with thy ruling faculty.[28]

Notes

1. The definitive biography of Aristotle's intellectual development remains Werner Jaeger, *Aristotle: Fundamentals of the History of His Development,* Clarendon Press, Oxford, 1934. Recent scholarship, however, has raised objections to Jaeger's interpretive biography.

2. Ibid., Ch. 1.

3. Ibid., Ch. 2.

4. Aristotle, *The Protrepticus,* translated by Anton-Hermann Chroust, University of Notre Dame Press, Notre Dame, Ind., 1964.

5. Ibid., p. 9.

6. Ibid., p. 44.

7. Ibid., p. 31.

8. Jaeger, *Aristotle,* Ch. VII, Sec. II.

9. Aristotle, *Metaphysics,* Book A (I). Parenthetical numbers refer to the line of text in the Greek. These numbers are conventionally employed in all translations.

10. Ibid. Book XIII, 4, 30–33.

11. Ibid.

12. Aristotle, *De Somnis,* Ch. I, 459b; Ch. 3, 462b.

13. Aristotle, *Posterior Analytics,* especially Book II, Ch. 19.

14. Aristotle, *Categories.*

15. Aristotle, *History of the Animals,* Book VI, Ch. 3, translated by A. L. Peck. Harvard, Cambridge 1970.

16. Aristotle, *On the Heavens,* Book II, Ch. 14.

17. Plato, *Theaetetus* (186), in *The Dialogues of Plato,* translated by Benjamin Jowett, Random House, New York, 1937.

18. Aristotle, *Nicomachean Ethics,* 1153b.

19. Aristotle, *Rhetoric,* 1389a–1389b.

20. Aristotle, *Politics,* 1252a.

21. Bertrand Russell, *History of Western Philosophy,* Simon and Schuster, Clarion Books, New York, 1945, p. 202.

22. F. M. Cornford, *From Religion to Philosophy: A Study in the Origins of Western Speculation,* Harper and Row, Harper Torchbooks, New York, 1957.

23. A collection of Epictetus's *Fragments,* reprinted from the Vatican Fragments, appears in *The Stoic and Epicurean Philosophers,* edited by Whitney Oates, Random House, The Modern Library, New York, 1940.

24. Ibid., Epictetus.

25. Lucretius, *De Rerum Natura,* translated by Palmer Bovie, New American Library, Mentor Books, New York, 1974.

26. Marcus Aurelius, *Meditations,* in *The Stoic and Epicurean Philosophers.*

27. Lucretius, *De Rerum Natura,* Book VI.

28. Marcus Aurelius, *Meditations.*

4 Patristic Psychology: The Authority of Faith

Dark Ages?

The period covering the decline of classical Roman civilization and the emergence of the Christian world in Europe includes within it an allegedly "dark" episode of unbridled superstition and paralysis of intellect. We must be wary of that Renaissance habit of treating everything separating their own time from classical Greece and Rome as "dark" and "medieval." Whom do we find when we thumb the pages of these centuries? Epictetus, Origen, Plotinus, Porphyry, St. Jerome, St. Augustine, Boethius, St. Anselm, Peter Abailard (Abelard), Robert Grosseteste, Roger Bacon, Thomas Aquinas, William of Ockham—this is the short list. And amidst their writings and achievements we also note the veritable invention of the university as we know it; the construction of more than a hundred "Gothic" cathedrals, each an architectural masterpiece; and the codification and promulgation of Roman law, modified to meet local conditions and laying the foundations for a unified culture and civilization. Not too dark, then, after all.

True enough, from roughly 500 to 800 A.D. the lamps of comprehension and genius are dim, the record of enduring accomplishment almost vanishingly thin. If there is a culprit to be named here, however, it is not the orthodoxies or superstitions of a Christian church but the utter collapse from within, and assaults from without, of all that was Roman civilization. During the early Christian era the cities were politically corrupt, financially and morally wayward. The now deified emperors were of a generally sorry character. Their official biographers provided the *Scriptores Historiae Augustae,* a lasting testimony to those special inanities bred by absolute power. For much of the time Athens was still the target of the philosophical pilgrim and Greek was the language of the learned. By the reign of Justinian, however, it had become a dead language in Rome, another sign of cultural collapse. A surer sign, however, was the contagious *nostalgia* expressed by the better minds of the period.

From the Hellenistic to the Patristic

The Epicurean universe of the "atoms and a void" was a nearly senseless din orchestrated by matter in conflict and by an indifferent destiny. Man's only hope was for a life reconciled to knowable natural law and one passed among friends. While preaching the indestructibility of matter, Epicurus still denied the material soul the ability to retain a meaningful existence once the body "is broken up." Epictetus, reflecting on this Epicurean materialism, had this to say:

> Epicurus understands as well as we do that we are by nature social beings, but having once placed our good not in the spirit but in the husk which contains it, he cannot say anything different.[1]

Epictetus was committed to a useful philosophy. He scorned those whose lessons were contrary to common sense or private experience. Just as he insisted upon philosophers living the philosophical life and not merely talking about it, he also insisted upon a philosophy that formally acknowledged what every sentient being knew to be true. One such fact, of course, was the reality of the will and its essentially spiritual nature. Unlike Epicurus, who reduced prophecy to charlatanism, Epictetus was willing to accept spiritual commerce with the future but condemned those who sought it as weak and fearful (Book II, VII).[2] The Epicurean injunctions against marriage and children were ridiculed by Epictetus as removing the very foundation of social life (Book III, VII).[3] The Platonist suspicion regarding the evidence of the senses was viewed with equal contempt, since Epictetus had little to say to one who "has sensation and pretends that he has not; he is worse than dead" (Book I, VI).[4] In all, Epictetus reasserted the central position of reason in the affairs of life; the indispensability and accuracy of perception when guided by reason; the power of the individual will to withstand the commands of princes and kings; the true happiness of a life devoted to harmony with nature and to rational principles; the transitoriness of wealth, popularity, and worldly power; and goodness as the end of human life. To these he added monotheism, convinced, like Aristotle, that heaven must have one ruler as every flock has one shepherd and every family one father.

Marcus Aurelius was greatly influenced by the *Discourses* of Epictetus and, if only for this reason, Rome's official philosophy was Stoical until well into the third century. However, the Romans themselves had moved some distance from the reserved and resourceful lives advocated by Zeno, Cleanthes, and Epictetus. We are on surer footing to read Epictetus as an example of what philosophy was fighting, not what it was observing. The literati among Epictetus' contemporaries were moved more by Horace, Martial, and Juvenal. Indeed, by a deft corruption of Epicurean teaching, the upper-class Roman of the second century traduced philosophical materialism into a rationale for sexual promiscuity, moral laxity, and general debauchery. Nero's reign (54–68) had set the tone and, within a half century, Romans forgot that he had been judged an enemy of state. They recalled instead only that degeneracy was a sign of power.

As the empire proceeded along the course of its destruction, two radically different remedies were available: Stoic resignation, virtue, and independence and skeptical indifference and ridicule. The former, for obvious reasons, was unacceptable to the free-living aristocrats whose station in life depended upon the good will of the emperor. Skepticism, unless reduced to comedy, is always too subtle and too removed to be adopted by the ordinary run of citizens. In declaring man to be basically a rational animal (Plato), a social animal (Aristotle), and a moral animal (Epictetus), the philosophers of antiquity recognized only generally that we are *psychological* animals at base. Thus, to declare our perceptions to be always illusory or correct is to tell us what we know or firmly believe to be false. To say we are merely matter is, in terms of daily realities of life, to state the useless. To assure us that the laws of nature will see us through the hard times when a Caesar's caprice ever threatens life itself, when Vandals threaten to destroy our culture, or when plague succeeds in killing the children is to remove all philosophy from serious consideration. Thus, at a time when life violated Stoic orderliness in every conceivable way, one could subscribe to Stoic doctrine only by denying that reality was real, or, following the Skeptic, by suspending all judgment entirely. Neither alternative was any more compelling in the second and third centuries than it is now.

The conspiracy of circumstances that led to the rise of Christianity can only be rather generally fathomed. Of the overall political and moral conditions it is important to note the waves of migration by Germanic and Frankish tribes into Rome and its once imperial world. The immigrants, now in power, brought their own traditions and cultish beliefs, these mixing with Roman and Christian attitudes in various and sometimes exotic ways. As it became ever more important for the leaders of this new world to make peace with Christianity, changes were worked in both directions. More will be said of this in the next chapter.

With the breakdown of traditional Roman practices and perspectives, there came a mounting and ever more public cynicism among the more tutored classes. The scorn of Tertullian's *De Spectaculis*[5] had long since threatened carnal Rome with a Day of Judgment. The persecution of Christians under Nero, Trajan, and (even) Marcus Aurelius provided the poor and hopeless non-Christians with a daring and romantic model and with what would later come to be grounds for reprisal. The eloquence of such early religious philosophers as Plotinus and Origen merged with the Stoic mandate that the true philosopher live an exemplary life. Origen's devotion to an existence devoid of lust is documented by his self-inflicted emasculation! Christianity went beyond the Stoic resignation in the face of death to the (Platonic) celebration of death which would free the soul from its bodily prison and reunite it with its Creator. To a worn and weary peasantry, the Gospels offered eternal life hereafter and rewards that would beggar the vaults of Croesus.

From Stoicism the Patristic philosophers borrowed the conception of nature as law. From the Platonists, they received and embellished the ultimate reality of idea over sense. From Aristotelian teaching, the way of Cicero, Seneca, and Epictetus, they assimilated the concept of the *logos:* the underlying rational principle of the universe. Against the classical philosophies, they rejected materialism and its gloomy and pointless implications. As Plotinus put it,

> Matter is not Soul; it is not Intellect, is not Life, is no Ideal Principle, No Reason-Principle; it is no limit or bound, for it is mere indetermination; it is not a power for what does it produce?[6]

These are the words of a founder of neo-Platonism, perhaps the major philosopher of the early Patristic period, though not a Christian himself. But in the period under examination, the term "Christian" was not as unambiguous as it would become centuries later. Roman imperial expansion and imperial collapse brought together a great diversity of beliefs and cults, each somewhat transformed by the others. Plotinus (205–270) and his most influential disciple, Porphyry, are illustrative. The former was born in Egypt and studied in Alexandria under Ammonius Sacas, who was also Tertullian's mentor. Here an intellectual pedigree was shared by a major critic and a major defender of early Christian teaching. Porphyry was an Athens-educated Syrian who settled in Rome, where he discovered a congenial group of Plotinists. The point, of course, is that the world of thought was now becoming scattered. The old and tight philosophical systems were being divided up into sects and infiltrated with notions their classical authors would have found bizarre. What gives the Patristic period the quality of a distinct age is not that its major figures all defended the claims of Christianity, but that they all sought to reestablish the authority of those cardinal tenets developed by the classical philosophers— tenets that were expressive of a greater culture. In a word, they were searching for a philosophical unity that might restore a form of social unity. Thus, the same Plotinus who at times might challenge Christian teaching is found at other times defending monotheism. The defense would extend to the immateriality and immortality of the individual soul, not to mention the essential "evil" of mere matter, deprived as it is of an intelligent principle. This is the sense in which Plotinus and Porphyry, as non-Christians, are as representative of the Patristic period as are such famous Fathers of the Church as Tertullian and Origen.

How should acts of self-mutilation or lives of celibacy or abject asceticism be understood? The pervasive psychology of the early Christian period was an amalgam of previously developed but now rather oddly comprehended phi-losophies—oddly comprehended in that their pagan origins required them to be purged, often of the very elements that gave them their initial coherence. Renunciation of the flesh, so common among the early communities of Chris-tians, was predicated on a radically *dualistic* ontology rich in moral implica-tions. There are two worlds, the world of matter (believed by many to be ruled by Satan), and the world of Spirit. The soul is degraded by the former. A body uncleansed is unfit to house a worthy soul. Even food can "block" the inner dwelling-place converting the body from what might be a temple to some-thing more akin to a kennel. The flesh-eaters have made their inner world an abatoir. The sensualists were the devil's playground. Not unlike today's Islamic fundamentalist, early Christians of this persuasion required women to be veiled and carefully monitored, discouraged socializing among unmarried men and women, and imposed other strict observances. Life as a trial or test that is failed by the many, with eternal damnation in the wings, cannot proceed on the grounds of whim and impulse.

The reach of these practices was limited by a number of factors. In general, the lower classes of society were not regarded as worthy of such concern and vigilance. Attitudes toward the conduct of a barmaid were quite different from those directed at the wife of the establishment's owner! Was this the lingering influence of Platonistic and Aristotelian class-consciousness? Could it be rec-onciled to belief in universal brotherhood and natural equality? It is worth not-ing that some of the larger enclaves of asceticism and monkish retreats were in just those places that had been centers of Greek culture and teaching. Antioch comes to mind immediately. Such centers had their local hermits and gurus, saintly persons known for good works and self-denial. Some of them, follow-ing Paul's teaching, insisted that even the institution of marriage tied one to the business of this world, this realm of corrupt matter. As a result, something of an aura of spiritual superiority surrounded the celibate, the virgin, the unat-tached widow, the community of prayerful men.

The teachings of the early Patristic scholars were simple and undogmatic, at least with respect to details. Plain folk from all over the Empire had to find resonant elements in the new faith and, from Paul of Tarsus to the mystic Ori-gen, the missionaries of Christianity were willing to bend and accommodate. On fundamental principles, however, there was no compromise, and these prin-ciples have remained central to Christian belief ever since: that every person is the child of God; that there is only one God; that we are made to serve God in this world and, through good works, to live eternally in His light; that the soul is the essence of human life; that neglect of the soul is a sin to be punished; that no force on earth can affect God's plan nor earthly wisdom fully reveal it; that God's goodness is the cause of all things; that in His goodness He sacri-ficed His only Son to take on the human coil and then die for human redemption; and that in that sacred death the soul's hope was reborn, so that we once might aspire to God's good grace.

With such brief, bold and uncomplicated declarations, the founders of the Church of Rome, Jewish Christians at the outset, offered an alternative. To an empire torn by fickle tyrants and ignorant chieftains, they offered brotherhood. To a mass facing hunger, plague, and the violent death of warfare, they offered eternal life and the virtue of self-denial and renunciation of the flesh. To the oppressed, the slave, and the exile, they offered the reassuring genealogy of God as the Father of all. To the pagan and criminal, the wanderer and the corrupt, they offered redemption. The poor were consoled by the meaninglessness of earthly riches; the aristocrats, by the good works the prosperous could do for their brothers. The army inherited by young Constantine was already mostly Christian, as, indeed, was his mother. His most impressive victories were won under Christ's flag. Thus, as the Stoics retreated to the serenity of philosophic speculation, as the fools at court submerged themselves in the lusty favors of the emperor, and as the Visigoths planned their assaults on Rome itself, Christianity was laying claim to the hearts and minds of the empire.

Christianity did not begin with a firm foundation in philosophy. Many of its earliest followers were drawn from the poorly educated classes of Egypt and the Nile Valley, Greece, Asia Minor, Syria, and Spain. Its spokesmen were no match for the urbane rhetors of Rome and Alexandria. Indeed, Tertullian had contributed a sneering anti-intellectualism to the movement that remained with it in various forms for almost five hundred years. Philosophy, after all, had failed. Those who converted to Christianity had, we were to assume, tried the way of the philosophers and found it wanting. Moreover, a philosophical foundation required argument, public discourse, and that openness to interpretation which comes only after the religion is itself secure. At the outset, and long before this security was attained, the faithful formed together in one or another study-group of men and women, a *didaskaleion,* often maintained for years. Their leader, rather than a set of texts, would come to be the authority on questions of doctrine. But with numerous such groups and the expected variety of spiritual leaders, the community of disciples in the first two centuries was scarcely united around a set of dogmatic teachings. If unification were ever to be attained, it would be at the price of interpretive license. At least *some* teaching had to be rejected as *heresy.* Furthermore, if there was to be a recognized Christianity, it would require the clearest distinctions between itself and both Pagan and Jewish laws and beliefs. In the first two centuries of the Christian era, movements in this direction were frequent but unsuccessful. Soon, however, nothing less than an emperor would participate, and political success was on the horizon. Although we can examine the psychological and social factors that guided the evolution of early Christianity, a sharply defined Christian *psychology* before the fifth century is more difficult to make out. With each *didas-*

kaleion framing its own special conception of human nature and its participation in the divine, with each region still laboring under one or another set of ageless pagan beliefs, and with the radical class-structure of the settled communities, there is too little that might stand as a defining perspective. By the fifth century, however, with the confidence that comes from success, the Church Fathers had created that studiously rational framework by which the Christian view of human nature might be taught in psychologically meaningful terms.

The Problem of Knowledge

It has become a tradition in discussions of the Patristic period to describe Christian belief as neo-Platonist and to defend this description with quotations from Origen, Plotinus, Augustine, Boethius, and others. This can be misleading on several counts. First, despite the breadth and penetration of Plato's dialogues, the passages containing the Platonic cosmology will not yield a Christian theology. Second, the spirit of Platonism is fatalistic; that of Christianity, optimistic. The soul that abandons the dead body of the Platonist wanders through the universe in search of True Forms and a next incarnation. The Christian soul confronts God directly and takes up eternal life with Him. Platonism, taken on the whole, recommends a contemplative, introspective life of quiet virtue. The Christian life is one of action and reform. Platonism is rationalism par excellence: through dialectical reasoning, the mind can be led to the latent knowledge of the soul. The early Christian belief, however, is that all knowledge begins first with faith, which leads to a transcendental awareness of God, who is the creator of all and who, therefore, is in all things. For the Platonist, reason is the light. For the early Christian, faith is the light and the path. In short, the Platonist would not know what one meant by "taking a truth on faith," and the early Christian could not imagine knowing any significant truth without it.

There is no doubt but that the early Christian theorists, when exploring the available philosophical alternatives, found many natural bonds between their beliefs and Plato's teaching. Predictably, they turned to these when it became useful to place Christianity in a context that could invite the attention of the more intellectually inclined. This reading of early Christianity as neo-Platonist is correct. That which would have Christian thought evolving from Platonism is not. This is movingly documented in St. Augustine's *Confessions*,[7] in which he credits the now lost *Hortensius* of Cicero with first whetting his philosophical appetite.[8] He insists that Aristotle's *Categories* did him more harm than good[9] and, while praising the Platonists for their appreciation of truth-as-incorporea,[10] he lauds Paul still more for identifying this truth with the grace of God.[11]

Augustine was the most influential philosopher in the history of Christianity, at least until St. Thomas Aquinas. He set the tone of Christian intellectual life for the better part of eight centuries. In limiting our analysis of the problem of knowledge to Augustine, then, there is little risk of rendering an incomplete account. Our understanding of the problem of knowledge in Augustinian terms begins with the realization that God is the ultimate truth and that to know God is the ultimate goal of the human will. Thus, inquiry of any sort must come to rest on this realization; otherwise, it is mere vanity and doomed to error and corruption. We examine human nature, for example, only because in the process we will reaffirm the existence of God.[12] The inquiry must avoid the pitfalls of the senses and must reject, a priori, the godless materialism of the Epicureans. In brief, the formal properties of the inquiry must be patterned after the Platonists.

When Augustine visited the venerable Simplicianus, he reported his devotion to the Platonists, and Simplicianus praised him "for not having fallen upon the writings of other philosophers full of fallacies and deceits, after the rudiments of this world, whereas in the Platonists God and His Word are everywhere implied." [13] However, the Platonist merely implicated God, whereas the Christian, with reason led by grace, is able to explicate God's reality and command.[14] Now, what is this reason led by grace? It is, in Augustine's phrase, an interior sense. It is that nonsensory inner awareness of truth, of error, of the moral right, of personal obligation, and of personal identity. This interior sense is the judge of perception and, therefore, is not reducible to perception. Unlike the five senses, it perceives itself perceiving as it perceives each of the separate senses perceiving.[15] In modern parlance, this interior sense is no less than consciousness itself, but, indeed, it is something more than consciousness. It is a moral consciousness whose character is outlined, in a boldly Freudian way, in the *Confessions:*

> You commanded me to abstain from sleeping with a mistress. . . . But there still live in that memory of mine . . . images of the things which my habit has fixed there. These images come into my thoughts and though, when I am awake, they are strengthless, in sleep they not only cause pleasure but go so far as to obtain assent and something very like reality. . . . [H]ow does it happen that even in our sleep we do often resist and, remembering our purpose and most chastely abiding by it, give no assent to enticements of this kind? [16]

For Augustine, it is not through the deliberations of the mind alone that one comes to know the truth of God. Rather, it is that there is mind which informs us of the divine agent. That the mind furnishes itself with number, with time, with memory-facts not discernible through perception alone convinced Augustine of the wisdom of the Platonists. They did not go far enough only because the Son of God had not seen fit to present Himself to the Hellenes. The life of Jesus, however, changed all that and thereby revealed those truths that mere philosophy could never discover.

What makes this early Christian philosophy a transcendental psychology is its insistence that human beings, as children of God, share in the divine wisdom and that through this fact and simple faith in it we can elevate our comprehension of the universe to a truly cosmic level. Put quite directly, the Christian pledge is that the faith conferred by grace and spiritual labor will equip the believer with the answer to the most vexing question: "Where did we and the universe come from and what is our destiny?" The Platonists, whose rationalism provided the broad philosophical guidelines of Christian theological discourse, had carefully segregated the true forms and the Republic. That is, Plato at his idealistic best, never complicated the affairs of State with those ultimate verities that only death could illuminate. He yearned for a philosopher-king, but only because he sought a government organized around defensible, rational principles rather than one immersed in petty squabbles and self-indulgence. Philosophy, after all, was a way of life recommended because it would make people happy. It would conduce to civility, justice, fairness, and virtue. In a word, it would allow aristocratic personages such as Plato and his circle to live out their years untrammeled by the bellicose strivings of "men of brass." To be sure, the *Phaedo* and the *Crito* are rich in their intimations of immortality, in their promise of an ultimate enlightenment. To this extent, the Platonic teaching qualifies unambiguously as transcendental philosophy. It asserts the existence of extrasensory, immaterial truths of a finer quality and graver meaning than any accessible to earthbound humans.

Christianity, however, went further. It required not a philosophical life but a religious one, which, if neglected, led not to ignorance and its attendant unhappiness but to sin and the ultimate retribution. It replaced the Platonic true forms with the all-seeing vision of the timeless architect of all truth, an architect whose infinite love was carefully balanced against infinite justice. Where the Republic was the enlargement allowing a clearer view of human nature, the Christian was the miniature in whom God's reality could be established. This shift in emphasis provided early Christian scholarship with a decidedly psychological cast. God was believed to be infinitely good, yet there was evil in the world. If this was not God's doing then there must be matters beyond God's control, or else, and in contradiction, God must be the author of both good and evil. To reconcile this seeming paradox, Augustine addressed the apparent tension between free will and determinism, a tension that has been discussed vigorously ever since. If all human beings are God's children and, therefore, are potentially able to see the light, those who do not must be converted; their souls must be saved even at the peril of their bodies. This issue became the basis of theories of the just war and of justifiable homicide, both psychological principles of conversion.

The early Christian's problem of knowledge was not one of uncovering the truth but of transmitting it, one of readying the pagan for the light of faith. In conceiving the problem in these terms, Christian scholars inquired more deeply into the psychological, as opposed to the purely rational, factors governing human judgment and conduct. The *Confessions* are particularly indicative of this shift in orientation. Augustine perfected the practice of public disclosure, confession of guilt, and the expression of piety and resolution. The point here is not that he discovered the psychoanalytic principle of *catharsis,* but that his attention was given over to that side of self and others that the Socratic dialectic had ignored, even condemned. Where Socrates merely counseled against the rule of passion over reason, Augustine laid bare the genuinely personal and psychological dimensions of the conflict. The Platonic dialogues, then, were transcendental without ever attaining the character of a *depth*-psychology. Aristotle's works were studiously psychological, in most respects assiduously non-transcendental. Augustine described human nature in an unblushing, otherworldly idiom.

Central to the Augustinian epistemology is the distinction between knowledge and wisdom, the former being "a rational cognizance of temporal things" [17] while the latter is "an intellectual cognizance of eternal things." [18] On this account, intellect and reason are different faculties. Reason is, indeed, a guiding light by which we might navigate through a confusing world, but reason alone could not equip us with that sublime knowledge of the eternal. And, of course, "it is not difficult to judge which is to be preferred or postponed to which." [19] In Book X of his work on the Holy Trinity (*De Trinitate*), Augustine reviewed the essential character of mind and emphatically declared it to be incorporeal.[20] He argued dualistically that the mind, while not a substance, was able to direct the material senses to "find out" what was of interest to it. However, the mere fact that it was so able to have commerce with the world of things did not indicate to Augustine that it was a thing itself. He rebuked former philosophers, and especially the Epicureans for confusing the objects of the mind's interest with mind itself. He rebuked the great run of individuals similarly for allowing their minds to become confused between the opposite poles of sense impression and eternal wisdom. Set free of sensory deceit, the mind can know itself, reflect on itself, love itself. Only this way can it find God.

The philosophy of Plato, for all the inspiration it provided, was not without dangers and liabilities when examined from the early Christian perspective. Recall, for example, the Platonic theory of knowledge which requires the soul to remember or retrieve those truths which it possessed in prelife, before it became tied to a body. Clearly, this sort of two-way eternal life was at variance with the Christian theory of God's creation of each individual soul, and the related theory of a common descent from Adam. The problem then, is one of explaining how the soul (mind) can possess those universal truths both Plato and Christianity granted. If every soul is, as it were, minted anew, having no existence until God brings it into being for this person, the soul must have a personal identity, and not simply that ghost-like and vaporous existence allotted to souls in the realm of the true forms. But would not this personal or individuated soul, by the very fact of individuation, be incapable of knowing the universal, the un-individuated?

On Augustine's account, this would be so were it not for the voice of the Divine that speaks directly to and through each soul. It is by the grace of God that the human mind receives this knowledge. Human freedom is such that one may choose not to listen, but the word is ever present. We see in this one of the differences between the Platonic and the Christian approach not only to the problem of knowledge but also to the question of personal responsibility. Hellenic philosophy judged human nature in conservative and not especially hopeful terms. Even the loyal lover of wisdom, the philosopher, could get only so far in this earthly life. The ultimate truths would forever evade the purely temporal being, and would become accessible only in that eternal time when being was no longer personal. This much, as we have seen in the previous chapter, survives in Aristotle's metaphysics and is quite discordant with Christian belief. But given the Hellenic perspective, it becomes a relatively straightforward matter to consider some persons as the "natural slaves" of Aristotle's *Politics* who must be led and controlled by those who have traveled further in the march toward truth. Metaphorically, they are like Plato's men of brass, etc., some of whom are destined to rule while others are to serve and be ruled. Just as natural slaves are not really responsible for their barbaric innocence, so the master, too, is guilty of no offense in ruling over them.

The Christian understanding is different from this in nearly every way. Every person is "a brother in Christ," every soul comprehended by God's mercy and love. What then of those who depart from the path of righteousness? Two answers are forthcoming from the Patristics: first, some such souls have been offered by God as examples to the rest of us, as proof of the fact that sin is sickness. Second, there are those whose vanity and lust are such that they refuse to hear the voice of God, and these are as worthy of contempt and punishment here as they will undoubtedly endure hereafter.

The Problem of Conduct

Let us now look more closely at the manner in which the problem of conduct was approached. If there was an element in Patristic philosophy setting it apart from virtually all preceding systems, even that of the Platonists, it was the explicit equalitarianism just discussed. The Platonists subscribed unwaveringly to psychological nativism in accounting for the differences among men. The "convenient fiction" of Plato's men of gold in the *Republic* served as the constant rationale for eugenics and as the occasional explanation for human corruption. Aristotle and the Peripatetics never abandoned the notion of the natural slave who, by constitution, could only follow reasoning but lacked reason itself. The rupture of this otherwise unbroken tradition is announced in *The City of God*.

> For, "let them," He says, "have dominion over the fish of the sea, and over the fowl of the air, and over every creeping thing which creepeth on the earth." He did not intend that His rational creature, who was made in His image, should have dominion over anything but the irrational creation—not man over man, but man over the beasts. . . . And this is why we do not find the word "slave" in any part of Scripture until righteous Noah branded the sin of his son with this name.[21]

How unfortunate for the animal kingdom that the line was drawn at this point, Augustine here being helped by previous generations of Stoic philosophers. Having drawn the line below human beings, however, he granted quite broad protections and promises to all those found above it. Moreover, given this insistence, scholars of the early Church subscribed to political philosophies, theories of justice, and principles of education which, it is fair to say, would have astonished the orthodox Platonist. The presumption of natural equality is a powerful one—one that perforce will color nearly every other aspect of a moral and social philosophy. Thus, whatever resemblances exist between the Platonists and Patristics, this one difference between their views is of overriding consequence. Augustine, Plotinus, Porphyry, and Simplicianus all acknowledged a variety of debts to Plato. They all admired the honorable life of Socrates. They all saw, in the theory of forms, the kernel of Christian transcendentalism. Hardly a line concerned with the immortality of the soul was written in the first four centuries of Christianity which did not refer to the authority of the Platonists for intellectual support. Still, when the time came for Augustine to share his vision of God's eternal city, it was not the *Republic* from which he sought inspiration, nor was it Plato's *Laws* that provided the maxims by which the Christian ordered his conduct.

It is to be noted that the task of explaining human error and limitation becomes far more difficult when egalitarianism is presupposed. For Augustine and his later disciples especially, the burden was increased by the transcendental character of their egalitarianism. Not only were all human beings presupposed equal, but they were deemed so by virtue of divine intervention. Because of this wedding of egalitarianism and transcendentalism, the problem of evil was the most vexing of all. The entity adopted to solve this problem was that of the free will. Without it, early Christianity would have been little more than Manichaeanism: the belief in a god of goodness battling eternally with a god of evil and with the fate of every soul hanging in the balance. Without free will, evil had to be attributed to divine authorship, which was a notion not only heretical but one utterly incompatible with the very concept of sin and personal responsibility. Only by asserting freedom of the will were the Patristic philosophers able to reduce evil in the world to human invention and, simultaneously, to elevate humans to the status of morally responsible agents. Although free will was able to achieve these desired results, however, it created still another and potentially more telling problem: if the human will is free, how can God be said to have knowledge of things to come? And, if God does not have such foreknowledge, how can He be said to be omniscient? This dilemma had been settled once by Cicero who, in his attacks on those Stoics who believed the future could be foretold, granted free will to man only at the expense of denying omniscience to God.[22]

Augustine's attempted solution began with an analysis of Cicero's conclusion. Praising Cicero for his reason but reproaching him for a lack of that wisdom which faith bestows, Augustine argued that the proposition need not be and, in fact, is not either-or:

> [T]he religious mind chooses both, confesses both and maintains both by the faith of piety. . . . God knows all things before they come to pass and that we do by our free will whatsoever we know and feel to be done by us only because we will it.[23]

Conditions may conspire to limit our range of possible actions, but we can always strive to want to do what is right:

> Wherefore our wills also have just so much power as God willed and foreknew that they should have; and therefore whatever power they have, they have it within most certain limits.[24]

The problem of conduct, on this analysis, is the problem of will: getting someone to recognize obligations to self, as a child of God, to God as the creator, to others as brothers. Failure of the will is sin and is unnatural.[25] Nature does not counsel us to seek evil. When we do, when we intend evil, we are making ill use of a good nature.[26]

Conversion to Christianity solved the problem of conduct as it solved the problem of knowledge. The Roman citizen could evade Caesar's notice, but all was visible to the all-seeing God. Caesar's guard could reward the brave act and punish the coward, but only God could know the true motive behind each and every act, and His rewards and punishments were of a very different sort:

> If a man does not pay his debt by doing what he ought, he pays it by suffering what he ought.[27]

The Patristic Legacy

Augustine was not the Church's only voice, but he was surely its most authoritative. His *Confessions,* taken as a whole, attacked intellectualism so broadly as to number too many casualties. His *City of God* inspired hope in an afterlife but necessarily reduced that great interest in a daily civic life that is the mark of every truly "classical" period. His *De Trinitate* and *De Libero Arbitrio,* which made "the goodness of God the cause of all things," relentlessly drew attention and energy away from Stoic science, Aristotelian logic, and Platonic rationalism. To the extent that we consider an unornamented and disinterested search for truth to be a noble one and a positive enterprise, Augustine's influence must be judged harshly. His teachings and the eagerness and talents of his followers induced fear and humility of a sort antithetical to creativity and culture.

Viewed against the background of fourth- and fifth-century alternatives, however, the growth of Christianity seems by far the better course. Despite the fear and trembling, man was first introduced to a psychological and theological theory of natural equality which has been a guiding force in the Western world ever since. Despite what may appear to be spiritual excess, this Patristic psychology, through its transcendental elements, rescued the mind from the blind alley of skepticism and the nihilistic prophesies of unbridled materialism. In Book XIX, Chapter XVIII, of *The City of God,* Augustine pauses to contrast the uncertainties proclaimed by Varro and the other philosophers of the New Academy with the certainties of the Christian faith. Skepticism regarding what the mind truly apprehends is denounced as nothing less than madness. The very grounds of skepticism are the misplaced confidence we have in fallible and deceiving senses which, after all, are material things sharing in the general limitations and corruptions of matter. The rational mind, unburdened by all this, can apprehend certainties and know the truth.

In focusing on the will and intention, the Patristics must be credited with greatly advancing the psychology of motivation. Augustine's attack on the heresies of the Pelagians affirmed freedom of the will and established the grounds on which moral blame and praise are justly ascribed. Pelagius had argued that the Augustinean position on freedom of the will was incompatible with divine omnipotence and concluded that this freedom was independent of the God's grace. The rejoinder has Augustine taking moral freedom to be a natural (God-given) endowment, itself expressing the power of divine grace to bring about something merely material nature could not attain.

It is true that other Patristic philosophers, if only to rebuke the philosophers of antiquity, preserved the older tradition, held it safe in monasteries from the western shores of Ireland to the flatlands of Syria. Thus, for every many years, a darkness settled over the mind of Europe, a darkness sustained by hunger and fear, by the failure of men and of law. We must be mindful of Christianity's contribution to this long silence, but we must also recognize how long it might have lasted had Christianity failed, for it was during this same five hundred years that new empires formed, new barbarian kings rose to power, and new cults emerged. Had these kings and cults no adversary, what would Europe's fate have been? Without these Christians, who would have labored to retain the Hellenic record or Roman Law? What would have stood between the king's caprice and the dignity of every human being?

For five hundred years the secular and the clerical powers grew. Conflict and competition raged with predictable ferocity. The citizen was torn for centuries between the king's command and God's. The modern world waited until a great king sought lasting peace with a good pope, when in 800 Charlemagne was crowned by Leo III. During this long pause, thought did not cease, but it was less public, less assertive, even less relevant than it had been for any time since the classical period. It was not until the reawakening of the philosophical mind in the tenth and eleventh centuries that these public and assertive elements returned. Whatever ill effects the authority of Augustine may have produced, and there were ill effects, this same authority formed the foundation on which was constructed a veritable monument to reason. We are, perhaps, so ready to note the Augustinian emphasis upon faith and upon forces of an utterly transcendental nature that we forget his central philosophical maxim: "Reason should be master in human life." [28] Indeed, rather than reinforcing the superstitions and innocent fears of his fellow Christians, he sought to dispel these very fears by relegating emotion of any sort to a rank lower than reason's. He railed not merely against lust but against terror as well. Far from attempting to foist belief upon the unsuspecting or controlling the conduct of the brethren through reckless propaganda, Augustine labored to know the truth and to free people through its power. This power was reason itself: that which separates not only man from beast but the fool from the wise.[29]

On specific psychological processes and functions, Augustine also carried on the tradition launched by Aristotle of examining distinct faculties and attempting to integrate them into a general theory of mind. That some of his views, though reasonable enough, would be provably wrong is less important than his keeping alive the relevance of theory and analysis to an understanding of human nature. If Scripture was in some sense the final word on such matters, it was surely not the only word.

In the matter of visual perception, Augustine was a representative of what might be called the "tactile-vision" school. The eyes must emit some sort of effluvium that makes contact with visible things, all this happening with nearly instantaneous speed. Perception is influenced by concentration and attention and is thus part of an overall *active* process, not simply passive stimulation by external objects. Our perceptions are integrated in that they have a duration during which time they are held together. This means that some *memory* process must maintain the continuity. Again, the general processes of integration are invoked to explain how the specific perceptual events become organized into what we now might call cognitive *wholes*.

Memory figures also in the storage of previous experiences, this time in the form of imagination, which may be either merely reproductive or fully creative. It figures centrally in the integrated and continuing sense we have of our very *selves*. In this Augustine not only pays attention to a problem that remains alive in philosophy—the problem of *personal identity*—but offers an early version of the theory Locke will defend thirteen hundred years later. Anticipating Descartes, too, he removes the grounds of skepticism regarding one's *self* by insisting that doubt itself entails being (*De Trinitate,* Book X, Ch. 14).

On the question of the relationship between the material and spiritual aspects of human nature, particularly as this bears upon matters of sensual pleasure, Augustine ushered in a different conception of Genesis and of marriage and family life. The more authoritative of the early Fathers (e.g., St. Jerome, Gregory of Nyssa, St. Ambrose) understood Genesis as describing two entirely different grounds of relationship between Adam and Eve before and after the fall. The prelapsarian couple was wed in friendship, whereas after the fall they were intimate and carnal. Sexuality, including that of marriage, is thus a correlate of the fallen nature of human beings. Augustine, however, interpreted the account in Genesis differently, concluding that marital bonds are bonds of friendship to which is added a quite natural sexual bond. If there is a defect to be found in this domain it is a defect of the will, a *willfulness,* that substitutes sexual gratification for what finally are the goods of marriage, the *bonus coniugali.* As with Aristotle, though it would seem essentially independently of Aristotle's works, Augustine takes the emotional side of life as natural and then seeks to define the conditions under which emotionality is to be encouraged or condemned. It is finally not the sentiment or emotion but the contexts in which we are disposed to have such feelings that determine the moral standing of the feelings themselves.

How, then, may we fairly summarize the Patristic contribution as embodied in the works of Augustine? In an all too general way, we can cite certain propositions of a distinctly early Christian shade which, over succeeding centuries, became an integral feature of psychological speculation but with more and more of the theological aspects stripped away:

1. Over and against earlier nature-philosophies, and in keeping with later developments in Stoic philosophy, the Patristics accorded to man a position unique in the world. This bias not only prevented the establishment of an ethological or evolutionary perspective but also discouraged the application of scientific principles of any sort to questions about human knowledge, conduct or will. This contribution was, on the whole, negative, all the more so for the animal-world thereupon abandoned to the domain of the merely useful and providentially applied.

2. Because of the unique position of humanity in relation to the bal-

ance of nature, the Patristic scholars insisted that each person was individually responsible for his actions. This attention to individual responsibility picks up where Plato and Aristotle leave off, but with otherworldly implications of overarching importance. *Conscience* is more or less put on the map of psychology.

3. In their antimaterialist convictions, the philosophers of the early Church insisted upon a psycho-physical dualism according to which the psychological characteristics of human beings were forever beyond physical analysis. In orthodox religious terms, the dualism was between soul and matter; later this became a dualism between mind and matter. In our own time, the issue survives as the mind-body problem. The Greeks invented the problem, and Augustine solved it in a way that was satisfying for over one thousand years. This contribution, it would seem, was mixed.

4. In carrying religiosity into every sphere of human concern, the Patristics often regarded daily experiences as trivial. They continued that Platonic tradition which held all perception suspect and sanctified this tradition with some powerful scripture. Occasionally this led to a modest, sometimes virulent, anti-intellectualism. This was its most negative effect, made worse when the political authority of the Church was finally able to render the position official.

5. Rationalism, when colored with mysticism, produced a psychology of what we might call "intuitionism," the belief in the power of the mind to achieve transcendental awareness of truth. The Patristics called this by several names: the "interior sense," the "light of faith," and "grace." It was the agent that gives intention to our actions and thereby makes us accountable. This contribution survives in the form of theories of the unconscious, notions of unconscious motivation, and theories that assert the innate origins of our moral sensibilities. In considering the failure of this intuition to be evidence of a defect, the Patristics—inadvertently, we might suspect—were presenting a theory of psychological deviancy as disease. More significantly, they were advancing the otherwise subtle notion that it is not enough to consider mere conduct (behavior) in one's attempt to understand individual psychology; only when an action is judged in light of the intention behind it can that action be said to be known. The Patristics were not behaviorists.

Galen (ca. 130–200 A.D.) and the Scientific Alternative

We noted in Chapter 2 that while philosophers concerned themselves with the eternal imponderables, the Hippocratics continued to assemble a collection of clinical observations and therapeutic outcomes. That is, they continued to develop Greek medicine as an essentially empirical science, indifferent or even hostile to the speculative excesses of the Socratics. A similar division of labor occurred in the Patristic period. As the fathers of the Church struggled to integrate pagan philosophy, barbarian ritual, and Christian teaching, Galen and his followers contented themselves with the more immediate problem of curing the sick, relieving pain, and understanding the causes of death and disease. Galen not only kept the Hippocratic system alive for subsequent historians; he also kept the idea of experimental science alive for subsequent scientists.

His most important psychological work was *On the Natural Faculties,*[30] in which he attacked not only the untested hypotheses advanced by philosophers concerned with biology but, more particularly, the very notion that an untested hypothesis has any place in biology. We are not to treat Galen as a radical empiricist, however. He was a practical man devoted to unearthing the facts of clinical medicine, and he was willing to employ any method promising success. With respect to rational deduction versus empirical induction, he had this to say:

> [I]t is not our habit to employ this kind of demonstration alone, but to add thereto cogent and compelling proofs drawn from obvious facts . . . [which] can actually be recognized by the senses.[31]

There are also nativistic elements in Galen's theories. Whole sections of the treatise are directed against the extreme Epicureans and Stoics who contended that man is to be understood as a material entity treated to experiences by way of external stimulation. By natural faculties, Galen meant those that exist by nature and, therefore, do not come into being in the empirical sense at all. Included among these natural faculties are those of the soul and especially those that ultimately reveal themselves in the form of reason and intellect. His theory, then, is not far removed from that of the Platonic school:

> Some of these people have even expressly declared that the soul possesses no reasoning faculty, but that we are led like cattle by the impressions of our senses, and are unable to refuse or dissent from anything.[32]

> Nature, however, knows better than these radical empiricists, for, she skillfully moulds everything during the stage of genesis and she also provides for the creatures after birth, employing here other faculties again, namely, one of affection and forethought for offspring, and one of sociability and friendship for kindred.[33]

Thus, with respect to the emotions, the social instincts, the maternal drive, and the affective dimensions of life in general, Galen's position is uncompromisingly nativistic. He rejected specifically that brand of empirical materialism according to which organisms enter the world as *tabula rasa,* whose knowledge and behavior must await the mechanical instructions given by experience. To the extreme Epicureans who wished to believe that we begin life as amorphous clay, gaining wisdom and virtue only through experience—those who think our essential character is but the consequence of certain channels having been etched into a human form from without—Galen offers this:

> Thus, every hypothesis of channels as an explanation of natural functioning is perfect nonsense. For, if there were not an inborn faculty given by Nature to each one of the organs at the very beginning, then animals could not continue to live . . . let us suppose they were steered only by material forces, and not by any special faculties. . . . if we suppose this, I am sure it would be ridiculous for us to discuss natural, or, still more, psychical activities—or, in fact, life as a whole.[34]

Galen is not to be viewed as a philosophical rarity in the Patristic period since, on the fundamental question of human nature, his system makes ample provision for native (God-given) forces. This system is even less a psychological rarity since, in its principal tenets, Galenism is neo-Platonist in its own way. However, as a scientist, he is unusual on at last two counts: he insisted on accepting the data of experience over the force of logic when the two were in apparent disagreement, and he dismissed scientific hypotheses that were devoid of empirical content. In accepting the psychological aspects of human and animal life as the consequence of hereditary influences and, as a result, in remaining skeptical toward the possibility of environmental factors having much of an effect upon the human condition, he located himself quite comfortably in an age soon to become Dark. Nonetheless, centuries later, when medieval philosophers would introduce once more the benefits of observation and experiment as methods by which the truths of nature might be uncovered, the debt to Galen would be recognized and noted as a major one.

Galen did not present himself as a philosopher. Quite the contrary. But his system of medicine was rife with philosophical implications. In opposing such radical materialism as had been advanced by the atomists he found himself constrained to propose a life principle by which the organic world was to be distinguished from mere matter. He termed this principle, *spiritus anima,* a principle that would reappear repeatedly in subsequent centuries, most notably in the "animal spirits" of Descartes and in biological theories based on *vitalism.* From Galen on, there would be a theoretical and philosophic tension between materialists and vitalists, the former insisting that the laws governing the physical world were sufficient to embrace not only living things but human life and human mind as well; the latter urging that life and its psychic attributes could not be entirely explained without recourse to an extraphysical, life-giving *vital* principle. A fair share of the controversies waged throughout the history of psychology is based either directly or derivatively on this tension.

There is still another feature of Galenism that would be authoritatively consulted in later ages, the use of living animals in surgical experiments. From his surviving treatises and reported findings, it is clear that Galen's work caused great pain to the animals thus vivisected. In light of prevailing religious beliefs and the high wall that the Stoics had erected to separate the human community from the balance of the animal kingdom, it is not surprising that Galen's conscience was untested by any of this.

Notes

1. Epictetus, *Discourses,* translated by P. E. Matheson, in *The Stoic and Epicurean Philosophers,* edited by Whitney Oates, The Modern Library, New York, 1957 (Random House edition, 1940).

2. Ibid., Book II, VII.

3. Ibid., Book III, VII.

4. Ibid., Book I, VI.

5. Tertullian, *De Spectaculis.* In *Corpus Scriptorum Ecclesiasticorum Latinorum,* H. Hoppe, Vienna, 1939. The most accessible unabridged essays of Tertullian's in English appear in *The Fathers of the Church: Tertullian (Apologetical Works) and Minucius Felix (Octavius),* translated by Rudolph Arbesmann, O.S.A., Fathers of the Church, New York, 1950. It is in his letter to Scapula (*Ad Scapulam*), the governor of Africa, that Tertullian insists on the right of every man to worship as he sees fit. In the *Apology,* he chastens the Romans for infanticide, proclaiming that the followers of Jesus will not even take the life of a fetus. Tertullian was one of those rare critics able to combine wit and judgment. He ridiculed the pagan excesses of the Romans and, in the same breath, judged Origen's self-emasculation with the question: "If God wanted eunuchs, could He not have made them?"

6. Plotinus, *The Enneads,* Book III, 6, 7. An English translation has been published by Pantheon Books, Random House, New York.

7. St. Augustine, *The Confessions,* translated by Rex Warner, Mentor, New American Library, New York, 1963.

8. Ibid., Book III, Ch. 4.

9. Ibid., Book IV, Ch. 16.

10. Ibid., Book VII, Ch. 20.

11. Ibid., Book VII, Ch. 21.

12. Ibid., Book V, Ch. 5; Book VII, Ch. 12.

13. Ibid., Book VIII, Ch. 2.

14. Ibid., Book VII, Ch. 21; and in *On Free Will,* Book II, Ch. 15.

15. Ibid., Book X, Chs. 12, 13, 14; and in *On Free Will,* Book II, Ch. 4.

16. Ibid., Book X, Ch. 30.

17. St. Augustine, *De Trinitate,* Book XII, Ch. 15, in *Basic Writings of St. Augustine,* 2 vols., edited by Whitney Oates, Random House, New York, 1948.

18. Ibid.

19. Ibid.

20. Ibid., Book X, Ch. 6 and 7.

21. St. Augustine, *The City of God,* Book XIX, Ch. 15, in *Basic Writings of St. Augustine.*

22. This argument appears in Cicero's *De Divinatione.*

23. *The City of God,* Book V, Ch. 9.

24. Ibid.

25. Ibid., Book X, Ch. 17.

26. Ibid.

27. St. Augustine, *On Free Choice of the Will,* Book III, Ch. 15, in *Basic Writings of St. Augustine.*

28. Ibid., Book I, Ch. 8.

29. Ibid., Book I, Ch. 9.

30. Galen, *On the Natural Faculties,* translated by Arthur John Brock, Putnam, New York, 1916. I call this his most important psychological work notwithstanding his authorship of an essay titled, *On the Affections of the Mind.* This latter work is finally no more than an attempt to understand delirium, fits, drunkenness, etc., in terms of the Hippocratic theory of the humours and the effects of an imbalance in the humours. It is not really a psychological treatise at all.

31. Galen, *On the Natural Faculties,* Book III, Ch. 2.

32. Ibid., Book I, Ch. 12.

33. Ibid.

34. Ibid., Book II, Ch. 3.

5 Scholastic Psychology: The Authority of Aristotle

Portraits of imperial epochs such as those of the pharaohs, of Alexander, and of the caesars can be painted with relatively broad strokes. Empires are regulated by a set of laws. They are governed by a visible coterie of powerful figures. They possess that unmistakably imperial tone of life. Their economies are based on a specific currency and depend on a few basic goods. They have rather transparent policies toward neighbors and rather well-articulated strategies for defense and conquest. They speak in one tongue that must be mastered by all who would share in the bounty and protections of citizenship. Portraits of "empire," then, are geometric. No matter how great the particular empire, one always knows how much canvas is required to contain it. On the few occasions when an empire succeeds, it becomes the focus of an age. That is, its entire culture becomes its principal export, and this culture is of such a nature as to transform and dominate alternative cultures. The empire that succeeds best and thereby creates a historically identifiable age is one that places a durable mark on every feature of social life. Accordingly, when the empire fails, nothing that defined it remains unchanged. Art, law, letters, economy, homelife, religion, politics, all display the signs of change as that which once animated each of them begins its decline. The historian's problem, in studying that almost fantastic period represented by the deceiving title "medieval," is that it is a post-imperial period, one occurring at the end of an empire that was great even by inflated imperial standards.

The problem, then, is not that the medieval epoch lacks distinguishing features but that each of its features is a feature-in-transition. As it is post-imperial, the medieval epoch is also pre-national. Indeed, the nations of Europe as received by modern times were medieval inventions. Empires have an imperial character and nations a national character. But what is the "character of a transition" if not a contradiction? It is only after we appreciate the complexity of this question that we can be protected against those misleading commonplaces employed in describing and dismissing the "Middle" Ages. According to these commonplaces, the period from Rome's conquest to the Renaissance was filled with a tame and dull homogenization of thought and belief; with a feudal system that reduced the faithful to servility; and with political, moral, and intellectual hegemony exercised by a clerical elite in league with dukes and princes.

With respect to the putative homogenization of thought and belief, it is only necessary to point to the difficulty of writing medieval history to expose the error. It is far easier to describe that which was Hellenic or Roman than to summarize that which was distinctively medieval. Indeed, it is precisely because the medieval epoch was neither imperial nor national that the search for its character must be so laborious. Thought and belief, always ambiguous terms when applied to more than one person, were probably more heterogeneous from the sixth to the eleventh centuries than during any equivalent span before or since. That superstition was rampant is beyond question, but this superstition, as a non-rational phenomenon, is a veritable symbol of heterogeneity of outlooks. That is, in order to render belief uniform, it is necessary to reduce it to a set of teachable principles, to put it on a rational and even quasi-philosophical level, and to argue it into an apparently unimpeachable form. Formal religions take hold only after dislodging the superstitious mind. They replace superstition. In this respect, we cannot even discuss Catholicism as the orthodox and dominant belief of the period until well into the eleventh century, and we must date its intellectual hegemony in the thirteenth.

The demise of a recognizably *Roman* world was brought about by internal corruptions and external force, the latter applied by waves of tribes from Gallic and Germanic Europe. These people had their own venerable customs and laws, most of them radically different from Rome's. The Salic laws were usually no more than a set of simple sentences indicating the penalty for specific infractions. Twelve solidi in the currency of the age was the amount it would cost to buy a slave, so we begin to see the importance attached to various persons and actions by noting the cash-penalties assessed: 200 solidi for killing a female child, 600 for killing a woman of childbearing age, 900 if she was pregnant. The same laws assessed fines of various amounts for touching different parts of the body of an unmarried woman, or for destroying various portions of an adversary's anatomy: a greater fine for the loss of an eye than a finger, etc. Needless to say, by the refined and developed standards of Roman law, all this was no more than a form of barter.

As for the feudal system, it was in fact a variety of systems evolving continuously over a period of centuries. It was never merely an alternative to anarchy and the daily threat of a violent death; it was typically the only alternative. As an essentially economic system, feudalism had much in common with Roman life from the second to the fifth centuries; that is, the feudal lord or baron was the economic equivalent of the Roman governor or provincial. But what was behind the system was a privatization of life and property quite unlike anything seen in classical Rome or Greece. Conquering hordes had lent their strength to leaders who promised the booty of victory and who enjoyed loyalty as long as their own personal strength and the success of their assaults continued. From the very outset, then, the conquest of Rome had private interests as an essential motive. These interests evolved into Feudal lordships, bounded properties, and small armies of laboring vassals who in simpler times might have been slaves.

As a system of classes, feudalism gave the lord or baron a position quite akin to that of chieftain in prefeudal tribal communities. Scarcity requires economy, and economy requires binding agreements which, in turn, demand authority. Whether this authority is vested in archons, consuls, chieftains, barons, or kings depends principally upon the size of the community and its historic sources. Contemporary Americans must be willing to sacrifice their lives in the national defense, an obligation quite in line with the serf's. The same Americans have a servile relationship to laws, including those that tax possessions, proscribe treason, and otherwise regulate public conduct.

Of course, American citizens share in the authorship of these laws and their amendment. But feudal life was also based on agreements by both parties, and these agreements had the force of law.[1] The serf entered into service through a ritual that included physical contact with his lord. Behind the symbolism stood the belief that something actually passed from the latter to the former; a form of sanctuary, an oath of protection having nearly corporeal properties. R. W. Southern, in *The Making of the Middle Ages,* recounts submission by one freeman for the purpose of gaining another vineyard. As Professor Southern notes in his analysis of the documents of the period, servitude was considered the lot of every man, if only servitude to God. Thus, some men were "serfs to the serfs of God," finding in their servility only the proper conduct of a Christian, since "all men labour and serve, and the serf is a freeman of the Lord, and the freeman is a serf of Christ."[2] Note also that the same Greeks who reviled tyranny benefited from slavery. Plato's men of gold, brass, silver, and iron are models for those hierarchic classifications of citizens that until modern times suffused every political community of the Western world. The slave in Greece was the one whose side lost the battle, whose birth was non-Hellenic, and whose debts could not be paid in any other way. The advent of the Dominate in the Roman Empire introduced to the Western community the precedent of man-as-god and, therefore citizen-as-servant. This is all worth considering not in order to absolve medievalism of wrongful deeds—if, in fact, it is a function of posterity to absolve its ancestry of anything—but to prevent the needs in question from being judged unique to a given age or religious temperament.

These generalizations, however, have limited validity when considering so various and evolving an epoch. It is only in the High Middle Ages (from, say, 1100 to 1350) that we can find written justifications for the policies and arrangements that applied to a substantial fraction of the Western world. Only at these later dates is the term "medieval" supported by those documented and reflective elements necessary for more precise historical analysis. From 500 to 1000 A.D., we can examine certain people, even certain groups. But we are not able to add these diverse settlements together to arrive at the character of an entire age, let alone its psychological attitudes. We can only try to capture, in a series of snapshots, how the mind of the Middle Ages grappled with itself and with the awesomely unpredictable world around it.

Fear and Magic

The changes that overtook the West between the fifth and tenth centuries were colossal by standards both ancient and modern. There is so little in common between the Frankish kingdom of Charlemagne and the late empire of Justinian that even comparisons are difficult to establish. Of all the prevailing forces working to change the world, none was more significant than Islam. The transformations imposed by the disciples of Mahomet (571–632) were progressive, cumulative, and overwhelming. Successively, Mohammedanism overcame Persia (651), Syria (636), Egypt (642), north Africa (698), and Spain (711), and even led to the blockade of Constantinople (717). The resulting economic and cultural consequences arose most immediately from the fact that the Mediterranean itself no longer served Western interests. The searching thesis that the Middle Ages are to be understood principally in terms of Islamic control of the Mediterranean Sea was advanced long ago by Professor Henri Pirenne.[3] His own description of what had occurred cannot be improved:

> The familiar and almost "family" sea which once united all the parts of this commonwealth was to become a barrier between them. On all its shores, for centuries, social life, in its fundamental characteristics, had been the same; religion the same; customs and ideas, the same or very nearly so. . . . But now, all of a sudden, the very lands where civilization had been born were torn away; the Cult of the Prophet was substituted for the Christian Faith, Moslem law for Roman law, and Arab tongue for the Greek and the Latin tongue. The Mediterranean had been a Roman lake; it now became, for the most part, a Moslem lake. From this time on it separated, instead of uniting, the East and the West of Europe. The tie was broken. . . .[4]

The part of the world which we now identify as Europe had become enclosed. To the north, the Danes and Saxons presented a constant threat to the vestiges of Roman civilization. From Spain in the south, the invasion and piracy of the Saracens were irresistible. In the East, of course, Islam was supreme. That life, which once had been Mediterranean, now collapsed toward a safe European Center, but it could not be moved northward intact. Instead, it degenerated into that form of tribalism which would evolve into feudalism. Even the secondary effects of the Mohammedan conquest were telling. Syria, for example, had been the major source of papyrus, and from 650 until the eleventh century the supply was virtually eliminated. The Arabic number system, which would animate European science and mathematics only after the fourteenth century, remained isolated from the Western mind.

Technology is the achievement of a settled people. It arises when an abiding problem, an abiding conflict with nature, insists on a solution: for example, the Egyptian farmer's need for water and the need to protect crops from an overflowing Nile; the Athenian desire to erect a temple with massive columns. Technological undertakings on a grand scale are pointless to those living a nomadic life. By the time the problem is solved, conditions have forced the nomad to a new but ever temporary home. Thus the threat and the reality of invasion, of piracy, and of crop failures not only robbed Europeans of their culture but denied them the very geographic permanence that might make a new culture possible.

Time itself was, for the medieval mind, far less metrical than it is to the modern mind. Indeed, the medieval sense of time was as exceptional as so many of the practices and beliefs we have come to identify with the entire epoch. For the medieval Christian, time had two major divisions: the brief and insignificant one in which one's sinful life proceeded, and the cosmically enduring one in which the suffering or the joys of the soul would occur. Thus:

> Every day, every hour, thus without ceasing
> I must finish my life, and recommence
> In this death uselessly alive.[5]

Christian belief and its essential elements of baptism, death, and resurrection fostered a perception of time devoid of scientific meaning and one not seen in philosophy since the age of the pre-Socratics. Each day was a kind of rebirth unconnected by that cognitive thread we know as history. Medieval life passed in a series of otherwise disconnected moments. Its events were sudden, as sudden as the imminent apocalypse that would end all life all time, all fear. Marc Bloch put it aptly:

> These men, subjected both externally and internally to so many ungovernable forces, lived in a world in which the passage of time escaped their grasp all the more because they were so ill-equipped to measure it. . . . The truth is that the regard for accuracy remained profoundly alien to the minds even of the leading men of that age.[6]

Without a sense of time, a sense reinforced and validated by empirical regularities, the concept of natural causes fails to overtake the unharnessed imagination. The medieval mind, this mind without a clock, was scientifically backward. In any case, neither humans nor the world was to survive very long. The end would come in the twinkling of an eye. What sign would there be?—Who would be so bold as to reckon the moment God's will would assert itself? "What is man?" asked Alcuin, the resident scholar in the court of Charlemagne. And he answers, "The slave of death, a passing wayfarer." And then, "How is man placed? Like a lantern in the wind." [7] These are lines provided by a man who described himself thus: "Alcuin was my name: Learning I loved. Oh thou that readest this pray for my soul." [8] The men and women who lived in this period endured the harshest circumstances. The power of the Church was political, but neither its moral nor its religious lessons could penetrate that large and shifting body of tribes that remained trapped by the timeless rites and visions of the pagan. With gruesome weariness these men and women struggled to rationalize the effects of their woeful lives. Not only was death a release, but the decaying body of the dead parent was praised for its sweet smell. The soul of this body now enjoying the presence of God Himself may still come back to sanctify its former shell. Thus a lock of hair, a rotting bone, the bristle of a beard, may have the power to heal or to protect its possessor against the devil and his demons. In one immense though bounded universe, man and his planet, the stars, the moon, each blade of grass, all shared in God's all-seeing vision. As late as the thirteenth century, St. Francis would repeat this spirit of medieval life in words whose beauty has survived the tragedy of it all:

Praised be my Lord for brother wind
And for the air and clouds and fair and every
kind of weather.
Praised be my Lord for sister water.
Praised be my Lord for our sister, mother earth,
The which sustains and keeps us
And brings forth diverse fruits with grass and
flowers bright.
Praised be my Lord for our sister, the bodily death.
For the second death shall do them no ill.[9]

It would be misleading to reserve to the Middle Ages the mind's domination by fear and magic, for every period of the human experience finds good sense in short supply. There can be no doubt but that the pervasive fear of death did have a paralyzing effect upon the few who might have made lasting contributions of an artistic or intellectual nature. Scripture was interpreted as predicting the world's end in 1000 A.D. or thereabouts, and to those who believed this (such credulity remains in abundant supply) the notion of launching a major project of any sort would have seemed chimerical. Even for those who were unaware of the exact date of the apocalypse or who entertained a different date, the crucial fact was that there was some date and that not too distant. That a fear of the end, both a personal and a worldly end, was generally held seems confirmed by the incessant rationalizations invented to control it. Death as freedom, as liberation, as rebirth, as escape, even *death as a good* are the themes that abound in the literary evidence of the period.

The run of competent and occasionally great philosophers in the West begins with St. Anselm (1033–1109). We are tempted to connect this resumption of intellectual pursuits with a fact that must have surprised a good many Christians; the fact that they were still alive after 1000 A.D.! There were, of course, less airy considerations responsible for the reawakening, and we shall examine several of them. Nevertheless we are advised not to dismiss too eagerly what, to the modern mind seems utterly incomprehensible: that scholarship, art, enduring institutions, and culture itself suffered neglect by a people generally persuaded that the world would not endure.

Accepting this, however, we must be careful not to judge these people and their other characteristics in terms foreign to the rest of the story of civilization. If we are to segregate the Middle Ages on any basis that might satisfy the demand for rigor, we must avoid variables such as poverty, plague, fear, superstition, zealousness, faith, and the like. We find these generously represented throughout the course of history. What is different, however, is the ease with which the literate and even intellectually gifted medieval citizen abandoned so completely those critical faculties that are the stamp of a tutored mind. What was unique about the Middle Ages, then, was not the record compiled by the struggling masses but the record attained by those very figures who presumed to lead, and by those who might pass for the best minds. In this respect, the achievements by scholars between Boethius (480–524), who was the last of the great classical thinkers, and John the Scot (ca. 810–ca. 877) were negligible. This interval, coinciding with the expansion of Islam and the scattering of the Western community, may be described legitimately as a "dark age." It may, however, also be viewed as a period of regrouping and assimilation during which monks in the west of Ireland preserved the language of the Hellenes; when kings and popes sought economic and even cultural foundations for peaceful coexistence; when the teachings of Islam forced the philosophically inclined believer to articulate arguments for Christianity. It was also the period in which Arab and Jewish scholars could reflect upon the ancient wisdom of Greece and Rome and thereby begin to invigorate the growing empire intellectually.

Charlemagne and the Prelude to Renewal

When Charlemagne allowed himself to be crowned by the pope, he displayed the rare talent of the political genius who knows what the future will demand.

More than any religious connotation that might have attached to the event, the coronation symbolized the rebirth of the Western world, a world now organized around a set of religious principles shared by both the temporal and the spiritual monarchs of that world. The result was at least the possibility of European unification against an enemy who, now, was a heretic as well. The Western king could now claim to defend a competing truth and one as great in its implications as that which impelled the soldiers of Islam.

We would overestimate Charlemagne's personal contribution to credit him with the creation of a European community of nations. But his alliance with the Church of Rome did establish at least a sense of community in a part of the world which, for several hundred years, was marked by provincialism. He served as king of the Franks for thirty-two years (768–800) and, until his death (814), as Charles I, emperor of the Holy Roman Empire.[10] He was devoted to learning and committed to the education of his people. He imported Alcuin from England to establish a palace school and strongly encouraged the bishops of the empire to incorporate schools within the churches and abbeys. These schools and even the curriculum introduced by Alcuin[11] were the wellsprings of the medieval university, an invention of the twelfth century. He was no Pericles, nor was Alcuin an Anaxagoras, but Charlemagne did restore respectability to learning and thus cut a seam in the heavy curtain that had darkened European scholarship for four hundred years. He restored some stability to the economy, provided at least the hint of safety for those who might otherwise have wandered, and planted the banner of Christianity in every major center of Europe from the Ebro River in Spain to the southern edge of Denmark. To the extent that the modern Western world retains institutions, ideas, and a general perspective that are Greco-Roman in their broadest features, then, to that extent, Charlemagne can be said to have been a maker of the modern Western world. He did not eliminate fear or magic—though he legislated against both—but he created a climate congenial to their antagonist, reason.

The Revival of (Aristotelian) Rationalism

There was a twelfth-century Renaissance which, in purely intellectual respects, compared favorably with its more vaunted successor in the fourteenth and fifteenth centuries.[12] The most vivid signs of this general rekindling of energy and hope are, of course, the majestic Gothic cathedrals. Even today, in an age that has made bigness a virtue, these churches seem to be and are of gigantic proportion. Still more impressive is the fact that nearly all of them (at Paris, Chartres, Amiens, Laon, Beauvais, Rheims, LeMans, Tour, Orleans, Mt. St. Michel, at Canterbury and Oxford, at Prague and Cologne, and throughout Europe and the British Isles) appeared in less than two hundred years. Coincident with their appearance were the revival of classical Latin, the resumption of serious philosophical inquiry, and, most significantly of all to intellectual history, the passage from Saracen Spain to Christian Europe of the Arabic translations of Aristotle's works.

The major figure here must be Avicenna (980–1037), whose Islamic rendition of Aristotelian thought would reacquaint the Western philosophical community with a far wider range of Aristotle's work than had been available in the west for centuries. His commentaries on Aristotle came to pose any number of questions for the thirteenth-century Christian theologians. Consider only Aristotle's metaphysical arguments for the eternity of the world—arguments directly contradicting Genesis and obviating the need for the God of Christianity. Consider, too, Aristotle's entire ethical system in which the concept of sin plays no part and according to which the ultimate sanctions of morality are grounded in considerations of the here and now.

Avicenna was a physician and a prolific commentator on the scientific and philosophical works of Aristotle. He anticipated such vexing psychological issues as our awareness of our selves as continuing entities, even amidst bodily changes. He explained the sense of "self" as something immediately known by a thinking thing (the *res cogitans* Descartes will offer six centuries later) independently of any and all experience. His hypothetical "flying man," suspended in space and blindfolded, retains a vivid awareness of his very *self* even while shielded from all external stimuli. Thus is self-knowledge based upon the very cognitive nature of a rational being.

It is in the Sixth Treatise of his *Deliverance* that he develops his psychological theories more fully along Aristotelian lines, but with a robust infusion of Platonic theory as well. In Chapter IX of this work he argues that the basis of all rational concepts must be immaterial, for such abstractions cannot be materially represented and thus cannot enter into causal relations with the material senses. The senses respond to the particulars of stimulation (*this* tree), but the rational faculty is able to abstract from this the *intelligible form* of "tree" and thus form a universal concept. By the same token, this immaterial grounding of rationality points to an immaterial soul which is therefore saved from degenerative changes. The soul is immortal.

First slowly and haltingly but soon quickly and continuously, the authority of Aristotle challenged and overtook the traditional authority of neo-Platonism. St. Anselm (1033–1109), who may be said to have initiated the philosophic revival in the West, titled his major work *Faith Seeking Understanding,* and argued patiently for the role of perception and reason in Christian life. His principal authority, however, is St. Augustine. Next, we confront the *Four Books of Sentences* of Peter Lombard (1100–ca. 1164), the most influential work in religious philosophy of the time and one that proclaims even more forcefully the position of reason in the affairs of faith. Still, the authority remains Augustine, who directs us to find God with that quality of ours, "than which our nature has not better, which is the mind." [13] But at the same time we discover Peter Abailard (1079–1142) placing the greatest emphasis on the authority of Aristotle, whom he calls "our prince" and who, for the next two centuries will be reverentially and simply dubbed "the Philosopher."

The revival of rationalism cannot be reduced to a single or even small set of causative agents. Nor is this the place to identify the many factors involved. Charlemagne's contribution has been noted, a contribution sustained and reinforced by King Otto, whose German kingdom enjoyed a mini-renaissance in the tenth century. St. Peter's, which had been assaulted by the Saracens as recently as 846, now commanded the brave loyalty of all of Christian Europe, and the Roman Church was able to administer more effectively the efforts of her monks and missionaries. Rome itself was now secure enough to serve as a center for grammar and law as the empire sought to restore its classical character. Not only was the empire secure against Islam, it was even able to reclaim territories previously lost. By 1085 the recovery of Spain had proceeded as far as Toledo, and by 1118 it included Saragossa. The notorious First Crusade (1096) was, we must note, a mere expression of the revival and not an antecedent. [14]

By the end of the twelfth century we find Europe in a state somewhat similar to that of Athens under Alexander: war, plague, and discord giving way to order, security, and growing prosperity; political instability submitting to the will of the great leader; and a visible but now conquerable foe providing the stimulus to cohesion. The nightmare had ended, men and women and the world itself were still in one piece, and the future, if only because there was a future, seemed far brighter.

Scholastic Psychology

In *The World of Medieval Learning,* Anders Pilz reproduces and discusses a quite remarkable illustration taken from a work by Peter of Spain published in 1514. Although the publication falls well within the period of the Italian Renaissance, the illustration is quintessentially medieval. As Anders Pilz says, it is the one illustration "that can sum up the referential system of the medieval scholar." [15] It is "Porphyry's Tree." Here Socrates (expressed in the diminutive "Sortes") and Plato are sketched at the base of the tree, with Socrates standing and Plato in a chair. Between them as the very roots of the tree are the names of particular persons (Henricus, Petrus, et al.) who, as individuals, fall outside any classification. It is only a genus or species that can be classified, not any detached item, including a specific person. But these particulars arise from the species *homo,* that universal class of which all individual persons are members. Combined with nonhuman creatures, *homo* arises from the larger class *animal,* of which there are categories of the *rational* and the *irrational.* These in turn

Porphyry's Tree

are derived from the more basic category of the *corpus animatum,* or *living body* which may be endowed with sense (*sensible*) and thus qualify as *animal,* or be without sense (*insensible*) and thus be *plant.* Ever more basic is the category of *substance* (*substantia*) formed by the most fundamental of all ontological distinctions, that between the *material* (*corporea*) and the *immaterial* (*incorporea*). To turn the tree upside down is to trace the ontological pedigree that culminates in identifiable and particular persons. The tree culminates in the universal class *substantia* into which everything having real existence falls. This entire apparatus is forged from Aristotle's (ten) categories. The most fundamental disputes of the medieval period pertained to the nature of the universals, their ontological standing, and the role of mental processes either in comprehending them or being their very source. Porphyry's tree continues to be a useful framework within which to comprehend the issues and theories spawned by Scholastic psychology.

Among those who write general histories of Western thought, there is a tendency to reduce the works of Augustine to "neo-Platonism" and to reduce the works of Thomas Aquinas to "Aristotelianism." The respect in which this practice is arguably valid is, alas, the respect in which we might reduce *all* philosophy either to Platonism or Aristotelianism. It is true that Thomas Aquinas (1225–1274) has been the intellectual voice of the Roman Church since the fifteenth century, and it is also true that his two major works, the *Summa Theologiae* and the *Summa Contra Gentiles,* derive their inspiration from Aristotle. But derivation is not duplication. One could not reconstruct Thomistic thought merely from a knowledge of Aristotelian thought. The two philosophers undertook their works in vastly different intellectual climates, with vastly different orientations and with vastly different objectives. If we are to comprehend the medieval view of psychological man, we must focus on Thomistic psychology. Thomas Aquinas was not the only or even the most influential figure to raise the important questions, but his were the only works that included all the perspectives prevailing in his age. And if we are to comprehend the portion that is distinctly psychological, we must first appreciate the problems facing Thomas and his Church in the thirteenth century. Let us, then, review briefly the problems confronting Thomas Aquinas in regard to human psychology and then proceed to his essentially Aristotelian solutions.

1. *Can man know God?* This is the central epistemological problem of the High Middle Ages. Peter Abailard, in his *Sic et Non* ("Yes and No"), listed some 158 theological questions that had been answered in a contradictory way by Scripture and by the early Christian Fathers. He was also one of several influential twelfth-century spokesmen to reject the real existence of universals and to adopt nominalism: that is, to assert that only individual entities are real

and that so-called universals are merely class-names (*nomines*) referring to concepts in the mind. By the time of Aquinas' floruit,
there was a creeping skepticism toward doctrine, a growing demand for rational as opposed to scriptural authority. The question
of whether or not man can know God, then, was just the most pressing member of the larger set, *Can man know anything?*

2. *What are our duties to God?* This question assumes that the general epistemological question has been answered; that is, we know
what man can know of God and God's will. Now what remains
to be determined is man's obligation to self, to the State, to his fellow man.

3. *What is sin?* In the ninth century, the Church was so disorganized
that the pope was unable to provide Charlemagne with an official
liturgy for the celebration of the Mass. By the thirteenth century,
intense efforts were directed at rendering Christian practices uniform. The Holy Roman Empire was indeed an empire, requiring
laws, agreements, and an understandable set of first principles. The
historic civil law of Rome, again taken to be the law of nations (the
ius gentium), was revived in the centers of legal scholarship and
incorporated into the decisions of royal and ecclesiastical courts.
But this had to be reconciled with a body of belief far more detailed
than the belief existing at the time of Justinian. Secular and clerical
powers were now so intermingled that the citizen-believer desper-
ately needed guidance for a lawful and Christian life. The concept
of sin had to be refined in such a way that all kings, bishops, and
farmers alike could be made aware of the specific obligations they
had to their Creator.

4. *What is the nature and status of the human will?* Scripture had
granted freedom of the will while insisting that God caused all
things. In the previous chapter we reviewed the Augustinian resolution of this apparent conflict, a resolution not complete enough
for the more enlightened and critical minds of the thirteenth
century.

5. *What is the right form of life for man?* Certain interpretations of the
classical works tended toward the conclusion that the soul dies with
the body. Thomas Aquinas needed to fashion an explanation of the
soul's mission, and it needed to be one that would not be caught in
the logical traps set by such Islamic authorities as Averroes (1126–
1198) ("The Commentator"), whose commentaries on Aristotle
reached unwanted antirealist and even materialistic conclusions regarding the essence of mind and thought. Averroist disciples in the
West (e.g., Siger of Brabant, ca. 1240–1284) were tempted by the

Averroist position that the soul perishes with the body, since it is the soul that grants individual (personal) identity and since that which is individual is, necessarily, destructible. Only abstract reason (*nous*) survives, and this is the same in all people. Thus, personal survival is impossible.

Many other questions were taken up by Thomas Aquinas, including others as fundamental to theology as these were. However, we need explore only his treatment of these to outline his broad perspective on human psychology. We might begin this outline with an observation by Maurice de Wulf on Scholastic psychology in general:

According to the medieval classification of the sciences, psychology is merely a chapter of special physics, although the most important chapter; for man is a microcosm; he is the central figure of the universe.[16]

Until the thirteenth century, the medieval view of human nature was essentially Augustinian, which is to say Platonistic in its most defined features. The Holy Trinity served as a metaphor of human consciousness viewed, accordingly, as the trinity of sense, reason, and intellect. Each of these faculties was able to provide knowledge of a certain sort, but only the last (*nous, intellectus*) could discern truth itself.

As early as the ninth century, John Scotus Eriugena could offer a more or less complete system of psychology by combining Platonic idealism with the tenets of Christian faith. In his *De Divisione Naturae*,[17] a work that is astonishing given the sorry intellectual climate surrounding its authorship, he reaffirms the Platonic distinction between attribute (*accidens*) and essence (*substantia*) and argues that the senses can apprehend only the former. That is, the senses as material agencies can be affected only by the material aspects of the world. These, because they are of an ephemeral and crude sort, have little direct connection with the ultimate and sublime reality of God. So far, this is orthodox Platonist teaching. Through the gift of reason (*logos, ratio*) the perceiver is able to assimilate these crude physical facts in such a way as to comprehend the supra-factual order and design of the universe. Even this higher degree of cognitive power is limited, however, because the universal order so disclosed is only an order among things; it is an ordering of effects but not an awareness of their true, nonphysical causes. It is only when the passive senses and the active reason deposit their contents into that spiritual realm of intellect that the fundamental truth of nature can be discerned. Since these truths are above and before things, and since they are, alas, ideas, that which discovers them must, itself, be immaterial. As for human beings, they, too, are a kind of intellectual idea held eternally in the mind of God.

A distinguishing feature of the philosophical revival that took place in the twelfth and thirteenth centuries is the rejection of such extreme idealism. John Scotus Eriugena rejected the facts of sensation as reaching basic truths and thereby spoke in defense of an idealism that could never advance beyond the point at which Plato left it. The Scholastic philosophers, though never abandoning the spirit of idealism, were willing to deal with the perceptible realities of nature as facts and as facts that expressed truths. We discover this willingness in St. Anselm's *Dialogus de Veritate,* which credits the senses with an accurate reflection of the facts of nature while condemning the "interior sense" for deceiving itself by creating false opinions about sensation.[18] For Anselm, truth finally is perceived by the mind alone but, in this act of creative perception, the mind makes use of the accurate neutrality of the senses. It is this line of reasoning that led him to his celebrated *Ontological Argument*[19] for the existence of God. We need not analyze this complex and vexing thesis, but we might note its psychological orientation. Anselm advanced the theory that what the mind is capable of entertaining is, by that fact, real. In order for the mind to be impressed with an idea, it must enjoy a faculty that is compatible with that which might impress it. For example, if we are able to see color, it is not only because there is color but because nature had equipped the human faculties to be responsive to this feature of the world. A faculty does not exist for which there is no corresponding item or process in nature. Now, the mind, as an enlarged faculty, is able to comprehend "that than which there can be nothing greater" and this, finally, is God. Since we are able to comprehend the possibility of that "than which there can be nothing greater," the possibility must refer to what exists in reality.

The ontological argument is neither trivial nor "subjective" in the prosaic sense. God's existence on this account is not based simply on the fact that someone might conceive of it. Nor does the argument require, as some have suspected, that an island "than which none can be greater" exists because we conceive of one. First, there cannot be an island than which none can be greater because such an island would be the universe or would be infinite and, therefore, would not be an island. We can, according to the argument, have the conception only if the capacity for the conception were imposed upon our understanding. But, to be imposed, an agency commensurate with the conception itself must exist. This at least, is the first step in approaching the subtle complexity of the ontological argument.

Anselm was not proposing to replace faith with reason, nor was he suggesting that God's existence in any way depended on the idea one might have of it. He was not advancing a Platonic idealism either. Instead, and with the influence of a great and revered teacher, he was permitting faith to rest upon a rational foundation such that mind and spirit need not battle any longer. It remained only for Peter Lombard's best-selling *Four Books of Sentences* to persuade the medieval faithful that God is known in his works, known through an intellect informed by perception. With this final spadework accomplished, the thirteenth century was able to host that grand synthesis of thought stretching back to Plato and Aristotle and, along the way, include Augustine, Boethius, John Scotus, Anselm, and Peter Lombard, in the synthesis that is Thomistic Scholasticism.

Although the ideas and theories comprising Scholastic thought are varied in both tone and origin, the synthesis itself can be attributed to two men, Albertus Magnus and his pupil, Thomas Aquinas. It is the latter's writings that present the Scholastic approach, indeed the Scholastic solutions to the problems bequeathed by ancient philosophers, problems that had to be solved in a manner compatible with what were taken to be the truths of the Christian religion. The approach to the problem of knowledge is illustrative.

Consistent with Aristotle, Scholastic psychology distinguishes between the factual knowledge of the senses, a knowledge that the sensory powers of animals can provide, and that knowledge of principles that only reason can embrace. Thus, although experience can inform about things, it cannot provide us with a knowledge of laws. Put in another and more classical way, experience is and must be of particulars only, whereas reason comprehends the universals. It is from the particulars of sense that reason abstracts the universals. Reason is a faculty of the soul. In fact, the soul itself is an intellectual principle.[20] As with Avicenna, the Scholastics discover in Aristotle grounds on which to deny the corruptibility of the soul.[21]

Animals, allegedly lacking the intellectual principle, survive by instinctual patterns of responding. Human beings, through reason, apprehend universals, and this allows them to fashion an infinite variety of solutions to all sorts of problems. They are not limited to perceptual data in reaching judgments, nor are they driven to accept natural phenomena instinctively.[22]

Against Plato, the Scholastic system admits not only the true and factual status of sensations but also places importance on the physiological mechanisms by which the senses inform the intellect.[23] On the occasion of physical death, the nutritive and sensitive faculties of the soul cease, for these require a body for their expression. But the will and the intellect survive.[24]

Still in the patrimony of Aristotle, medieval psychologists distinguished between different species of intellect. There is first that power of the soul by which thought is possible, called "potential intellect." Operating on the actual perceived facts of the world, the potential intellect gives way to the "actual intellect." By reflecting on its own operations and contents, the soul functions as an "acquired intellect," one that comes about only when the mind has been supplied with data and with reflective powers. Behind these operations must be some causal power that drives the actual and acquired intellects, and this is what the Scholastics referred to as the "agent-intellect" (*agens intellectus*). By way of the agent-intellect, what was only in potentiality becomes realized. It is the actual intellect that abstracts the form of a thing from its appearance. The senses can embrace only the attributes, but the intellect can discern the form.[25] Now this is not a Platonic process: that is, the form is surely not the idea by which the thing comes about. Rather, the form is in the matter as a principle, but the senses can respond only to the matter and not to the principle. Thus what we come to know is not what the senses have responded to, for knowledge is an abstraction. It is based on principles of matter, but the principles themselves are of course immaterial. The psychology of knowledge, then, is a cognitive psychology, not an empirical psychology.

For Aristotle, the agent-intellect was a kind of light, making manifest what would otherwise be latent. As the sun makes apparent that which exists whether seen or not, so the agent-intellect makes actual that which remains only a potentiality. Thus, it is an interior light, illuminating particulars in such a way as to allow reason to discover their principle. For Aquinas, this interior light is the gift of a providential God by which human knowledge can reach the fundamental principles of things.[26] The abstractions made possible by this interior light are not (Platonically) removed from the world of matter (though they themselves are not material), but they reveal a nonsensory aspect of matter. We can be sure these features or principles are inherent in things, or else we could not derive the abstractions from the evidence of sense: *Abstrahentium non est mendacium* ("The abstraction is no lie!"). The connection between matter and principle or object and form is discovered, not invented, by the mind. As de Wulf noted in his analysis of the Thomistic theory of knowledge:

> We perceive directly reality itself, and not our subjective modification of it. We perceive it thanks to a close collaboration between sense and intellect. . . . Truth is the correspondence between reality and mind.[27]

Human knowledge, in human earthly life, is imperfect, and this is because human reason is imperfectly equipped to grasp the divine essence. Faith has been made available to men so that the imperfections of reason will not cause them to stray from God. However, faith and reason do not conflict and cannot conflict, since both seek the same truths:

> Science and faith cannot be in the same subject and about the same object; but what is an object of science for one can be an object of faith for another.[28]

Scholastic rationalism is of a limited sort. Reason can take us only so far in the search for truth. Even with the aid of faith, the rational mind cannot know all in this life. Not only are we all limited as a genus, but not every individual enjoys the same rational faculty as every other. All human beings possess the agent-intellect but not necessarily in the same degree.[29] There will be differences, therefore, in achievement and comprehension. Moreover, since reason's operations are performed on the data of experience, those whose experiences are limited (such as children) or distorted (as in the case of the sick) will have an impoverished intellect.[30] During earthly life, which finds the soul united with the body, it is impossible for our intellect to understand anything actually except by turning to "phantasms," which are the mental representations arising from sensations. As Thomas Aquinas expresses the point "man is impelled in the consideration of intelligible things by being preoccupied with sensible things." [31] These phantasms are the images that perception makes of external objects. The child, the delirious, and the enfeebled, all fail to record reality aptly and thus fail to understand. But even for them, there is an afterlife in which the soul no longer requires the senses, and in which truth no longer must be abstracted from sensible things.

In light of the foregoing, differences among individuals as regards their capacity for knowledge leads inescapably to governance by some and servility by others. Before the sin of Adam, we lived in a state of innocence. Even in this state, some would lead and others follow, but only by common consent; in a state of innocence, slavery is repugnant.[32] However, ours is no longer a state of innocence. Through original sin and weakness of the will we fail in our duty either by failing to perceive duty or by failing to act once we have perceived it. In either case, the sin is a departure from the rule of reason.[33] It is the rule of reason that establishes the eternal law in the mind. As one may know of the sun through its rays, so, with respect to the eternal law, "every rational creature knows it according to some reflection, greater or less." [34] Those privileged to know it clearly must lead those who are less fortunate. The will of the prince is indeed the law of the land, but only insofar as the prince discharges the responsibilities for which the office was created.

In addition to advancing a general epistemological system that combined empirical and rational elements, the Scholastics also examined more specific psychological functions such as learning, memory, habit, emotion, and language. Throughout the thirteenth and fourteenth centuries the guiding maxim, "Nothing is in the intellect which was not first in the senses," remained, and to this extent Scholastic *applied* psychology was empirical. Medieval students relied on similarity and vividness to help their memory for facts[35] and domestic pets were trained by the disciplined dispensing of sweets and cuffings! There was also a more than a casual interest in dreams, at least as much as every other epoch has displayed. But more important than any of these to the future of a scientific psychology was the medieval discovery, although a limited one, of the experimental approach to nature. This approach was initiated by Grosseteste and developed by Duns Scotus and William of Ockham.

Experimental Science

[Robert] Grosseteste appears to have been the first medieval writer to recognize and deal with the two fundamental methodological problems of induction and experimental verification and falsification which arose when the Greek conception of geometrical demonstration was applied to the world of experience. He appears to have been the first to set out a systematic and coherent theory of experimental investigation and rational explanation by which the Greek geometrical method was turned into modern experimental science.[36]

When we list the philosophical achievements of the twelfth century, we include high on the list the effective integration of rationalism and empiricism, of Aristotle's logic and Aristotle's experimentalism. In this connection, we must recall that Aristotle's principal failing as a scientist was rooted in an attachment to idealism inherited from Plato and never fully rejected, an attachment that led him to require a kind of rational ("final cause") justification even for the ordinary facts of science. It led him also to a model of scientific explanation that was demonstrative, logically structured, and formal. It was as a naturalist that he is found so carefully observing the habits and appearances of the animal kingdom, urging Alexander to order his troops to bring back specimens, reflecting on lunar eclipses, explaining why a greater variety of animals is found in Libya, etc. In these spheres he not only captured the spirit of modern science but did much to lay foundations for it.

Grosseteste and his colleagues at Oxford and, a little later, Albertus Magnus and others at Paris succeeded in joining these historically dissociated elements of the Aristotelian legacy. Aided by recent translations of Greek studies in optics and of Euclid's mathematical essays, they set out deliberately to establish scientific proofs through experimental procedures—in other words, to bring logical analysis to bear upon nature only after recording natural events with precision. This program can be viewed either as a natural extension of Aristotle's own program or as an improvement on it, but, in either case, it served as an anticipation of the more fully developed conceptions of scientific understanding that would flourish in the late Renaissance.

The religious orthodoxy of the twelfth and thirteenth centuries had a salutary effect upon these scientific enterprises. Peter Lombard had offered convincing arguments for finding God in His works, and this notion was a veritable theological directive. Even before Thomas Aquinas formally distinguished between the truths of science and those of faith, arguing that one did not conflict with the other, Peter Abailard (like Avicenna before him) had posited a dualistic psychology that placed perception and its objects exclusively in the material realm. Even more significantly, an orthodox believer such as Grosseteste had no doubt about the Final Cause, which is God, and therefore could examine efficient causes without fear of heresy. His credentials as an unrelenting enemy of Epicurean materialism were commendable. He never entertained the notion that nature was only material and so he felt and perhaps was freer to explore those respects in which it was also material.

It was in the twelfth and thirteenth centuries also that translations of Greek medical works became widely available. The Hippocratic school and its descendants down to Galen were uniformly practical in their approach to disease. Diocles, the nutritionist, never paused to assess the logical structure according to which cucumbers from Antioch soothed the bowels. Nor was it necessary to analyze the syllogistic terms applicable to muscle, nerve, and artery. Greek medicine was observational, correlational, practical; in short, it was *clinical,* as medical practice is to today. It remained for the medieval scientists to recognize that the methods that advanced medicine were useful in any branch of inquiry, including, for example, optics, physics, and astronomy. Of course, the resulting factual knowledge could not be of scientific consequence unless it was (rationally) tied to a general theory. What was necessary, then, was to enter the observational domain with a hypothesis, to record events predicted by that hypothesis, and then to deduce the chain of effects that must take place if observation and hypothesis are properly compatible.

We can appreciate the novelty of approach and outlook in this age through two passages, one from Roger Bacon and the second from Duns Scotus. Here is Roger Bacon in his *Opus Majus:*

> I now wish to unfold the principles of experimental science, since without experience nothing can be sufficiently known. For there are two modes of acquiring knowledge, namely by reasoning and experience. Reasoning draws a conclusion and makes us grant the conclusion, but does not make the conclusion certain, nor does it remove doubt so that the mind may rest on the intuition of truth, unless the mind discovers it by the path of experience. . . . Aristotle's statement then that proof is reasoning that causes us to know is to be understood with the proviso that the proof is accompanied by its appropriate experience, and is not to be understood of the bare proof. He therefore who wishes to rejoice without doubt in regard to the truths underlying phenomena must know how to devote himself to experiment.[37]

We must not review the many actual experiments conducted by Roger Bacon in his attempt to confirm the value of this new science: one, for example, involved the use of a stone prism to separate sunlight into the spectrum. The freshness of his thought is apparent in the words themselves. Although not rejecting Aristotle's rationalistic approach to proof, he tempers it with the undeniable, immediate facts of controlled observation. Reason, the interior sense, works with perception (the external sense) and forges certain knowledge. Note that this is not the certain knowledge of the syllogism; that is, a purely logical certainty of the kind entailed by the Aristotelian notion of demonstration.

Rather, it is empirically certain knowledge, something that neither Plato nor Aristotle was willing to endorse with enthusiasm.

The same idea is advanced by Duns Scotus:

> As for what is known by experience, I have this to say. Even though a person does not experience every single individual, but only a great many, nor does he experience them at all times, but only frequently, still he knows infallibly that it is always this way and holds for all instances. He knows this in virtue of this proposition reposing in his soul: "Whatever occurs in a great many instances by a cause that is not free, is the natural effect of that cause." This proposition is known to the intellect even if derived from erring senses.[38]

Robert Grosseteste, Roger Bacon, and Duns Scotus, who made up the "Oxford school," had successfully begun to delimit a new approach to epistemology that would be made official by William of Ockham (1300–1349) and would secure for Ockham one of the more celebrated positions in the history of science. Scotus pleaded for the validity of empirical knowledge. Ockham went even further by reserving to observation the very basis on which any universal concept might be formed. In strict nominalist fashion, he argued that the universal is the name invented by the perceiver to represent that class of particulars learned by experience. In the absence of sensory commerce with particulars, no universal could be conceived. Logic will take us just so far in our quest for knowledge. At some point, we must temper or even abandon the formalisms of logic in favor of evident truths.

Ockham's razor, the maxim that has guided scientific explanation for six hundred years, was an instrument honed by many hands. Grosseteste, Thomas Aquinas, and Duns Scotus were as disinclined to "multiply causes unnecessarily" as was Ockham. Where he exceeded his predecessors was in examining more carefully the nature of the mind itself, in his attempt to assess the reality of universals and, for that matter, particulars. It is not surprising that Renaissance scholars would be able to identify more readily with the writings of Ockham, not because Ockhamism was liberally heretical, but because it was fundamentally psychological.

Ockham's Psychological Synthesis

Ockham was more deliberately psychological in his approach than any of his Scholastic predecessors or contemporaries. Both Thomas Aquinas and Duns Scotus, as well as Albertus Magnus, had addressed the issue of human reason and had posited psychological principles of perception, memory, and the will. Their efforts, however, were pre-theoretical and were never far removed from the larger religious context which was their central concern. Ockham, on the contrary, proposed specific psychological principles in a rigorously theoretical manner, and brought these principles to bear upon religious questions. Where others had taken the truths of theology as the starting point in their exploration of human nature, Ockham chose to employ human psychological dispositions in an attempt to discern the way in which theological conceptions came into being. His approach is not revealed in any single treatise on psychology proper but suffuses the entire range of his scholarship.

Oswald Fuchs has prepared a most useful compilation of Ockham's psychology, drawn from his major works.[39] At the core of Ockham's theories is the concept of habit (*habitus*) which, in the Latin of Ockham (and in the technical, philosophical Latin of the Romans as well), is understood as an acquired disposition, a perfected state or condition. It represents a qualitative change in the individual such that one is now able to do easily, or perceive readily, or behave effortlessly where, initially, these acts and experiences were difficult or incomplete.

Since the habit disposes the individual to conduct of a certain sort, it must either be present as a condition of birth or it must be acquired. As dissatisfied with the Platonic solution given in the *Meno* as was Aristotle, Ockham argues that habits must be acquired, since fundamentally they are dispositions toward specific objects. We cannot be born with the habit of dressing neatly, since nothing in our intrauterine life is able to anticipate fashion. Now, if our commonest habits require experience and practice, is there any reason to assume that any habit results from something different? Ockham's answer is a resounding *No:* All our habits are the results of experience; none is innate, including our moral habits or "virtues." Although Aristotle's associationistic theory of learning was similarly empirical, as is shown most clearly in his description of the new-born soul as a *tabula nuda* in his *Nicomachean Ethics,* Book III) and also in his *Posterior Analytics,* in which he argues that we know principles through experience and practice (Book II, 100a), Aristotle is not entirely consistent on the point. In the *Categories,* habit is rendered as essentially changeless (8a–9b), and in the *Metaphysics* habits of a certain sort are viewed as inherent (Book V, 1022b). Thus, Ockham is not to be judged merely as an Aristotelian. His theory is far less compromising in its empiricism. It is also more specific in distinguishing between instincts and similar physiological dispositions and habits that are intellectual, gradual in their appearance, and changeable through disuse or conflict.

Consistent with his nominalist position on the question of universals, Ockham insisted that the habit conduces to a given act, is disposed to particular objects, and is strengthened by the specific exercise of the act of which it is the habit. However, the mind, not nature, can (and does) come to create general categories in which are included acts and objects of a perceptibly similar nature. By a process of abstraction (or in modern parlance, what behaviorists might refer to as "stimulus generalization"), the mind will relate habitually to a number of objects that resemble each other. Cautiously, Ockham avoids speculating on how this is done, describing the process only as *natura occulta,* a process of a hidden or secret nature. There is no universal outside the mind (*extra animam*), but one produced by the mind after frequent experience.

Ockham also discoursed on the passions and appetites, which, through their association with habits, motivate behavior of an appetitive variety. While insisting that some acts of the will are moral, Ockham argued that these are subjectively moral in that they are acts of the will; that is, moral acts, like other habits, are acquired and therefore cannot be by necessity. Any absoluteness possessed by them must derive from God, not from our experiences.

William of Ockham is a fitting conclusion to the Scholastic period. His emphasis on human psychology serves as an introduction to the Renaissance, as does his focus on experience, experiment, and natural causation. Dante died in 1321, Ockham in 1349. Dante was, in all symbolic respects, the last spokesman for the distinctively romantic tone of medieval life, a life in which allegory was real, fact suspect, and nature threatening. Ockham, more than symbolically, is one of the first spokesmen of the coming age, an age of confidence, individualism, and grandness coexisting with "natural magic," witch-hunts, superstition, and attempts to forge a scientific worldview.

Chivalry, Honor, and the Ideal Life

Of the various institutions, beliefs, and practices that serve to separate the medieval epoch from other historic periods, none is as unique as chivalry. It is the embodiment of the popular psychology of the Middle Ages in the way that Scholasticism is the embodiment of the academic psychology of the period, And, like Scholasticism, chivalry resists easy classification. We begin to understand it as an idea only by recognizing how readily the medieval mind blended fact and metaphor so completely as to create a cognitive reality that was nearly otherworldly. In *The Waning of the Middle Ages,* Professor Huizinga has described this feature of the medieval mind thus:

> The conception of chivalry as a sublime form of secular life might be defined as an aesthetic ideal assuming the appearance of an ethical ideal. Heroic fancy and romantic sentiment form its basis. But medieval thought did not permit ideal forms of noble life, independent of religion. For this reason, piety and virtue have to be the essence of a knight's life. . . . The . . . revival of the splendor of chivalry that we find everywhere in European courts after 1300 is already connected with the Renaissance by a real link. It is a naive prelude to it.[40]

Committed with as much earnestness as the Scholastics to the idea of natural order, but with far less criticality, the medieval citizen perceived the world to be a set of established and immutable arrangements, arrangements of a hierarchic order with the Church supreme, its safety and prosperity in the hands of knights and noblemen. It made no difference whether the commoner was a plain farmer, a simple merchant, or a cultivated burgher. Unless one served God directly as a cleric or a knight, one's position was of diminished significance. Even as the lowest economic classes loomed in real importance, the members of these clerical and aristocratic classes refused to acknowledge the growth of their power. Riveted in the mind was the notion of eternal order, social level, and God's will animating both.

We need not pause to criticize here. We can have sympathy or scorn, for example, for the hopeless attempts to reclaim Jerusalem, the "Holy Sepulchre," at a time when the Balkans were on the verge of falling to the Turks, or for the oddly macabre tournaments in which life and limb were sacrificed in the name of honor. These are merely symptoms or correlates of a deeper set of values and a more fundamental perceptual bias:

> All realism, in the medieval sense, leads to anthropomorphism. Having attributed a real existence to an idea, the mind wants to see this idea alive, and can only effect this by personifying it. In this way, allegory is born. It is not the same thing as symbolism. Symbolism expresses a mysterious connection between two ideas, allegory gives a visible form to the conception of such a connection. . . . Embracing all nature and all history, symbolism gave a conception of the world, of a still more rigorous unity than that which modern science can offer. Symbolism's image of the world is distinguished by impeccable order, architectonic structure, hierarchic subordination.[41]

In parallel fashion, the Scholastic philosophers and the public at large interpreted nature according to a great scheme, with each piece fitting its intended slot, each act serving an ultimate purpose, each event fulfilling the prophesy of the symbol. The two abiding conditions of the human coil, love and death, were dressed in rituals of unimaginable complexity and formality. Pure sensuality, which has never vanished in any epoch, was disguised by the heavy curtain of courtly love that required honor before pleasure: that is, it required suffering and sacrifice as conditions for pleasure. In literature, the allegory reigned, notably in the influential *Roman de la Rose,* in which the union of lovers is preceded and attended by a veritable legion of virtues, saints, demons, and coordinators—right up to and including the moment at which the rose is finally plucked.

The chivalric ideal represented and preserved the medieval belief in man as divine creation in order to serve specific and different functions. Classes, since they existed, were intended to exist for the same reason that the number "7," chosen by God for the planets, was sacred. The world itself is worth noting only as a symbol or metaphor of God's work, and, when noting it correctly, one must find in it order, symbol, and the divine. Faith is the light; reason the guide; suffering, honor, and humility the way. With the pope as Plato's philosopher-king, with the knights as his guardians, and with a world in which symbols enjoy the ultimate reality of the Ideas, the medieval mind created a republic where, in fact, a civilization was crumbling.

In observing the parallel development of a Scholastic theory of conduct and the chivalric ideal, we do not mean to suggest that the latter was a conscious product of the former. Whatever praise might be heaped on the knight-errant, it surely would not include an award for scholarship. Rather, we describe the development as parallel, in that Scholasticism and chivalry both reflect the same features of the High Middle Ages but in different ways. The citizen recognized distinct classes, each with distinct responsibilities. Thomas Aquinas set forth these distinctions more formally:

> For everything is for the sake of its operation, since operation is the ultimate perfection of a thing.[42]

However, not only do species differ in their inclinations, owing to a different plan God has for them, but individuals also differ from other members of the same species, "and a sign of this is that they are not the same in all, but differ in different subjects."[43] It was not necessary for the nobles of the thirteenth century to read the *Summa Contra Gentiles* in order to learn of their own special obligations, nor was it necessary for Thomas Aquinas to study the social arrangements of the court in order to discern that God has intended each individual, uniquely, to participate in the divine adventure. We may ask, however, in the light of this encompassing perspective, how the High Middle Ages were saved from the doleful form of Stoicism that engulfed Rome in the late stages of the empire? In some respects, of course, the medieval perspective must be judged as Stoical. Every major Christian spokesman agreed that the universe and all its contents were under the influence of God's gaze and owed their existence and their destiny to His will. Except for the introduction of a personal and knowable creator, the medieval scholars in this respect are indistinguishable from several of the followers of Zeno. For that matter, they are not far removed from the rationalist astronomers of the seventeenth century or any radical determinist of our own time. But the inclusion of a personal and a knowable God makes a great difference here, a difference great enough to call for restraint in assessing the Stoic elements in medieval thought. For the medieval Christian, God has endowed human beings (and only such beings) with free will, thereby permitting autonomy and indeterminacy in human affairs. The price we pay for the freedom is, of course, responsibility. We can be said to sin only because we can be said to choose. Our actions are worthy of praise or blame only to the extent that they are voluntary. Scholasticism grants to animals only an instinctive disposition to move toward or away from objects depending on which of the animal's appetites are excited by the stimulus: animals, on the Scholastic account, are reflexive but not reflective. Man, as a rational creature, can control his appetites and thereby resist actions. Both human and animal actions occur according to a principle that exists within them; but, whereas the animal has only the appetitive principle, humans have, in addition, the intellectual principle.[44] Accordingly, when we sin it is not because the will has taken control of reason but because the latter has failed to note the evil that attaches to an otherwise pleasurable sensation.[45] This is entirely consistent with Socrates' insistence that actions committed to ignorance are involuntary. Thus, to the extent that we are rational, our will seeks the good, unless misled by sensation. To the extent that we reject reason, we may free ourselves from blame but only at the forfeiture of all that attaches to being human.

Scholastic Psychology and Its Implications

The two and a half centuries beginning in 1100 were among the most inventive in Western history. Not only did scholars such as Roger Bacon, Robert Grosseteste, Duns Scotus, and William of Ockham define and promote the experimental approach to knowledge, but even heads of state shared in the new learning. Prominent in the latter category was Frederick II, Holy Roman Emperor and King of Sicily (1194–1250). It was this "Renaissance man" who addressed philosophical and scientific questions to scholars throughout the realm and to the Arabs as well. It was he who, putting philosophy and religion aside, experimented with predatory birds, proving their ability to locate prey by smell by blindfolding them. It is said that he reared children in stark silence to determine whether Hebrew was the universal innate language. He also is said to have tagged fish and returned them to the lake to determine their longevity.[46]

Frederick II was not common in these respects among the political leaders of his time—in what age would such a king be common?—but he does appear in the earliest chapter of a new kind of learning, scientific learning. He did not reject the Scholastic system, nor did the Renaissance that followed in any but the most carping respects. Rather, he was a product of Scholasticism, as were Grosseteste, Scotus, and Ockham. Scholasticism was not a single body of truths or a rigid set of methods. It was not a cult or a version of Christianity. It was a movement and, as such, was as diverse as the other isms which fill the canvas of intellectual history.

In according to human life a central position in the creation, Scholasticism defended this priority on grounds as psychological as they were theological: man was the animal with free will, reason, and intellect; man was made in the divine image; man was born in sin and ever on the edge of error; and man, like God, was to be known by his works. These themes would be expanded and secularized over the next two hundred years and would take the form of Renaissance humanism. The chivalric ideal would be translated into the concept of human dignity. Experimentalism, under heavy pressures in a very practical world but one still spiritual in tone, would appear at first as natural magic, then as natural science. As children of the Divine, human beings would come to challenge the authority of pope and king alike. It was not for the Middle Ages to achieve these levels. Instead, it set the stage. As a period simultaneously discovering experimental science and the special place of human reason in the great scheme, the Middle Ages brought psychology closer to its modern expression by removing it further from its transcendental heritage. That the Scholastics did not reinvent modern psychology is to be understood in terms of their unwillingness to take the decisive step taken earlier by the Hellenic and Hellenistic Greeks: the step of viewing mind as object. The Renaissance fared no better in this respect. In addition, the Scholastics were concerned more with systematizing than with manipulating nature. As a result, they succeeded in recognizing the systematic character of experimental science but not in initiating actual experimental programs. They were, in this regard, more aloof than the Greeks, and they paid the same price: a somewhat impoverished technology.

Their fleeting tilts with empiricism aside, they were rationalists to the core; that is, a generally impractical sort. But the age they forged, unlike so many in history, commands more respect the more we learn about it. It is a measure of the intellectual originality of the Scholastics that we identify the "modern" period of scientific and philosophical inquiry with an age (the seventeenth century) that devoted itself to refutations of Scholastic arguments. It is more than curious, for example, that Newton, Galileo, Descartes, Hartley, Hobbes, Francis Bacon, and John Locke either explicitly cite Scholastic sources or offer their own theories against the background of Scholastic thought.

The Scholastic record speaks for itself, and can mount a formidable defense of at least the following claims: that the period from 1150 to 1300 hosted more original and significant contributions to logic than any comparable period since Greek antiquity; and that these contributions were not reviewed and extended until the late nineteenth and early twentieth centuries. Scholarship in psychology in the thirteenth and fourteenth centuries came closer to the sorts of issues that would define modern psychology than did any until late in the eighteenth century. Specifically, the Scholastic analyses of the competing claims of nativism and environmentalism, sensationism and conceptualism, free will and deterministic behaviorism, mind-body aterialism and dualism, and individualism and social determinism were sophisticated, deep, suggestive, and mature by the best standards of assessment that we have. The Scholastics recognized more fully than most schools except the Socratic the need for linguistic analysis as the first step in the evaluation of philosophical arguments. More completely than Aristotle, the Scholastics examined rival philosophies of science and sought to establish the respective functions to be served by observation, logic, and hypothesis. That so much of the scholarship of the period was devoted to theological details no longer troubling to the modern imagination simply reflects what they took to be the higher values—future historians will, perhaps, wonder why so much of our own scholarship is devoted to mechanical details. But in poring over their problems, they made lasting contributions to ethics, politics, law, and (even if negatively) science. This much is now in the record. Ours is not yet complete.

Notes

1. An excellent analysis of the legal and moral obligations existing between vassal and lord is provided in Marc Bloch, *Feudal Society,* University of Chicago Press, 1961, Ch. XIV. Translations of several original documents of interests appear in *The Records of Medieval Europe,* ed. Carolly Erickson, Doubleday, Garden City, N.Y., 1971, pp. 161–163. See also R. W. Southern, *The Making of the Middle Ages,* Yale University Press, New Haven, 1959.

2. Southern, *Middle Ages,* p. 104.

3. Henri Pirenne, *Medieval Cities: Their Origins and the Revival of Trade,* translated by Frank D. Halsey, Princeton University Press, Princeton, 1952.

4. Ibid., p. 25.

5. Cited in *Studies in Human Time* by Georges Poulet, Johns Hopkins Press, Baltimore, 1956, p. 10. Professor Poulet has provided a most valuable study of the manner in which different epochs conceived of time and how these varied conceptions are revealed in the literary works of the periods.

6. Bloch, *Feudal Society,* pp. 73, 75.

7. Cited in Peter Munz, *Life in the Age of Charlemagne,* Capricorn Books, New York, 1971, p. 119.

8. Ibid., p. 122.

9. *The Canticle of the Sun,* by St. Francis of Assisi, in *The Writings of St. Francis of Assisi,* translated by Fr. P. Robinson, Dolphin Press, Philadelphia, 1906.

10. Munz, *Age of Charlemagne.*

11. The trivium, which comprised all of the medieval curriculum, was Alcuin's invention.

12. The most authoritative and seminal study of the renascent features of the period remains Charles Homer Haskins, *The Renaissance of the Twelfth Century,* Harvard University Press, Cambridge, 1927.

13. St. Augustine, *De Trinitate,* Book XI, translated by Whitney Oates, Random House, New York, 1948.

14. Haskins, *Renaissance of the Twelfth Century,* p. 14.

15. Anders Pilz, *The World of Medieval Learning,* translated by David Jones. Basil Blackwell, London, 1981, p. 56.

16. Maurice de Wulf, *An Introduction to Scholastic Philosophy,* Dover edition, 1956, p. 125).

17. John Scotus Eriugena, *De Divisione Naturae,* Book 4, Chs. 7–9. A good trans-

lation can be found in *Selections from Medieval Philosophers,* Vol. I, edited by Richard McKeon, Scribner, New York, 1929.

18. St. Anselm, *Dialogus de Veritate,* in McKeon, *Selections.*

19. St. Anselm's ontological argument is presented and discussed in *The Ontological Argument,* introduction by Alvin Planting, edited by Richard Taylor, Doubleday, Anchor Books, Garden City, N.Y., 1965.

20. Thomas Aquinas, *Summa Theologica,* Question 75, Part I. All references to the work of Thomas Aquinas are based on the translations of Anton Pegis in the Random House edition of the *Basic Writings of Thomas Aquinas.*

21. Ibid., Question 75, Article 6.

22. Ibid., Question 76, Article 5.

23. Ibid., Question 77, Article 5.

24. Ibid., Question 77, Article 8.

25. Ibid., Question 79, Article 3.

26. Ibid., Question 79, Article 4.

27. de Wulf, *An Introduction,* p. 45.

28. Aquinas, *Summa Theologica,* Question 2, Article 4.

29. Ibid., Question 79, Article 6.

30. Ibid., Question 101, Article 2.

31. Ibid., Question 84, Article 7; Question 94, Article 1.

32. Ibid., Question 96, Article 4.

33. Ibid., Question 73, Article 7.

34. Ibid., Question 93, Article 2.

35. For an excellent and intriguing review of the techniques employed over the ages to improve memory see Frances A. Yates, *The Art of Memory,* University of Chicago Press, 1966.

36. A. C. Crombie, *Grosseteste and Experimental Science,* Clarendon Press, Oxford, 1953, pp. 10–11.

37. Roger Bacon, *Opus Majus,* Vol. II, Pt. VI, Ch. 1, translated by Robert Belle Burke, University of Pennsylvania Press, Philadelphia, 1928.

38. John Duns Scotus, *Concerning Human Understanding,* in *Philosophical Writings,* translated by Allan Wolter, Thomas Nelson and Sons, London, 1962.

39. Oswald Fuchs, *The Psychology of Habit According to William of Ockham,* The Franciscan Institute, St. Bonaventure, N.Y., 1952.

40. Johan Huizinga, *The Waning of the Middle Ages,* Doubleday, Garden City, N.Y., pp. 69–71.

41. Ibid., p. 205.

42. Thomas Aquinas, *Summa Contra Gentiles,* translated by Anton Pegis, in the *Basic Writings of Thomas Aquinas,* Ch. CXIII, p. 223.

43. Ibid.

44. Aquinas, *Summa Theologica,* Question 6, Article 2.

45. Ibid., Question 6, Article 4.

46. An excellent discussion of science in the Court of Frederick II is presented by Charles Homer Haskins in *Studies in the History of Medieval Science,* Frederick Ungar, New York, 1924, Chs. 12–15.

6 Nature and Spirit in the Renaissance

Was There a Renaissance?

Abrupt historical transitions exist far more frequently in the historian's mind than in the world of events. At the same time, we have observed that dramatic alterations in perspective do occur and that these have occasionally taken place on a relatively large scale. Two ready illustrations are to be found between the Augustan period of Rome and the seventh century, and between the Homeric period of Greek history and the Age of Pericles. A more recent, equally vivid change in perspective occurred in far less time: that between the early empire of Charlemagne and the philosophical revival of the twelfth century.

In comparing the members of any one of these apposed pairs, one is struck by the degree and number of differences in institutions, social organizations, political processes, aesthetic productions, and philosophical and psychological inclinations. This makes it all the more surprising that most general histories focus on the Renaissance and accord to it a status rare in the annals of historical analysis. It is a focus or emphasis that tempts the more critical reader or writer to overcompensate by denying anything original to the period. But this is a compensation well worth avoiding. Undoubtedly the Renaissance that flowered in Florence in the fifteenth century and spread throughout Europe by the sixteenth is a period of change, and the changes it brought either touched or completely altered the full range of existing social and political institutions. The medieval town gave way to the Renaissance city. Chivalry disappeared and patronage took its place. The scholarship of devotion gradually relinquished first place to humanistic studies and halting steps toward the natural sciences. The full partnership of Church and State became more competitive, occasionally combative. Secular authority grew, and economics came to challenge theology as the State's first science.

When scholars and aesthetes of the nineteenth century invented the term, they did not intend "renaissance" to imply a mere development of trends already in place in medieval Europe. Writers such as Walter Pater, T. A. Trollope, and John Ruskin sought to convey the notion of a rebirth and, specifically, of a rebirth of the classical approach to life. We need not cavil with this Victorian usage, for, although the nineteenth century created the term, it was the Renaissance scholars themselves who insisted that they were in fact re-creating the classical outlook. But what is to be made of such a claim? In asserting their role as new classicists committed to restoring a particular set of values and institutions, the spokesmen of the Renaissance clearly regarded themselves within a historical context. They rejected what they considered to be the perspective of their immediate predecessors. More than this, however, they recognized the evolutionary nature of the human condition, which is, after all, what a historical perspective entails. These very claims served notice on the medieval view of man and society, the view of eternal order with each person in his intended place.

Was there a Renaissance? If the term is to signify a renewed commitment to a contemplative and philosophical life, there surely was no Renaissance in Europe between 1350 and 1600. If it is meant to convey a renewed appreciation of the classical works, and especially the classical philosophical treatises, we need only recall that Lorenzo Valla (1405–1457), perhaps the period's most celebrated philosopher, was passionately contemptuous of very nearly every Greek and Roman thinker, with the possible exception of Epicurus.[1] Ficino's neo-Platonist "Academy" in Florence (established through the support of Cosimo de Medici), as well as Pietro Pomponazzi's Arisotelianism are sufficient in themselves to confirm the interest of Renaissance scholars in ancient philosophy. But, in light of the feverish philosophizing that took place from the eleventh to the thirteenth centuries, this interest can hardly be called a revival! Moreover, the growth of trade, the establishment of banking agencies, the undertaking of massive public and religious works, all are features of the Renaissance but are by no means sudden in their appearance. Each can be traced, uninterruptedly, to Charlemagne, to the Ottonian monarchy of Germany, to the Lombards in Italy, and to the Norman kings in France. By the time of the Renaissance, each had attained a certain stability and centrality, but none was new.

If one is to adopt the term, therefore—and it is one so grafted to the historian's lexicon that no argument will succeed in removing it—we inevitably impute features that were not unique to the period. It would be sounder to acknowledge that something was born, not reborn, in that period, and there is a theme in Renaissance literature that for all intents and purposes really is quite new to the world of letters, one that could scarcely have found approval in Plato's Academy or the Lyceum or the court of Marcus Aurelius. As expressed in the Renaissance, it is a theme that would not have been approved even in the universities of the Schoolmen (Scholastics). It is the theme of *the dignity of man,* with the associated insistence that the world was made for human beings, more or less to do with as they will. Perhaps there is no clearer expression of the view than these lines from Marsilio Ficino in 1474:

> Universal providence belongs to God who is the universal cause. Hence, man, who provides generally for all things, both living and lifeless, is a *kind* of god. . . . Man alone abounds in such a perfection that he first rules himself, something that no animals do, and thereafter rules the family, administers the state, governs nations, and rules the whole world. As if he were born to rule, he is unable to endure any kind of slavery. Moreover, he undergoes death for the common weal, a thing which no animal does. . . . [O]ur soul will sometime be able to become in a sense all things; and even to become a god.[2]

Humanism, Individualism, and the Renaissance City

Our present age, which, more than ever before, finds the public commitment to improve the lot of the less fortunate ranking near the top on the list of the affairs of state, tends to lull us into thinking that humanism has always referred to humaneness. Also, living as we do in an intellectual and cultural period in which religious orthodoxy and power are in scarce supply, there is a tendency to think of humanism as something of an anti-religionism and to assume, thereby, that the humanism of the Renaissance was a movement away from the Church. These notions are false. The humanism of the Renaissance is inconceivable without Christianity, but in no case is it to be confused with altruism, charity, or a widespread dedication to the welfare of the unfortunate. It was first and foremost a new individualism of the rugged sort. Earlier ages hosted heroic epics in which finely crafted personalities appear, but the Renaissance conception of the person and of what we call "personality" was different. In the Renaissance, personality is taken to be protean and organic, evolving as a result of changing conditions, and taking on a fixed form only when the individual has settled on a position in society and has essentially chosen a form of life for himself. Just as it is no longer assumed that the occupation of one's father determines the future course of one's working life, it is, increasingly, no longer assumed that one's origins bear determinatively on the balance of one's life. There is a new perspective on human nature, and a new place created for its development: the city.

The city was surely one of the great achievements of the Renaissance. Florence, the quintessential Renaissance city of the fifteenth century, was chiefly the creation of the Medicis, whose genius for finance was undiminished from the time of Cosimo the Elder (1389–1464) to Cosimo the Great (1519–1574). Not content with remaking the economic and architectural face of Italy, this family covered all its bets by siring two popes, Leo X and Clement VII. More than great kings and merchants, however, it was the Renaissance city itself that sustained the growing sense of individualism and its limitless possibilities. The medieval town had room for but a handful of leading figures. The Holy Roman Empire, like all empires, was of a scale to embarrass all but regal initiative or papal intervention. The city, however, was perfectly suited to the fierce economic and cultural competition that yields visible and often sudden heroes, and sudden reversals of fortune. It affords anonymity when needed, celebrity when earned.

This capacity of urban life to nurture individual excellence was recognized clearly by the luminaries of the Renaissance. Early in the fifteenth century, Leonardo Bruni, the chancellor of Florence, wrote his *History of the Florentines* in which he attributed much of the success of the city to the fact that it did not labor under the stultifying and regulatory influences of a Roman empire, or an empire of any sort. Charmed by Aristotle's *Politics* and Plato's *Republic,* Bruni led his contemporaries to a reverence for their cities and a commitment to attain the greatness that only urban life allows.[3] Bruni's praise of the city and the very Renaissance idea of the city was rooted in the conviction that we are born for a life of civility and cultivation, and the best of us for something bordering on the divine.

If the spirit of humanism was individualism, the term itself was intended to describe only the intellectual pursuits held proper for enlightened members of the urban centers of Italy. These pursuits revolved around the venerable humane letters (*litterae humaniores*), or "humanities," as they are now called. The subjects comprising this domain were the Greek and Latin classics, though Renaissance scholars were not much for careful, philosophical analysis, let alone philosophical originality. Renaissance humanism, then, was neither self-consciously humane nor, in any modern sense, especially humanitarian. When the first graduated tax was introduced in Florence, intended to provide relief for the poor, the affluent Florentines were stubborn in their opposition. The humanistic movement, if it may thus be called, evolved from and was something of the concluding state of chivalry through which the privileged classes might work toward noble and honorable ends. As an intellectual movement, humanism was devoted to a broader range and a different type of classical scholarship than that which had caught the attention of the Scholastics. As a moral instrument, it was intended (or at least used) to protect the cultural, artistic, and intellectual prerogatives of aristocrats and the circles that formed around them.

Petrarch (1304–1374)

The great Petrarch, called by more than one historian the very father of the Renaissance, goes to some lengths, in an essay entitled *On His Own Ignorance and That of Many Others,*[4] to prove the weaknesses and contradictions of those vaunted Greeks and Romans. Commenting on Aristotle's treatment of happiness in his ethical works, Petrarch offers this, which is enough to convey the flavor of his rebukes:

> He knew so absolutely nothing of true happiness that any pious old woman, any faithful fisherman, shepherd or peasant is—I will not say more subtle, but happier in recognizing it. . . . He saw happiness as much as the night owl does the sun.[5]

It is passages such as this that sharply distinguish the painstaking analytical efforts of the Scholastics and the larger agendas the Renaissance commentators were prosecuting.

No single work of Petrarch's stands as a psychological treatise, nor, for that matter, does the bulk of his scholarship qualify as philosophy. Like Voltaire after him and Diogenes before, his position within the most important intellectual movement of his age was earned more by the daily effect of his letters, his immediate effect on his friends, his influence among political and financial luminaries, and his winning rebukes of those who might stand in the way of progress, even if they were drawn from the ranks of Aristotelians. His "book" on his own ignorance was really a long letter to his friend Donato. In it he attacks the aloof rationality of Aristotle with a poet's passion, insisting that man can be happy only when in possession of faith and immortality, two properties conspicuously absent in Aristotle's treatises on *eudaimonia*.[6] Petrarch's own authorities are the Bible, St. Augustine, and Cicero, not to mention fishermen and pious old women. He rejects Aristotle not because the philosopher was wrong (which, by Petrarch's lights, he was), but because philosophy itself, and especially Aristotle's, failed to make men either good or happy.

> It is one thing to know, another to love; one thing to understand, another to will.[7]

We find here and elsewhere in Petrarch one of the principal elements of Renaissance thought, a skepticism toward intellectualism. Note that this is not a religious skepticism. Quite the contrary. Petrarch was concerned to re-create a healthier and purer climate of faith, one not cloudy with Scholastic analysis. He saw his opponents in the camp of the Aristotelians, whose contempt for Plato caused them either to ridicule Christian belief or to insist that such belief be reconciled with the tenets of logic. The opposition, then, was none other than medieval philosophy. By the fourteenth century, Scholasticism had invited many quasi-heretical factions who found the Aristotelian system congenial to the coy badinage the skeptic so readily heaps on those of simple faith and little learning. We should be less surprised, then, to learn that Petrarch, this father of Renaissance humanism, comes down emphatically against, of all things, freedom of speech:

So sweet does the word Freedom sound to everyone that Temerity and Audacity please the vulgar crowd, because they look so much like Freedom. Thus the night owls insult the eagle with impunity.[8]

And here is still another characteristic that will be present in so much of the thought of the age: suspicion, even hostility toward philosophical speculation and a reserved position on intellectual freedom. Petrarch and his contemporaries were of a decidedly practical disposition. The worth of a person or an idea is judged by the effects in practice. Given this orientation, it is less surprising that the entire Renaissance would fail to produce a philosophical system of the highest order. And the conservatism toward intellectual freedom, in its extreme form, would lead finally to the persecution of witches and the Inquisition. But the emphasis on achievement and palpable success animated the monumental creations in art and architecture for which the age will be remembered.

If the Gothic cathedral has that ambiguous and ethereal character of a Platonic idea, the Renaissance city displays the orderly reasonableness of Aristotle's categories. But there is no contradiction in identifying Petrarch with both. He sought to remove belief from the context of philosophical debate but, at the same time, to restore the citizen to a life of participation and achievement of the sort proclaimed by the ancients. In separating, or seeking to separate, philosophy from religion and in striving to give the former a practical function, he was instrumental in furthering the late medieval development that would ultimately flower as the experimental science of Galileo. His attacks on Aristotle were not of philosophical consequence; most of the anti-Aristotelianism of the Renaissance is less than commanding in its comprehension of Aristotle's works. They were, however, of general intellectual consequence. The authority of the Aristotelians, by the fourteenth century, had begun to retard and even repress philosophy and science. There is a general impression, though one not without severe limitations, that the modern world did not begin until this authority was overcome. To the very limited extent that this is true, Petrarch figures centrally.

Architecture speaks as directly for the spirit of an age as do its words. We need only compare the disciplined, line-and-angle proportions of the Renaissance palace with the free, "barbaric" imperfections of the Gothic cathedral. Without denying the beauty of either, or insisting that one is superior to the other, we still recognize a vanity, a pretense, even a showmanship in the Renaissance structures when examined against the herculean innocence of a Chartres or a Mt. St. Michel. This distinction was first examined by John Ruskin, who played so important a part in the "Gothic revival" in Victorian England. We might develop a more complete understanding of the departure the Renaissance represents and of the variety of conservatism for which Petrarch spoke by recalling Ruskin's analysis. In what he has to say about the Gothic and the late or "High" Renaissance, Ruskin summarizes the truisms of an age and conveys at once that sense of living freedom and the residuals of that earlier age of the living metaphor:

> Of servile ornament, the principal schools are the Greek, Ninevite, and Egyptian; but their servility is of different kinds. The Greek master-workman was far advanced in knowledge and power above the Assyrian or Egyptian. Neither he nor those for whom he worked could endure the appearance of imperfection in anything; and, therefore, what ornament he appointed to be done by those beneath him was composed of mere geometrical forms . . . which could be executed with absolute precision by line and rule. . . . The Assyrian and Egyptian, on the contrary, less cognizant of accurate form in anything, were content to allow their figure sculpture to be executed by inferior workmen, but lowered the method of its treatment to a standard which every workman could reach, and then trained him by discipline so rigid, that there was no chance of his falling beneath the standard appointed. . . . The workman was, in both systems, a slave. . . . The third kind of ornament, the Renaissance, is that in which the inferior detail becomes principal, the executor of every minor portion being required to exhibit skill and possess knowledge as great as that which is possessed by the master of the design; and in the endeavor to endow him with this skill and knowledge, his own original power is overwhelmed, and the whole building becomes a wearisome exhibition of well-educated imbecility. . . . But in the medieval, or especially Christian, system of ornament, this slavery is done away with altogether; Christianity having recognized, in small things as well as great, the individual value of every soul. . . . It seems a fantastic paradox, but it is nevertheless a most important truth, that no architecture can be truly noble which is not imperfect. And this is easily demonstrable. For since the architect, whom we will suppose capable of doing all in perfection, cannot execute the whole with his own hands, he must either make slaves of his workmen in the old Greek . . . fashion, and level his work to a slave's capacities, which is to degrade it; or else he must take his workmen as he finds them, and let them show their weaknesses together with their strength, which will involve the Gothic imperfection, but render the whole work as noble as the intellect of the age can make it.[9]

Marsilio Ficino (1433–1499) and Hermeticism

Ficino is of interest because he was selected by Cosimo de' Medici to lead the newly established Platonic Academy in Florence (1462). Renaissance Platonism, which was never a re-creation of pure, Platonic philosophy, was yet one more movement or attitude that served to challenge the authority of Aristotelianism. It is important to continue to refer to the contrary perspective or "the enemy" as Aristotelianism and not Aristotle, for much of the criticism and scorn displayed toward Aristotle was based on a failure to comprehend the philosopher's system. Moreover, the attempt to prove rationally God's existence and the ultimate nature of nature itself was a Scholastic attempt and one fashioned from Aristotle's methods and arguments; it was not undertaken by Aristotle in the same way, nor was Aristotle as confident that the realms of society, politics, biology, or psychology would fit neatly into a logical framework.

In addition to installing Ficino as director of his new academy, Cosimo de' Medici kept him busy by providing a steady infusion of Greek manuscripts to be translated. He was no more insistent regarding any of these than he was in the case of the *Corpus Hermeticum,* a collection of books thought to contain the pre-classical and mystical secrets of the disciples of Hermes. Tradition had it that Hermes descended directly from Zoroaster and that his genealogical successors included Orpheus, Aglaophemus, and Pythagoras himself. Thus, the "divine" Plato's philosophical inspiration derived initially from Egyptian mystery religions and only these possessed the purest theology. Ficino titled his translation of the *Corpus* "Pimander," and it became one of the most widely read and influential manuscripts of the period. Among its most devout patrons was Giovanni Pico (see below). It was even greeted tolerantly by Pope Alexander VI.[10]

The Renaissance love of antiquity was based in part upon the general belief that the purest and most philosophical eras had been those of Greece and Rome and that, from that time onward, civilization had undergone steady deterioration. Added to this linear perspective on virtue was the suspicion that if one could penetrate history even further back, beyond Athens and Rome to the very fountains of Greek wisdom, one would find the absolute essence of spiritual energy. Egypt always held a deep fascination for the Renaissance faithful, and the *Pimander* unleashed a wave of Egyptomania. The Hermes legend, which the seventeenth-century scholar Isaac Casaubon proved to be authored by various hands between 100 and 300 A.D., came to dominate the spiritual aspects of Renaissance life. By deftness of interpretation and through agile translation, Ficino was able to establish a nearly perfect compatibility between its central themes and those of the New Testament. More encouraging even than this surprising agreement was the possibility held out by the *Corpus Hermeticum* that the individual in this life could command supernatural powers and share in the Cosmic spirit. The Egyptian priests had unearthed all the necessary chants, had discovered all the secret astrological configurations, and had documented the most reliable number combinations and alphabetical series. They had divined all the essences, talismans, foods, and ceremonies. Those experienced in their ancient art could summon the forces of heaven and partake in the Oneness. It was Hermetism that placed the sun at the center of cosmic concern, allowing the earth to move, a principle duly acknowledged by Copernicus in defense of his seemingly odd speculations which were to the same effect. It was this same mystical wisdom that accorded the magus (a kind of magician) the status of a priest. Finally, it was a fundamental tenet of the Hermetic gospel that man is an agent of change, a rational-spiritual force capable of altering nature's course. That human dignity of which Pico wrote so movingly we now recognize as indebted to the Hermetic revival. It proved to be a boon as well for alchemy, for the new (Copernican) astronomy, and for a good share of what passed for science from the sixteenth to the seventeenth century. Giordano Bruno's vaunted defense of Copernicanism was, in the last analysis, no more than Hermeticism, bearing little relation to the scientific merits of the case. In addition, many of the most ardent assertions about the human will and dignity were made by those firmly rooted in the new Hermetism. Indeed, the Renaissance focus upon the will itself is, in considerable measure, inspired by lessons from the *Pimander.*

Ficino would write in a separate work that man is something of a god; one who rules over animals, who provides for both the living and the lifeless, who instructs others.

> It is also obvious that he is the god of the elements, for he inhabits and cultivates all of them. Finally, he is the god of all materials, for he handles, changes, and shapes all of them. He who governs the *body* in so many and so important ways, and is the vicar of the immortal God, he is no doubt immortal.[11]

Ficino distinguished between two kinds of magic; one practiced by those who merge with demons through religious rituals, and the noble sort practiced "by those who seasonably subject natural materials to natural causes."[12] Even among these there are mere inquisitive types who show off with ostentatious tricks. But there is also the *necessary* form of natural magic joining medicine with astrology and as compatible with scripture as it is essential to the development of knowledge.

As with so many facets of Renaissance life and thought, these lines from Ficino and the Hermeticism they express preserved the worst of what was old and promoted the best of what was to be the new. Hermeticism, with its brooding cants, its nonsensical astrological and dietary hocus pocus, and its ritual fever, kept some of the best minds of the period and most of the plain folks under the heaviest cowls of superstition and fear. The witchcraft of the sixteenth and seventeenth centuries is partly its legacy, as were the rites and tortures invented to solve the problem. Great artists were inspired by Hermetic tenets; others were driven to depravity by the same tenets. In his way, Ficino exemplifies these several and finally incompatible tendencies and certainties. The competing alternatives of Nature and Spirit were implicit in the Hermetic system. Part of this system, through the development "natural magic," led ultimately to an interest in experimental science and Francis Bacon's recipes for the manipulation of nature. The other part, through "spiritual magic," was matched with the macabre banalities of exorcism, witch-hunting, and self-mutilation.

Giovanni Pico della Mirandola (1463–1494)

One of the more durable characteristics of that form of humanism advanced by the Florentines of the Renaissance was optimism. The leading spokesmen of the period were all infected with the idea of progress. They saw themselves as agents of change, and this, after all, is the dominant element in the idea of freedom. Perhaps no one was more coherent in his assertion of this idea or more persuasive in bringing it to the foreground of Renaissance discourse than Giovanni Pico. His *Oration on the Dignity of Man*[13] establishes the human community as unique, even by heavenly standards! It is in this essay that Pico thanks God for granting to humans the freedom to change that even the angels do not possess. He notes that animals are tied to their instincts, mindlessly fulfilling the prescriptions of their species. The angels, enjoying a state of spiritual perfection, dedicate themselves eternally and immutably to the praise of God. Man, however, being neither instinctive nor perfected, is free to move in either direction; to degenerate to purely instinctual and sensual levels of existence, or to rise, through reason, to a nearly angelic station:

> If sensitive, he will become brutish. If rational, he will grow into a heavenly being. If intellectual, he will be an angel and the son of God. . . . Who would not admire this, our chameleon?[14]

Here, then, is one reply to Ficino's conservative neo-Platonism and hereditarianism. Indeed, man is "stamped" by God in such a way as to be predisposed, but it is freedom to which he is predisposed. God has not limited man in that dire, constitutional way suffered by animals:

> On man when he came into life the Father conferred the seeds of all kinds and the germs of every way of life.[15]

The notion expressed here has a modern equivalent in Haeckel's maxim, "Ontogeny recapitulates phylogeny": each advanced species progresses through the developmental stages of the less advanced on its way to maturity. Pico was not, of course, advancing an embryological theory but a psychological one: as individuals confront the challenge of life, they can use reason and intellect, thereby improving and developing these, or they can confine themselves to the more primitive apparatus of survival shared with lower forms of life. Pico, persuaded that all philosophical systems finally agree on certain principles, insists that the only really human life for human beings is one in which devotion, experience, action, and contemplation, are combined by the powers of reason. He does not reject Ficino; he assimilates his arguments. He does not shun the skeptic; he engages him in debate. He does not deny the Stoic immutability of natural law; he merely excludes man from its reach. To the cynic who points accusingly at philosophers' inability to agree, Pico insists that Plato and Aristotle, Scotus and Thomas, Avicenna and Averroes, when properly understood, were in substantial agreement and never contradicted each other on fundamental matters.[16]

In addition to his direct influence, Giovanni Pico indirectly influenced Renaissance scholarship through his nephew and biographer, Gianfrancesco Pico della Mirandola (1469–1533). Perceiving his uncle as hero and genius, Gianfrancesco was not, however, attracted by the *pax philosophica* so earnestly desired by Giovanni. In his *Examen Vanitas Doctrinae Gentium* (1520), he unleashed an attack on Aristotle which, in its tone and interpretive deficiencies, served as a model for the next two centuries of anti-Aristotelianism.[17]

Nature and Spirit

It is constructive to treat the two Picos together because they demonstrate the special, intellectual contradictions of the Renaissance and because these contradictions anticipated many of the psychological and social characteristics of the Reformation. The history of discourse on the human character may be summarized under two great headings: "Nature" and "Spirit." Beneath the former we find naturalism, Stoicism, materialism, and, ultimately, scientific determinism and logical positivism. Below the latter are the near opposites of these: spiritualism, idealism, transcendentalism, psychological indeterminism, and Romanticism. Every century or so the balance changes but the essential positions remain stubbornly constant. In the Hellenistic period, the controversy was over the reality of the Platonic Ideas. Among the Scholastics, this controversy surfaced in the form of the nominalist-realist antagonism. In the individualistic climate of the Renaissance, it becomes a battle between neo-Platonists and Aristotelians. In the eighteenth and nineteenth centuries the terms of conflict are "empiricism" and "idealism."

For the better part of five hundred years the method of analysis introduced to settle these disputes was logic and, more particularly Aristotle's forms of demonstrative proof. Nature was to be described according to the *Categories,* and the truths of nature were to be unearthed and preserved through the infallibility of the syllogism. The Aristotle of the *Historia Animalium* and other naturalistic works was either ignored or never found. Thus, Aristotle the rationalist dominated scholarship, while Aristotle the ethologist and empiricist enjoyed only the fleeting recognition of a Grosseteste, a Roger Bacon, a Duns Scotus.

From the earliest decades of the Patristic period theologians expressed concern for the intellectual hegemony Aristotle was coming to exercise over the new faith. Tertullian, with his Jovian impatience, dismissed the philosopher as a mere pagan. Plotinus read him as if he were something of an Epicurean. With Augustine's *Meditations,* Plato becomes the ancient of record, receiving first honors, but philosophy itself is relegated to a secondary position on matters of religious importance. The rediscovery of Aristotle in the twelfth and thirteenth centuries not only raised challenges to the authority of Platonism but led to such a veneration of Aristotle that his works soon served to arbitrate even questions of Scripture. By the fifteenth century the faith that Thomas Aquinas had sought to defend through rational analysis aided by Scripture had become, in many learned quarters, a mere branch of logic and rhetoric. Within the fifteenth-century Florentine circles, we discover several scholarly movements: (1) the revival of Platonism as a system, promising to overcome Aristotelianism; (2) a renewed Aristotelianism equipped to meet these neo-Platonist attacks; and (3) Giovanni Pico's *syncretism,* by which all philosophical perspectives are searched for truth and assimilated, peacefully, into a grand Christian scheme. To these we add a fourth movement which, while it employed the scholar's traditional tools, was not scholarly in conception or in its mission. This was the anti-intellectual movement away from books, away from argument, away from analysis, away from proof. In its sincerest expression it was led by Girolamo Savonarola's ceremonial "burning of the vanities." In its meanest form, it was led by the mob that burned Savonarola himself. Between the extremes of piety and hate stand the efforts of such sober commentators as Lorenzo Valla and Gianfrancesco Pico: the former found enough contradictions among the pagan philosophers to dismiss all of them in favor of simple Christian faith; the latter read Aristotle with such distorting lenses as to convert Aristotelianism into a Lockean form of empiricism and then set out to dismiss the philosopher because the senses are less than perfect!

It was in the fifteenth century that the Church was able to cement its position politically. Concurrently, it found it necessary to cement its doctrines as well. Scholasticism had never been more than the gospel of an intellectual elite. Within its fraternity, only Ockham toyed with heresy. By the Renaissance, however, the secular branch was not only more educated and powerful than it had been since the second century, but was increasingly at odds with the Church over economic and military issues. Philosophical disputes were no longer mere academic exercises. In a word, philosophy had become relevant. One moved through crowded Florence confronting a wave of street-corner philosophers peddling a variety of interpretations, all documented with quips and quotes, even if few bore any close connection to ancient wisdom. Ideas had become politicized and intellectuals had become warriors. The authorities, both secular and clerical, agreed to determine truth in their own special manner, and in 1478 the Spanish Inquisition was launched. For two hundred years Nature and Spirit would do battle: Nature, according to which man's dignity is a matter of record, his reason and power nearly limitless, his body and his artifices beautiful, his life and his city reasonable, and his mind inquiring; Spirit, with its extra-terrestrial vision, its sense of wonder and fear, and its murky and ascetic renunciations.

The Renaissance has left unambiguous records of this useless but ageless conflict. Nature and naturalism, in their historic marriage to empiricism and materialism, are expressed equally in the art, architecture, and political organization of the Renaissance city and in the ruthless and terrifyingly reasonable exploitation of the weak, the dull, and the poor. Renaissance naturalism simultaneously elevated geometry to the level of aesthetics and developed corruption and deceit to the level of a science. The Aristotle of the *Metaphysics* could be used to justify my success and your failure on the basis of that unavoidable final cause toward which all our actions proceed. The Aristotle of the *Politics* and Plato's philosopher-king now made Machiavelli's Prince almost syllogistically necessary.

The poor and the ignorant, important to any age because of their numbers and their labors, were now compressed into bustling and prospering cities; they now had firsthand knowledge of what they were missing. Not since the Augustan period had the lowest classes been so close and in such constant commerce with the educated and affluent elite. Serving at the Banquet of Reason, these plain people could not help but pick up many small but fascinating morsels: human dignity of man, freedom of the will, final causes, necessary truths. All heady notions, indeed. Unfortunately, these leftovers mixed poorly with Spirit and spirituality. Final causes and necessary truths, when digested by those used to simpler fare, create the sense of predestination and inevitability. Thus the doctrines of Spirit, which helped to mold some of the most sainted, pious, and loving people in all the world, also bred terrified souls, looking for a witch to burn, a prophesy to fulfill, an astrological sign that would disclose their fate. Renaissance spiritualism reached the summit of the Christian ideal and sank to the depth of fanaticism.

The contradictions within and between those movements we have identified as Nature and Spirit would come to determine the diverging paths taken by philosophers in the seventeenth and eighteenth centuries and by psychologists in the nineteenth.

Piety Reaffirmed: Luther and Savonarola

We are not concerned directly with theological claims and disputes and, if we were, Martin Luther and Girolamo Savonarola surely could not be treated under the same heading. Nor, to be sure, is it warranted to accuse either of them for those pietistic reactions to the new learning that led ultimately to the most scandalous forms of oppression and censorship. At one level, the two had several things in common. Both came from working-class families, both were priests, and both were gifted students and avid readers of the classics early in life. Both were passionately opposed to luxury, pomp, and privilege. Both were defiant and belligerent. Neither gained materially from the fame he won. Both were God-fearing Christians without a skeptical bone in their bodies. But there was, at a still deeper level of their being an additional similarity: Luther and Savonarola were overcome by and devoted their lives to promoting a mystical sense of life's meaning, and each, in his own way, instilled in the masses an intense hostility for all the trappings of what history recognizes as high civilization.

Savonarola's contempt for the Medicis was without limit or embarrassment. His rhetorical skill and his exemplary life escalated him to the position of de facto dictator of Florence, where in 1497 he led the citizens in the "burning of the vanities." The "vanities" were paintings, books, precious stones, and related symbols of now overly dignified man. He imposed upon his once comfortable clerical brothers a life of severe denial and insisted that this was to be the model for every Christian, popes and princes included. He justified his actions on the basis of Scripture and of direct, divine revelation, the latter permitting him to predict the fate of men and nations. His most vivid inspiration was the apparition of a burning sword commanding him to serve as God's gladiator, a vision amplified by dark voices, clouded skies, and by many other persuasive portents.

Luther, too, was directed to the clerical life by mystical influences, in his case, a terrifying vision. Poor origins and mystical experiences combined with an agile mind to locate him as a professor of theology at Wittenberg, whose Castle Church was to receive his Ninety-five Theses. Like Savonarola before him, he reviled the practice of taxing the poor to enrich the Church. The Theses, in fact were written in direct response to John Tetzel's sale of indulgences, the proceeds of which were to help defray the costs of Julius II's plans for St. Peter's, a work still being carried out in Luther's time under Leo X. But these disputes were only the most superficial aspects of a far more basic rift between Luther and those responsible for the affairs of the Church. At the core of a disagreement that would produce no less than the Reformation itself were competing conceptualizations of Christianity.

Where Luther's theology is rife with complex and subtle elements, his psychology is daringly straightforward. His conception of human nature is completely derivative, where this adjective is intended to contrast with inventive. All he seeks to know about mankind can be found in the New Testament; indeed, most of what he believes can ever be known is given by Paul. In his written debate with Erasmus on the question of free will, he summarizes the efforts of Greek, Roman, and Scholastic contributions to the issue as instances of the "plagues [Satan] has bred from philosophy." Scripture tells Luther that God's will is the cause of all things, and, therefore, the human will can count for naught. The fatalistic and predestinational refrains dominate his jeering reply to Erasmus,[18] and, in case anyone may be deceived into believing that Luther is just playing the scholar's game, he adds this:

> Let me tell you therefore—and I beg you to let this sink deep into your mind—that what I am after in this dispute is to me something serious, necessary, and indeed eternal, something of such a kind and such importance that it ought to be asserted and defended to the death, even if the whole world had not only to be thrown into strife and confusion, but actually to return to total chaos and be reduced to nothingness. . . . Stop your complaining, stop your doctoring; this tumult has arisen and is directed from above, and it will not cease till it makes all the adversaries of the World like mud on the streets.[19]

His attack on reason[20] is the rhetorical version of the burning of the vanities.

> What are the Universities, as at present ordered, but as the Book of Maccabees says: "Schools of 'Greek fashion' and 'heathenish manners' full of dissolute living, where very little is taught of the Holy Scriptures and of the Christian faith, and the blind heathen teacher, Aristotle, rules even further than Christ." Now, my advice would be that the books of Aristotle, the "Physics," the "Metaphysics," "Of the Soul," "Ethics," which have hitherto been considered the best, be altogether abolished. . . . My heart is grieved to see how many of the best Christians this accursed, proud, knavish, heathen has fooled and led astray with his false words. God sent him as a plague for our sins.[21]

Luther's psychology, for which there are generous Patristic precedents, begins with the conviction that man is born in sin. Where the medieval rites of chivalry and the Renaissance opportunities for patronage left room for good works to count in heaven, Luther leaves none: good works cannot be performed by an evil soul, and a good one can be responsible for no evil. In short, until the spirit is clean it is in peril, and there is no public gesture of altruism or grandiosity able to remove this peril. (The so-called Protestant ethic traces back to Luther only in a very circuitous way.) Purification of the soul requires a renunciation of the flesh, and at this point the distinctions to be made between Luther and Plato are less than striking. What is central to Luther's psychology is the will, since our spiritual status can only be assessed in terms of our intentions, and our intentions will either be inspired by God or by Satan. God's inspiration is, alas, grace. Without it, there is no hope; with it, no fear or danger. It will come to those who yearn for it and who, through the mystical effects of Scripture, prepare themselves to receive it. Those in the state of grace are doing

God's work. The rest are sinners, and this latter category is sufficiently wide to embrace popes, bishops, priests, and kings.

On Christian Liberty was Luther's inadvertent call to arms.[22] It was in this work that he insisted on the essential political freedom of each and every Christian, answerable to God alone and working for God's glory only, a slave to none but the willing servant of all. His message reached a large and eager audience, but when the Peasants' War (1524–1525) erupted, an angry and amazed Luther quickly issued a polemic, *Against the Thievish, Murderous Hordes of Peasants.* If the Renaissance as a whole is an era of contradictions, Luther's life is an apt analogy. His works bristle simultaneously with Dante's conservative awe toward the legitimate monarchy and with Locke's dedication to liberal reform; with Pico's elevating recognition of the worth of every single human life and with that Augustinian sense of the triviality of temporal affairs; with Valla's contemptuous rejection of the Greeks and with his Thomistic penchant for phrasing his case along essentially Aristotelian lines.

Luther did not create the Reformation; he spoke its message. It was the message of an emerging class, one whose power was already hundreds of years old but only recently sensed. For all his intensity, we must conclude that he never realized how far his own spirit of rebellion and reform would take the world when taken up by tens of thousands. In opposing exploitation or fear-mongering he succeeded, like Savonarola, only because he induced even greater fear. He threatened the Church through the masses, and the Church, predictably, threatened the masses. Where his own guide had been reason, he disdained its powers publicly. As reason receded from the arenas of debate, the rack, the stake, and the ax took over.

Psychology and the Witch Trials

The fiercest chapters in the persecution and execution of witches occur between 1400 and 1700. This is *after* the so-called "dark" and "middle" ages and in centuries including the Renaissance, the Elizabethan Age and the Age of Newton! Nor was the dreadful experience local. There were fewer executions in England (though not Scotland) and in Venice (though not in Italy at large) than in the rest of Europe, but the phenomenon was widespread. The veritable handbook for the interrogation, trial, and punishment of witches was the *Malleus Maleficarum*[23] authoritatively promulgated in 1486 by the Dominican Fathers Heinrich Kramer and Jakob Sprenger, who carefully noted the special procedures that had to be adapted and applied to witch-suspects. But King James I provides an echo of this work in his own *Daemonologie* more than a century later.[24] Both works are based on what has come to be called the "witch-theory," a woeful amalgam of ancient legal principles and the developed psychology of the Christian West.

Witches have been a fixture in history and were dealt with legally as early as

Rome's *Twelve Tables*. Often widows or otherwise unattached women, the witches of history have read palms or tea leaves, have given prophesies or herbal remedies or just advice; but they have also threatened with curses, performed what appear to be miracles, and been charged with great harms, including death and the destruction of property. They are prominent in the art and literature of the world, where they have been treated with a mixture of awe, reverence, fear, and loathing.

Prior to the development of religious psychology in the West, the secular law distinguished between "white" and "black" magic as perpetrated by witches, reserving punishment only for the latter (*maleficia*). Witchcraft that was harmless or judged to have desirable consequences was either explicitly unpunished by law or was overlooked by the authorities. In the eastern reaches of the Roman empire and of Christian Europe after Constantine, there was such close and continuing contact with classical Greek science and philosophy that policies were even more enlightened. Witches in these parts of the world (e.g., Syria) were thought of as suffering from just another malady to which we are vulnerable. A "bout of witchery" was not unlike other infections or visitations, and the priestly or monkish *exorcism* was undertaken just as it also was for the treatment of indigestion.

Among the more analytical and metaphysical writers in the West, however, witchcraft posed a problem of great religious consequence: Is the will in fact free or can Satan claim it for himself? If the latter position is taken, then the very moral grounds of earthly and ultimate judgment are shaken, for no one can be regarded as fully responsible. Thus, accepting that witches do indeed have supernatural powers, this power must arise from the witches' voluntary acceptance. Moreover, if this power comes from outside of the order of nature, it can only have either God or the devil as its source. As an evil power, its provenance is no longer a mystery: it is the devil's work. The logic of the case led inexorably to the conclusion that witches derived their powers from an implicit compact, a *pactum implicitum*, with the devil and were therefore guilty not merely of a crime but of *heresy*, for which the punishment was death.

By the time the theory was in place, so, too, were those developed principles of Roman civil law over which all sorts of barbarian tribal customs had overgrown during earlier centuries. As a result, the witch trials featured quite developed procedures for the production and weighing of evidence, for the protection of defendants and the determination of guilt. In the case of witches, some of this was relaxed or suspended. For example, although the law required the identification of those who bring charges against anyone, in the case of witches the identity of accusers was protected lest the witch undertake reprisals.

Torture had long been allowed and even required in Roman laws, as when slaves were to be interrogated. But there was a scrupulous principle in place by

1400 such that there could be no torture in the courtroom itself, and confessions gained through torture had to be repeated at a later time in full view of the court. Nor was it permissible to torture a pregnant woman or to repeat torturous sessions. Virtually none of this, however, availed the women charged as witches. They were tortured repeatedly, and pregnant ones were, at least occasionally, also included. Their accusers remained anonymous. Their loved ones were often presumed guilty of the same heresy, owing to the well-known more or less "infectious" tendencies of witchcraft.

In the matter of evidence, there was first confession itself. It is worth recalling that in the vast majority of cases only confessed witches were executed, the execution judged to be not only the right punishment but also a liberation from the pact. Confession was not the only grounds on which verdicts were based, however. There was the famous flotation test: The defendant was placed in a tank of water, held in place by long shafts until she was positioned in the center of the tank, and then left either to sink or float. In the latter case, guilt was presumptive, for the one who is commanded by the devil has greater buoyancy. There was also the "tear-test," which examined the defendant's capacity to cry when regaled with statements about God's love for her and the sacrifice of His only Son for her soul, etc. Failing to cry, she made clear that she had lost the virtue of remorse through her pact with the devil, although James I would caution against being misled by a witches "crocodile" tears.

When, in the sixteenth century, Johann Weyer would offer medical alternatives to the witch-theory—though not denying there were witches—he would note simply that the lighter bones of older women are probably the cause of their floating. It was Weyer's *De praestigiis daemonum et incantationibus ac veneficiis*[25] that helped to launch the modern psychiatric perspective on abnormal behavior. The book saw six editions between 1563 and 1583. By this date, such defenders of orthodoxy as Peter Binsfeld, Bishop of Trèves, could still claim that only "a few medical men, advocates of the devil's kingdom" deny that witches have intercourse with the devil, fly in formation to their rendezvous, etc. But Weyer, too, occupies a position of characteristic uncertainty and wandering perspective. He never denied the reality of Satan, nor did he rule out the participation of demons in mental illness. Rather, he considered the weakened mind of the aged and the psychologically distressed to be something of a devil's workshop wherein all varieties of illusion may be crafted. Still, Weyer was concerned with taking the physical evidence into account. *Of course* women tend to float, given their lighter constitution. To Weyer, this was simply a fact of nature, one known since the time of Hippocrates and lacking in any and all spiritual implications. He observed that drugs such as belladonna, opium, henbane, and tobacco enhanced such effects and could produce them in their own right. Again, forms of mental disorder did not require the participation of occult forces or beings. The poor souls who regard themselves as possessed and admit to being witches are, he reasoned, generally suffering from melancholia. They need the attention of a physician, not an inquisitor.

There were still other scientifically oriented figures struggling to establish a counter to superstitions. Paracelsus (1493–1541) wrote his brief treatise *On Diseases That Deprive Man of Health and Reason*,[26] noting in the Preface that the European clergy attribute these diseases to "ghostly beings," whereas "nature is the sole origin" of them. In this work he would oppose the Galenic theory, adding a medical theory of his own; with all this, however, he still reserved a category, the *Obsessi*, for the devil's handiwork. Thus, even the enemies of the more radical superstitions were hostages to milder forms themselves.

In the thousands and then tens of thousands of persecutions for this special heresy, there would be many cases in which the powers of the witch or the maladies she allegedly caused were not unlike conditions that might also occur naturally. It was not uncommon in such cases for medical practitioners to be summoned to give advice as to whether the symptoms were natural or not. Among its other contributions to history, then, the witch-phenomenon played a part in introducing medical "expert testimony" into adjudicative deliberations. In the same vein, it became ever more important for those who were actually engaged in studies of nature and natural processes to make clear the difference between their activity and the shady doings of the witch. Thus, in these same centuries we find ever more rhetorical and even defensive claims as to the true nature of scientific study. The borders between natural science and natural magic were not so clear in 1400. If they were clearer in 1600, the inducements of the witch-panic had more than a little to do with the anxious progress in this area.

Reason in Waiting

Thomas Macaulay may well have had Erasmus in mind when he wrote in his History of England:

> Every where there is a class of men who cling with fondness to whatever is ancient and who, even when convinced by overpowering reasons that innovation would be beneficial, consent to it with many misgivings and forebodings. We find also everywhere another class of men sanguine in hope, bold in speculation, always pressing forward, quick to discern the imperfections of whatever exists, disposed to think lightly of the risks and inconveniences which attend improvements, and disposed to give every change credit for being an improvement. In the sentiments of both classes there is something to approve but, of both, the best specimens will be found not far from a common frontier.[27]

Erasmus can be grouped with Savonarola and Luther to this extent: he was a reformer, an enemy of clerical abuse. Moreover, like the older Savonarola and the younger Luther, he was a scholar whose inspiration was grounded in Scripture. Here, however, the similarities end abruptly. Erasmus was a man of universal genius whose rare capacities would have surfaced in any age and under any circumstances. Those affairs of life and of state to which he directed his genius are neither national nor seasonal. Luther himself noted as early as 1517, in a letter to John Lang, that Erasmus could not be counted on to further Luther's objectives because "for him human considerations have an absolute preponderance over divine." [28]

In many relevant respects the Renaissance was more of a "middle" age than its predecessor. It was the interim separating two millennia, one ancient and one modern. The burning question put to the best minds of the sixteenth century was whether one period could gently give way to the next or whether the new could arrive only upon the death of its parent. Erasmus, more than any other figure of the period, hoped and worked for the first of these possibilities, and failed. Failure was inevitable. His time was one of desperation, foment, intrigue, and unrelenting enmity. His temper was one of conciliation, wit, urbanity, and graciousness. We appreciate his merits more adequately by realizing that as he wrote witches were being burned, astrologers were revered, and elixers concocted to rid the body of daemonic possession. Among the scholars of the day, invective and solemnity had all but replaced the search for truth. Europe was dividing itself between king and pope, between reason and passion, between obedience and revolution. Erasmus tried to hold these together, knowing all the while that it simply couldn't be done.

If Luther may be said to have had a scriptural conception of human nature, Erasmus must be said to have had an Erasmian view. To avoid this tautology is to court inaccuracy, for the Erasmian view is sui generis. That his conception was Christian can be admitted only if we recall his devotion to "St. Socrates." That his position on fundamental questions of faith was orthodox may be granted as long as we note his sympathy for the Skeptics,[29] not to mention his charming reproach of theologians in *The Praise of Folly*.[30] There was none more learned in Greek or Latin nor any more eager to promote classical letters but, unlike the Florentine Graecophiles who had gone past respect to adoration, Erasmus was wholesomely critical even of the most celebrated of ancient writers. His admiration of Lorenzo Valla[31] was based in large measure on Valla's willingness to scorn the ancients when they earned it.

Erasmian psychology is common-sense psychology, at once practical and eclectic. It flies off every page of his *Colloquies,* the little works rewarded doubly by being condemned both by the Faculty of the Sorbonne and the Council of Trent.[32] Of special interest are those dealing with exorcism and alchemy. Not only are these twins of gloomy nonsense held up to ridicule but they are presented in such a way that to hold out any validity for either is to proclaim oneself an innocent. Perhaps the best of the lot is the one titled "Charon," from the Greek mythological ferryman who brought the souls of the deceased across the river Styx. The colloquy is between Charon and Alastor, the latter querying Charon on the nature of his cargo, and on the varied problems confronted by one with a job such as his. Erasmus stages the colloquy in order to display the banality of war, especially that so-called "just war," and to depict the shabby rationalizations invented by zealots to justify what is finally homicide. Charon complains that the forests have gone bare feeding wood to witch-burners. Alastor asks if it is true that the souls of Frenchmen and Spaniards are rather light! He then observes that many a bishop, whose life would not have amounted to much at all, grows rich and famous in time of war: "They make more profit from the dying than the living." [33]

In *The Alchemist,* the master of the art, "who knows no more about it than an ass," is no more than a swindler who wins his way into the greedy heart of a great University scholar. In *A Pilgrimage for Religion's Sake,* another simpleton is presented, this one believing that the Virgin Mary has sent him a note complaining of the requests she receives from gamblers, loose women, and so forth. He knows the letter is authentic because the handwriting matches that which wrote "venerabilis" on Venerable Bede's epitaph. And so, one by one, Erasmus pricks vanity, exorcises exorcism, gleefully laments superstition, and otherwise reminds us of our duty to the simple lessons of saintly lives and of the extent to which all the trappings invented to decorate such lives serve only to obscure them.

Erasmus was neither scientist nor philosopher. He was not a dramatist or a statesman as these ranks are customarily claimed and awarded. He was a judicious observer, and, by this very fact, he pleaded the case for nature. His letters, colloquies, essays, and adages spoke directly of the observable world—of its problems, its charlatans, its warmongers, its vanities, its follies. His works reflected not only the labors of a polished mind and sympathetic heart but the precision and objectivity earned by judgments when good sense, opened eyes, and freedom from cant address the universe. He was the noblest side of Renaissance humanism.

Leonardo da Vinci (1452–1519)

In his *De Divisione et Utilitate Scientiarum* ("On the Division and Utility of the Sciences"), Savonarola accorded first place to philosophy and, revealingly, the lowest rung to ethics, economics, and politics. But can we believe that this was his genuine position? Can we believe, that is, that one of the most potent political forces in Renaissance Florence, a man who made a mockery of the classical writings, actually held philosophy in the highest regard and politics in the lowest? In the same work, we are reassured to discover that it is theology that is the ultimate science and that all the others are subsidiaries. Thus, when we read Savonarola's little treatise carefully, we note that philosophy is first among the servants of theology and that the latter, notwithstanding the rigors of Scholastic analysis, is fundamentally spiritual.

Of the records left by the Renaissance, perhaps it is the paintings that offer the clearest evidence of the beliefs and attitudes of the era. The texts of the age are helpful, but, owing to the censorship exercised over the written word, they are often of uncertain credibility. Contrast Savonarola's confusing classification of philosophy with Raphael's fresco *The School of Athens,* painted about 1509. Center stage is given to Plato and Aristotle, around whom are arranged seven liberal arts. This work captures the distinctively Scholastic flavor of Renaissance scholarship at least as effectively as any tract written by Valla, Pico, or Ficino. Raphael leaves no doubt in the viewer's mind: human creations fit neatly into categories; the categories complement each other but do not overlap; human invention appears in different degrees and one degree is more valued and true than a lower one; and philosophy (metaphysics), as Aristotle said, is the Queen.

Art historians have properly found significance in Raphael's inclusion of painting among the notables in this fresco. The inclusion is prima facie evidence of the status achieved by the artists of the Renaissance and of their clear awareness of this status. But what is all the more apparent in this work, and there are literally hundreds of others that make the same point, is that the Renaissance mind, when it searched for the eternal truths, produced Scholastic solutions. It is precisely because of this that the only alternative for many of the Renaissance anti-Aristotelians was to reject all philosophy. Metaphorically we may say that the Renaissance scholar could examine the world through one of only two spectacles: either of philosophy, and this meant Scholasticism; or of theology, and this meant spiritualism. To escape the categories meant to leave nature altogether, for what was "nature" other than quantity, quality, substance, state, and position. If we are to contrast the truly exceptional philosophical minds of the Renaissance with the large Scholastic background, we begin by noting that Erasmus and Leonardo da Vinci, through superficially different media of expression, escaped the categories and did not abandon nature.

It is always hazardous to rest a case on a single work, let alone a single paragraph. Nonetheless, if one were called upon to locate a brief statement within the corpus of Renaissance thought, a statement that completely anticipated the modern era, the following from Leonardo's *Book on Painting* could hardly be surpassed:

> Many will think they may reasonably blame me by alleging that my proofs are opposed to the authority of certain men held in the highest reverence by their inexperienced judgments; not considering that my

works are the issue of pure and simple experience, who is the one true mistress. These rules are sufficient to enable you to know the true from the false—and this aids men to look only for things that are possible and with due moderation—and not to wrap yourself in ignorance; if this has no good result, you would have, in despair, to give yourself up to melancholy.[34]

And, again, in his essay on physiology:

Though human ingenuity may make various inventions which, by the help of various machines, answer the same end, it will never devise any invention more beautiful, nor more simple, nor more to the purpose than Nature does.[35]

With respect to the soul, he retains this same, modest naturalism, leaving "to the imagination of friars, those fathers of the people who know all secrets by inspiration," [36] the question of the soul's spiritual composition. For his own part, he will eat no meat, cause no animal to suffer, his ethics being as basic as his natural philosophy and clearly tied to it.

Leonardo wrote on the full range of scientific and engineering topics that would be the focus of attention for centuries. Many of his ideas and inventions would not be improved upon until our own time. His anatomical sketches can still be used for instruction in biology, just as his drawings of various devices can still inform the mechanical engineer. His genius as inventor and painter is so widely recognized and discussed that we need not tarry to revere it. In matters that came to be of importance to modern psychology, his greatest contributions were in the field of visual perception. He explored and developed the geometric principles of perspective and, in the process, defined the conditions necessary for depth perception. He studied as well the factors conducive to illusions, to distortions of apparent size, to color contrast, to brightness-enhancement at boundaries, and to the perception of motion. His outlook was one of undaunted empiricism and his methods of inquiry followed accordingly. He shared with Aristotle the conviction that all knowledge begins with perception. To use his words, "wisdom is the daughter of experience." [37] He was duly respectful of reason, a capacity he enjoyed bountifully, but was unalterably opposed to the form of rationalism that required a rejection of the facts of experience. Reason and sense, for Leonardo, were conversant with different features of the knowable world, the former providing the certain truths of mathematics, the latter yielding the probable causes of natural events. He distrusted any science devoid of mathematical substance, and he argued that any science, in its final stage of development, would be mathematical in form. This smacks of Cartesian science and also allows us to locate Leonardo, as a theorist, in the rationalist tradition. However, in requiring that the scientific work with the clear and immediate evidence of sense before attempting to "mathematize" nature, he was an empiricist of the non-ideological sort.

On the question of human moral dispositions, Leonardo was less than modern, subscribing to an essentially Platonic position. Although willing to attribute failure and vice to bad experience and a lazy attitude, he also believed that people differed in their constitutional makeup: nature made some individuals golden, some brass, and so forth. However, "The greatest deception men suffer is from their own opinions," and this seemed to be a sufficient injunction against excessive theorizing about human psychology. To know man, one must study him the same way one would go about studying anatomy, geography, and mechanics. The method is observational, the perspective naturalistic, the descriptions quantitative. Long before Newton would immortalize experimentalism with the maxim, "Hypothesis non fingo!" ("I frame no hypothesis!"), Leonardo had set the cornerstone.

Was There a Renaissance?

We end this brief analysis of Renaissance philosophical psychology with the question that began it. It is true that the period bracketed by the years 1350 and 1600 yielded no new philosophical system; that the dominant theme in intellectual circles was the one set by the Scholastics; that witchcraft, alchemy, necromancy, persecution, fear, and mysticism were abundant; that the chief concerns and controversies still centered around theological matters; and that the main variable governing political life was the ageless quest for power. But it is also true that a strong and growing middle class began to assert itself; that skepticism was diffused more widely; that literacy and education extended to greater and greater numbers; that art and architecture of unprecedented appeal were created; that science and engineering attained a status not enjoyed since antiquity; and that man and his planet were, in increasing fashion, assimilated to that system of thought we have called naturalism. Intellectual epochs, as we have said so often, do not begin and end abruptly. While Leonardo preached the value of empiricism and experimentalism, Calvin was contributing predestinationism, and Calvin's following was a good deal larger than Leonardo's.

To the extent that there is an irreconcilable antagonism between naturalism and spiritualism, the Renaissance is important as a period in which the options and obligations held out by each were made clear and pressing. Reading Calvin and Erasmus transports one back to Athens, to the spare and righteous life recommended by the Pythagoreans and the reasonable, cheery gospel of Anaxagoras. As the Catholic Church began to suffer from divisive pressures, various sects splintered away and ironically strived for a condition of spiritual life much like the one founded by the early fathers of Christianity itself. We do not underestimate the sincerity or intelligence of either the reformers or the orthodox in the fifteenth and sixteenth centuries to note the similarity between their convictions and those that once divided Plato and Aristotle, or Augustine and Aquinas, or Pythagoras and Anaxagoras. Nor do we trivialize their concern by observing the frequency with which it has been expressed and defended historically. In every age there are those who fine excesses in the material conditions of life and who seek to rescue us from these by recommending or imposing blind obedience, discipline, denial, and humility. If they succeed, it is not long before still others perceive that human life must amount to more than these; that people must do what they can do; that happiness, a sense of dignity, and a means of personal growth and individual achievement are as essential as food and shelter; that nature is what it appears to be; and that what does not appear, alas, is not.

The Renaissance featured a rebirth of intellectual independence, too, which allowed secular knowledge to develop and be promulgated. The birth of public universities as early as the twelfth century in Italy not only extended higher learning beyond the walls of the abbey but also established practical subjects as appropriate topics for study and debate. The University of Padua was central to this movement. It was here that Pomponazzi rekindled Aristotelianism not as the foundation of theology but as the essential ingredient of naturalism and of any systematic approach to the human condition. As Professor Wade has observed:

> In analyzing his [Pomponazzi's] works . . . the modern historian is struck to see that the five problems with which the Paduan of the Renaissance was preoccupied (existence of God, immortality of the soul, nature of matter, free will, and good and evil, i.e., providence) are the same with which Voltaire struggled at Cirey . . . and strikingly similar to those which engaged the attention of Sartre, Camus, and the other existentialists in the twentieth Century.[38]

The secular universities, the invention of the printing press, the migration of classicists from fallen Byzantium, the growth of nation-states and their kings as a check on papal authority, the mystical naturalism of the Hermetic truths, and the preceding two centuries of Scholastic intellectualism blended to form a truly original concoction. Where the quest for personal power had once been confined to an aristocratic few, the Renaissance citizens openly sought to make nature work for them: if orthodox, to make it work religiously; if Hermetic, to make it work magically; if humanistic, to make it work in their behalf politically and intellectually. Pragmatism had never been seen on so great a scale, not even in Augustan Rome. Even the fear of the period, its superstitions and mysticism, were pragmatic at the roots. Medicine, which was essentially Arab medicine, reacquainted the Western mind with Averroes in contexts largely devoid of theological (Scholastic) content. The latent message of Averroist medicine was the message of materialism, and those who absorbed it added to the ranks of naturalism. In the hands of a Pomponazzi, the Averroist form of Aristotelianism led straightaway to a universe of "two truths": the truths of reason nurtured by perception, and the truths of faith revealed by intuition. The latter are beyond proof and therefore need not play a significant role in the affairs of the former. The leading humanists of the period—Erasmus, Thomas More, Montaigne—did not need Averroist science to adopt the same position.

The Renaissance did not succeed in separating Nature and Spirit in a way that might have dignified each. In requiring the two to coexist, Renaissance scholars too often laced each with tenets of the other or talked themselves into believing that each was only a different form of the other. It remained for seventeenth-century philosophers to divorce the two and to promote a notion of irreconcilabilty which survives to the present day.

The City (Again)

Of the several depressing spectacles allowing the twentieth century to call attention to itself, perhaps that of the modern city is the grayest. Millions of bodies are hurled about daily as they seek shelter and success in the faceless blocks of glass which pass for architecture. It is hardly surprising, then, that each year the historic cities of the Renaissance become more appealing to the pilgrim. Moreover, the appeal is based on something in addition to the exquisite beauty of any given building or any collection of art. It is based as well upon the unmistakable sense of order and proportion conveyed by the city as city. In an important way, it was the Renaissance that invented the idea of a city, at least as that idea survives, though weakly, in our time. The founders of Florence set out to plan a city around definite aesthetic, practical, and political considerations. Not until the fifteenth century did an essentially philosophical vision materialize into conscious urban planning.[39] The vision was Leon Battista Alberti's (1404–1472) and its materialization the result of Nicholas V's genius. The latter (1397–1455), as pope, remade Rome. Professor C. W. Westfall reviews the relation between Alberti and Nicholas V this way:

> Alberti's theory of architecture and Nicholas's program for papal government flow together. In his testament the pope outlined a purpose for buildings, and in his treatise the architect described the principles of design for such buildings. Buildings that were perpetual monuments to doctrine would also be architectonic structures conjoined into a city that would facilitate the activity of citizens whose love for God moves them to strive for the good through *virtu*. The pope, while governing the Church, would be a stinging *exemplum* of *virtu*, mov-

ing the lower members of the hierarchy to seek God actively. Rome
and its buildings would be proper settings for Nicholas's exemplary
actions.[40]

The leading architects of quattrocento Rome and Florence were not mere
designers. They were, in a manner of speaking, manual philosophers conditioned by Augustinianism to appreciate the city as teacher, the city as a moral
agent and moral engine. Here, in the beauty and order of a structurally planned
space, men and women and children would be constantly reminded of the obligation to impose beauty and order upon their lives and their wills. In this
regard, the Renaissance city is not a sudden invention but the logical extension
of the chivalric animus. When communities were scattered and fragile, it was
the knight who kept lonely vigil over the human potential for sacrifices and
honor. The military and economic security of the Church of quattrocento Italy
transformed the idea of city as fortress to that of city as agent. Even with this
enlarged set of responsibilities, the city retained an essentially chivalric character. That is to say, even under the more modern press of finance and political machination, the city retained the quality of a symbol—a largely medieval
symbol. Where the silhouette of the knight, painted against a drear and porten-
tous sky, once chilled the designs of a luckless felon, once fired the moral
passions of children, and once served to imbue peasant and priest alike with
the spirit of God's justice and mercy, there now stood the outline of the city.
Within its precincts churches and houses, meeting halls and municipal offices,
and gardens and markets would be placed according to a plan. Streets would
be laid in order to join related functions. Size would be employed as the symbol
of significance. Each frieze and corbel not only ornaments a building but tells
a story, teaches a lesson, recalls a value, intones an obligation, evokes a noble
sentiment. Here is architecture as applied psychology and civics. Florence and
Rome in the Renaissance were cities on a grand scale even by modern standards. They were places in which one could fine anything known or done by
the civilized world. They possessed most of the wealth, nearly all of the art,
and the sum of the talent of the age.

There is a dark side to all of this, too. Ancient Greece hosted a symbiotic
arrangement between aristocrat and farmer. At the height of Athenian influence, the agricultural class still played a salient role in the affairs of state. The
medieval township or fortress-community preserved this tradition. Even the
Roman patrician revered the land and sought the pastoral reaches of the exur-
ban territory for peace and reflection. The Renaissance city, however, as a valid
city, began to change this, and the change has progressed almost without a
pause. By the fifteenth century, the needs of major cities have already begun to
be insatiable. The countrysides become mere suppliers; their inhabitants, near-
barbarians. Now, urbanism, urbanity, is the culture, and the city is at once a

symbol of some higher mission and a reality in its own right. That is, the possibilities the city once announced are now obscured by the realities it presents to its citizens. Living becomes coping, and the quest for perfection gives way to the struggle for success. The chivalric ideal becomes perverted into the gaudy ritual of self-conscious heroism as the knight is displaced by the banker. The city takes on its own life and begins to use and abuse the citizen in return for which it offers mere services. Values give way to fashion, valor to competition, restraint and sacrifice to unimaginable poverty. Understood sociologically, the modern world begins with the Renaissance city, which, ironically, was to reawaken the classical ideal of life.

Notes

1. E. Cassirer, P. O. Kristeller, and J. H. Randall, eds., *The Renaissance Philosophy of Man,* University of Chicago Press, 1948, p. 147.

2. Marsilio Ficino, "Platonic Theology," translated by J. L. Burroughs, *Journal of the History of Ideas,* April 1944.

3. Eugenio Garin, *Science and the Civic Life in the Italian Renaissance,* translated by Peter Munz, Doubleday, Garden City, N.Y., 1969, pp. 21–48.

4. Petrarch, *On His Own Ignorance,* translated by Hans Nachod, in Cassirer et al., *Renaissance Philosophy of Man.*

5. Ibid., p. 74.

6. Ibid., p. 75.

7. Ibid., p. 103.

8. Ibid., p. 121.

9. John Ruskin, *The Stones of Venice,* Vol. II, Reuwee, Wattley and Walsh, Philadelphia, pp. 159–171.

10. Marsilio Ficino, in Cassirer et al., *Renaissance Philosophy of Man.* Regarding its reception, see Frances A. Yates, *Giordano Bruno and the Hermetic Tradition,* University of Chicago Press, 1964.

11. Ibid., p. 147.

12. Ibid.

13. Giovanni Pico della Mirandola, in Cassirer et al., *Renaissance Philosophy of Man.*

14. Ibid., p. 225.

15. Ibid.

16. Ibid., p. 245.

17. Charles B. Schmitt, *Gianfrancesco Pico della Mirandola and His Critique of Aristotle,* Martinus Nijhoff, The Hague, 1967.

18. Yates, *Giordano Bruno,* p. 121.

19. Ibid., pp. 128–130.

20. Ibid., pp. 185–186.

21. Martin Luther, *Twenty-Seven Articles Respecting the Reformation of the Christian State,* in *Introduction to Contemporary Civilization in the West,* Vol. I, edited by J. Buchler et al., Columbia University Press, New York, 1946, p. 630.

22. Ibid., pp. 634–647.

23. Henry Kramer and Jacob Sprenger, *Malleus Maleficarum* (1486), translated by Rev. Montague Summers, Benjamin Bloom, New York, 1970.

24. James I, *Daemonologie,* in *Minor Prose Works of James VI and I,* edited by James Craigie, Scottish Text Society, Edinburgh, 1982.

25. Johann Weyer, *De Prestigiis Daemonum* (1563), translated by John Shea, in *Witches, Devils and Doctors in the Renaissance,* edited by George Mora, Medieval and Renaissance Texts and Studies, Binghamton, N.Y., 1991.

26. Paracelsus (Theophrastus von Hohenheim), *The Writings of Theophrastus Paracelsus on Diseases that Deprive Man of Health and Reason,* translated by Gregory Zilboorg, in *Four Treatises of Theophrastus von Hohenheim Called Paracelsus,* Vol. I, edited by Henry Sigerist, Johns Hopkins University Press, Baltimore, 1941.

27. Thomas B. Macaulay, *History of England,* Vol. I, Harper and Row, New York, 1849.

28. Heinrich Boehmer, *Martin Luther: Road to Reformation,* Simon and Schuster, New York, 1957, p. 160.

29. *Luther and Erasmus: Free Will and Salvation,* Library of Christian Classics, Vol. XVII, Westminster Press, Philadelphia, 1969.

30. Erasmus, *The Praise of Folly,* translated by Hoyt Hopewell Hudson, Princeton University Press, Princeton, 1941.

31. See *The Epistles of Erasmus,* translated by Francis Morgan Nichols, Vol. I, #26, Longmans, Green, London, 1901.

32. *Erasmus: Ten Colloquies,* translated by Craig R. Thompson, Bobbs-Merrill, Indianapolis, 1957.

33. Ibid., p. 116.

34. *The Literary Works of Leonardo Da Vinci,* edited by Jean Paul Richte, Vol. I, Phaidon Press, London, 1970, pp. 116–117 (#12).

35. Ibid., Vol. II., pp. 100–101 (#837).

36. Ibid.

37. Ibid., p. 240 (#1150).

38. Ira O. Wade, *The Intellectual Origins of the French Enlightenment,* Princeton University Press, Princeton, 1971, p. 63.

39. On the origins of city planning in the fifteenth century, see Carroll William Westfall, *In This Most Perfect Paradise: Alberti, Nicholas V, and the Invention of Conscious Urban Planning,* Pennsylvania State University Press, University Park, 1974.

40. Ibid., p. 62.

2 From Philosophy to Psychology

7 Empiricism: The Authority of Experience

We must not corrupt our hope. To prostitute our past-cure malladie to Empiricks.
> William Shakespeare, *All's Well That Ends Well,* 1602

An empirical law, then, is an observed uniformity, presumed to be resolved into simpler laws, but not yet resolved into them.
> J. S. Mill, *A System of Logic,* 1846

A mere Rationalist (that is to say, in plain English, an Atheist of the late Edition).
> Sanderson, Preface to *Ussher's Power of Princes,* 1670

According to rationalism, reason furnishes certain elements, without which, experience is not possible.
> Fleming, *Philosophical Vocabulary,* 1857

The empirical philosophers are like to pismires [ants]. . . . The rationalists are like the spiders.
> Bacon, *Apothegms,* 1626

Coming to Terms . . .

There is a largely undefended convention adopted by historians of psychology of presenting rationalism as the nearly mechanical "cause" of empiricism and materialism. On this construal, empiricism is to be understood as a long reply to the claims of the rationalists, and materialism as a refinement of or inference from empiricism. Thus we learn (over and over again) that the "father" of modern empiricism (allegedly John Locke) set out to overturn the theory of "innate ideas" advanced by that patron of modern rationalism, René Descartes. It does not seem to trouble those who subscribe to this thesis that Locke never discusses Descartes in *An Essay Concerning Human Understanding* and that, as we shall see, Descartes specifically denied allegiance to the "innate ideas" theory with which he was charged. Nor is confidence in this convention weakened by the undaunted rationalism we discover in Locke's treatment of "intuitive" knowledge, or his theory of the "original acts of the mind" or of morals. By and large, all three of these *isms* are generally found side-by-side throughout intellectual history. It is only rarely the case that any one of them is utterly destitute of and uniformly hostile to features of the other two. In important respects, therefore, it makes no difference which is treated first simply because none can claim temporal priority.

What, then, is the advantage of beginning the present account with empiricism? First, it is the dominant voice in contemporary psychology, and its principal claims are likely to seem more familiar to the reader than those of rationalism or (even) materialism. Secondly, if its "father" is to be named, the honor goes to Francis Bacon, whose target was not rationalism per se, but Scholasticism. As we shall see, Bacon's thought reveals traces of Renaissance hermeticism and Scholastic empiricism. Accordingly, it makes sense to follow the two preceding chapters with an examination of Bacon, and this, of course, means an examination of the new empiricism. By the "new" empiricism I mean to contrast Bacon's metaphysics with the older Scholastic psychology. Had Bacon argued for nothing more than the thesis that our knowledge of the world comes from immediate experience, he would have done no more than echo Thomas Aquinas, William of Ockham, and numerous other commentators from the thirteenth and fourteenth centuries. But Bacon was to propose something beyond a theory of human knowledge; he was to propose an official method by which knowledge is to be pursued. This, too, was not new, since Grosseteste, Roger Bacon and other members of the Oxford school had arrived at much the same conclusion centuries earlier. But Bacon enlarged the scope of methodological empiricism to include subjects that the older empiricists either failed to address or placed outside the arena of perceptual knowledge.

Rationalism and empiricism, as we know them today, have the inevitable Platonic and Aristotelian ancestry. But they have more immediate origins in Renaissance humanism, hermetism, and skepticism. When historians tell us that the modern era of science did not begin until the authority of Aristotle was overcome, they are generally referring to the emergence of an experimental attitude and the suspension of purely logical approaches to discovery. However, this very experimental attitude, one that Aristotle possessed in abundance, is already highly developed in the "natural magic" of the sixteenth century. Unlike Aristotle's own empirical perspective, which was that of the detached, disinterested observer of nature (an attitude reinforced by the simple desire to discover the makeup of the physical and animal world), Renaissance empiricism and experimentalism were typically based on the motive to control and manipulate nature, to make nature conform and obey, to change the world. It is this difference that accounts, in large part, for the inextricable connection between science and technology in the Renaissance, and for the otherwise surprising indifference of Greek science toward technology. In this we have one more instance of the failure of the Renaissance to achieve what it proclaimed to be its central mission: the re-creation of the classical outlook.

Renaissance rationalism, too, was given more to the "spiritual magic" of the Hermetic legacy than it was to that classical version of rationalism best represented by Anaxagoras and Aristotle. In their contempt for the Scholastics and, therefore, for the Aristotelian foundations of Scholasticism, Renaissance rationalists such as Giordano Bruno looked beyond the tidy veracity of the syllogism for still grander truths. These were to be found first in the eternal Platonic Ideas and, on closer inspection, in the (presumably) pre-classical teachings of the *Corpus Hermeticum*. The modern era, then, did not begin either with the science of the Renaissance or with its suspicion of Aristotle. It did not, indeed, begin in the Renaissance at all except to the extent that the quiet skepticism of Erasmus and the insolent skepticism of Gianfrancesco Pico are precursors to the *Novum Organum* of Francis Bacon (1561–1626).

The modern era began, it would seem, with a rejection of the Renaissance, which is to say, with the careful partitioning of naturalism and spiritualism. To the extent that it may be called modern it is to be seen as an era in which small projects replaced grand ones, an era in which scientific humility replaced the limitless reach of magic. The Renaissance was philosophically conservative and scientifically radical. The seventeenth century reversed this when it did not reject both the conservative and the radical molds. Leonardo announced the modern era and Francis Bacon attempted to define it. As can be discerned in his remarks quoted at the beginning of this chapter, Bacon was no admirer of the empiricism practiced in the Renaissance. He rebuked those who would attempt to construct theories from limited material, or, as he said, "the narrowness and darkness of a few experiments."[1] While scorning the Aristotelian form of empiricism, according to which experiments are performed in keeping with an initial metaphysical bias and one that must be satisfied by the "experiment," Bacon was recommending an empiricism of his own: the direct, theoretically neutral observation of nature for the purpose of learning the physical facts of the real world. This form of empiricism, which Bacon construed to be as much at odds with Aristotle as it was with mysticism, is precisely the form that Gianfrancesco Pico construed as Aristotelianism's fatal flaw! Gianfrancesco's contention was that Aristotle based his epistemology on the faulty data of sense; Bacon's, that he did not limit his science to the facts of sense. What we learn from these contradictory complaints is, of course, that each age reads Aristotle pretty much as it pleases.

The empirical movement launched by Bacon is the scientific movement itself. Empiricism, understood within this movement, is an overarching philosophy that confers epistemological authority on direct experience. It takes the evidence of sense as constituting the primary data of all knowledge. It stipulates that knowledge cannot exist unless this evidence has first been gathered, that all subsequent intellectual processes must use this evidence and only this evidence in framing valid propositions about the real world. How does this differ from rationalism? Rationalism, as received by our own time, is the product of the philosophical systems created by Descartes, Spinoza, and Leibniz. It is no coincidence that two of them, Descartes and Leibniz, were distinguished mathematicians. Descartes founded analytical geometry, and Leibniz invented the calculus independently of Newton. The modern rationalist, from the seventeenth century on, has shared with the ancient Pythagoreans, and with the Plato of the "number theory," a vision of the real world as a system of mathematical, harmonic relationships. Persuaded by the proofs of mathematics that certain knowledge exists, the rationalist tends to be aloof toward the imprecise and ephemeral facts of experience. It is only when reason explores the universe that a small set of fundamental and irrefutable principles are clear and that, from these, the more detailed facts and fabric of nature can be deduced rationally. What modern rationalism retains of its Aristotelian origins is the contention that the very act of perception must assume a categorical framework if experience is to be anything other than a buzz of confusion. The mind must be so constituted as to segregate and organize sense data. It must be so equipped as to direct the senses, thus separating their illusory from their real content. One way or another, the rationalist position incorporates the concept of an a priori cognitive capacity. Without it, meaningful experience is not possible. Rationalism takes all settled knowledge to be the result of a rational analysis of the evidence of sense. Such evidence cannot be gathered except by a rationally directing principle. Accordingly, the primary "datum" of which our knowledge is comprised is that innate disposition called "the laws of thought."

We should also recognize a point of agreement between rationalists and empiricists which survived unimpaired until the advent of behaviorism in the twentieth century. This common feature may best be labeled "mentalism." The leading architects of empiricism all based their epistemologies upon what seemed to them to be the fixed dispositions of the mind. In other words, their philosophies were explicitly designed to account for the facts of mental life. Although all agreed that the mind is furnished by the senses, they agreed as well that philosophy's task was to determine how this occurred and what it implied. The empirical tradition, therefore, is in no sense anti-mental, notwithstanding its emphasis on perception. Rationalism, of course, is unabashedly mentalistic. With the terms more or less in place, then, we can turn to the first significant modern defender of empiricism and empirical science, Francis Bacon.

Francis Bacon (1561–1626)

Francis Bacon was born to a ranking family, but his own star rose higher than all of theirs combined. Successively, he was a leading and prosperous attorney, a member of the Privy Council, the Lord Keeper of the Great Seal, and the Chancellor of the Realm, and, in 1618, he was named Baron Verulam by his admiring king, James I. It is a sign of his time that he was forced to resign from office in 1621, charged with accepting bribes. That the charge was valid is not to be questioned, only that there could have been no Bench in all of England had simon-pure been a condition of service.

Luther's Reformation, which on the Continent was a popular cause and one fired by the poverty consuming the German states, was to be absorbed by England on a different basis. There, a terrible rift had been infecting the affairs of Church and Throne for centuries. As early as the twelfth century English kings had sought to gain independence from Rome by contending that in temporal matters the Sovereign's will was the law of the land. The famous showdown between Henry II and Thomas à Becket, Bishop of Canterbury, which ended with the latter's murder (1170), was followed by a wave of Roman Catholic zeal threatening to reduce the throne to a form of vassalage. Throughout the later Middle Ages and the sixteenth century all varieties of intrigue were fashioned to maintain a delicate balance of power among English, European, and papal forces. It was in the reign of Henry VIII that the European reform movement had so threatened papal authority in Europe that an English king might take an extreme action. In 1533 Henry declared his marriage to Ann Boleyn legal, despite the pope's refusal to annul his current and childless marriage to Catherine. Anne lost her head within three years of their wedding day, charged with adultery, although her crime seems to have been the failure to produce a male heir. The marriage had lasted long enough, however, to yield a daughter, Elizabeth, later Elizabeth I, the most kingly queen in the long history of that office.

If the reign of Elizabeth (1558–1603) had embraced no more than the works of Shakespeare, it would still demand the attention of scholars. But the Elizabethan era yielded a trove of cultural achievements. It was also the period in which Calvinism got its grip on the English mind, when Puritans began their holy vigil against pleasure, and when a throne that had long been cudgeled by papal Bulls found its moment of revenge. Catholics were no longer permitted to celebrate the Mass. Altars and icons were removed from the churches. Houses of suspected Catholics were entered and searched. The response from Rome was a decree from Pius V that announced Elizabeth an excommunicate and instructed English Catholics to ignore her laws. Recalling the circumstances of her birth, the oppressed Catholic majority now judged the Queen to be no more entitled to the crown than any other of Henry's bastards. Not unpredictably an Irish revolt was precipitated and was put down at a cost in lives and money almost unprecedented in England's history to that date. War with Spain followed and the defeat of the Spanish Armada began a period of oceanic rule which England enjoyed until our own century. But neither military success nor artistic accomplishments were able to mute the hostile cries of the religiously persecuted. The Catholic majority was outraged, and the small Puritan faction even more so. Elizabeth's cousin, Mary, had been the exiled "Queen of Scots" for years and posed a constant threat to Elizabeth's publicly perceived legitimacy, for Mary Stuart was the great granddaughter of Henry VII, and her pedigree was without blemish. Elizabeth had her beheaded in 1587, but by then Mary Stuart's son, James VI, was already over twenty-one, was king of Scotland, and was on record as a defender of Protestantism. At the age of thirty-seven, he succeeded to the throne of England as James I. Within two years appeared *The Proficiencie and Advancement of Learning* (1605), Bacon's first major work. Elizabeth was dead and a time for renewal and rededication had arrived. Blood was to be spilled again in the civil wars of 1642–1649, but in the reign of James I (1603–1625) there was to be an interval of peace and ground-laying that would start English scholarship on a voyage from which it has never turned back. Bacon was one of its great captains and the *Novum Organum* its chart.

Many ideas developed in the *Novum Organum* were initially proposed in the *Advancement of Learning,* a work serving rather as an outline. Bacon's project in the *Advancement* was manifold: to delineate the principal causes of enduring ignorance and disagreement; to identify true authority in matters intellectual; to outline those areas of inquiry about which little is known; to present methods able to accommodate the foregoing; and to defend the basis upon which any undertaking is worth the effort. On the question of method, the *Advancement* proves to be a skimpy text, but its deficiencies were ably overcome in the *Novum Organum.* On the final objective Bacon's position can be summarized quickly: the value of any undertaking is assessed in terms of the potential benefit to the human race. The ultimate standard is pragmatic. If the project can do people little or no good in the daily affairs of life, there is the strongest presumption that it is worthless. This position, we see, is but the bold version of the pragmatic spirit that first blossomed in the Renaissance. It is also a specific disavowal of the aspect of Reformation theology that would depreciate the value of human works. In Bacon's system, usefulness and worthiness are synonymous. This is made clear in a passage not yet fully divorced from the Hermetic tradition:

> But the greatest error of all the rest is the mistaking or misplacing of the last or furthest end of knowledge. For men have entered into a desire of learning and knowledge . . . seldom sincerely to give a true account of their gift of reason, to the benefit and use of men . . . for

the glory of the Creator and the relief of man's estate. But this is that which will indeed dignify and exalt knowledge, if contemplation and action may be more nearly and straitly conjoined and united together than they have been; and conjunction like unto that of the two highest planets, Saturn the planet of rest and contemplation, and Jupiter the planet of civil society and action.[2]

What of legitimate authority in the epistemic domain? The *Advancement* shows that Bacon was not the zealous and radical antitraditionalist so often conveyed by caricatures of his writing. He praises Aristotle often, as well as other leading philosophers. However, he condemns those (and especially the Scholastics) who consult antiquity with such diffidence as to be unwilling to add a line or a reservation to Greek or Roman teaching:

> For as water will not ascend higher than the level of the first springhead . . . so knowledge derived from Aristotle, and exempted from the liberty of examination, will not rise again higher than the knowledge of Aristotle.[3]

The complaint, then, is not with the ancients but with their disciples. Similarly, those following in the steps of Luther, who had revived many ancient works to defend his charges against the Church, have become more interested in the words and the grammatical style of the older writers than they have in the very matters which Luther addressed:

> Words are but the images of matter. . . . [T]o fall in love with them is all one as to fall in love with a picture.[4]

In addition to stultifying reverence for antiquity, there are human inventions that stand in the way of the advancement of learning, and these are inventions of the mind based on gullibility. Bacon notes three of these: astrology, natural magic, and alchemy.[5] As Aristotle was the "dictator" of the Scholastics, so too have miracles, spirits, illusions, and magicians come to master the common mind. A part of the source of these superstitions he attributes to too much concern for "final causes" and particularly to the inclusion of discussions of these final causes in the area of physics rather than metaphysics. Perhaps the most modern notion submitted in the *Advancement* is that which insists on the division of natural science into these two branches (physics and metaphysics) and the requirement that the issues in physics be treated with methods different from those purely logical devices normally employed in metaphysical analysis.[6] In addition to this division, Bacon argues that mathematics be reserved to metaphysical matters and, in this way, he stands four-square against that rationalist (neo-Platonist) tradition which seeks to comprehend nature in terms of number theories, harmonies, and the like.[7]

The portion of the *Advancement* of special importance to the history of psychology is that concerned with subjects that have not received their due in the deliberations of natural philosophers. It is in these sections that Bacon invents the disciplines of theoretical and experimental psychology without naming them. He is clear on these points:

> We come therefore now to that knowledge whereunto the ancient oracle directeth us, which is the knowledge of ourselves; which deserveth the more accurate handling, by how much it toucheth us more nearly. This knowledge, as it is the end and term of natural philosophy in the intention of man, so notwithstanding it is but a portion of natural philosophy in the continent of nature. . . . [W]e proceed to Human Philosophy or Humanity, which hath two parts: the one considereth man segregate, or distributively; the other congregate, or in society. So as Human Philosophy is either Simple and Particular, or Conjugage and Civil, Humanity Particular consisteth of the same parts whereof man consisteth; that is, of knowledges which respect the Body, and of knowledges that respect the Mind . . . *how the one discloseth the other and how the one worketh upon the other.* . . . [T]he one is honoured with the inquiry of Aristotle, and the other of Hippocrates.[8]

And then a little later in the essay:

> For Human Knowledge which concerns the Mind, it hath two parts; the one that enquireth of the *substance or nature of the soul or mind,* the other that enquireth of the faculties or functions thereof. Unto the first of these, the considerations of *the original of the soul,* whether it be *native or inventive,* and *how far it is exempted from the laws of matter.*[9]

We cannot read these passages without being reminded of Aristotle's *De Anima,* but we must recognize a telling difference: Bacon is presenting this subject as a branch of natural science and is reducing it to a set of questions of an experimental nature. This is not reheated Aristotelianism!

In the concluding pages of the *Advancement,* Bacon presents his own theories of humanity and these turn out to be surprisingly nativistic. He is persuaded that everyone can be identified in terms of several innate characteristics or temperaments; that some minds "are proportioned to great matters, and others to small," [10] and that while some of these characteristics "are inherent and not extern" others are caused by fortune.[11] Of the inherent determinants he includes sex, age, nationality, disease, constitutional deformity, and beauty. The environmental determinants number sovereignty, nobility or obscurity of birth, poverty and wealth, and such ambiguous agencies as magistracy, privateness, prosperity, adversity, constant fortune (luck?), variable fortune, rising per saltum (in a leap), and rising per gradus (in humbler steps). Some of these track Bacon's own youthful attempts to gain favor in Elizabeth's court through his aloof uncle, Lord Burghley.

At no point in the essay does Bacon abandon or even question the Hippocratic and Galenic theory of humours, but he is less rigidly nativistic on this score than either the Hippocratics or Aristotle. Choosing (out of context) the line of Aristotle's *Metaphysics* that argues that what is by nature cannot be changed by custom, he insists that even human nature is alterable through practice, reward and punishment.[12] He agrees with Aristotle that a stone hurled in the air a thousand times "will not learn to ascend," but he distinguishes between the fixed laws of mere matter and those covering human conduct. The latter are not "peremptory" but provide a certain "latitude." Since virtue and vice are, accordingly, only or largely habits, it is possible to create virtuous citizens by a proper regimen. It is in this context that he praises Machiavelli for having described what men do rather than what they ought to do. Bacon is concerned with human nature in both its inherent and its acquired respects. He recommends a psychology able to distinguish between the two and able to manipulate the manipulable feature toward the betterment of the two and able to manipulate the manipulable feature toward the betterment of the human condition. He is, in a word, demanding an "operative" science to complement the purely speculative one bequeathed by Aristotle and left unimproved by the Schoolmen.

Bacon's empiricism was of a limited sort. He was committed to the view that human psychology bears the indelible stamp of hereditary influence. Given this, he was forced to conclude that in regard to human conduct and human achievement, external conditions could produce changes of only a limited variety and degree. His was decidedly not a behavioristic empiricism. It was rather a methodological empiricism. This is amplified in the *Novum Organum,* to which we now turn.

If the *Advancement of Learning* is the work of a young man testing the waters of a new monarchy, the *Novum Organum* is the churlish scolding of a seasoned and respected scholar writing from the lofty position of a Baron of Verulam. Thus, though the effort of an older man, the work is less cautious, less temperate, and less balanced. And it was incomparably more influential than its predecessor. It was written as two books of *Aphorisms,* 130 in the first and 52 in the second. It is in Book II that the heralded method is disclosed; however, after the brilliant, often witty, and always challenging Book I, Book II proves to be a rather pretentious, even half-baked collection of veiled theories, Hermetic innuendos, Scholastic distinctions, and labels. Had only Book II survived, more than one Ph.D. dissertation would have been needed to establish precisely what the "Baconian" method was. Fortunately, in Book I Bacon exhumes it from the crypt:

> Now for grounds of experience—since to experience we must
> come—we have as yet had either none or very weak ones. . . . Nothing
> duly investigated, nothing verified, nothing counted, weighed or mea-
> sured, is to be found in natural history: and what in observation is
> loose and vague, is in information deceptive and treacherous. And if
> any one thinks that this is a strange thing to say, and something like
> an unjust complaint, seeing that Aristotle, himself so great a man, and
> supported by the wealth of so great a king, has composed so accurate
> a historia of animals . . . it seems that he does not rightly apprehend
> what it is that we are now about. For a natural history which is com-
> posed for its own sake is not like one . . . for the building up of phi-
> losophy. They differ in many ways, but especially this: that the former
> contains the variety of natural species only, and not experiments.[13]

Bacon's criticism of Aristotle is that the latter busied himself merely with noting and naming rather than searching for causes. Overwhelmed as he was by
the importance of his final causes, Aristotle the naturalist mixed metaphysics
with physics or, worse, failed to develop natural physics by treating its subject
matter in a metaphysical way. Had Aristotle discovered the experimental approach, he would not have spent his time on a merely historical assessment of
the animal kingdom, identifying the types of animals found in different places
at different times. Instead he would have inquired into the physical (efficient
and material) causes of the variety. The experiments envisaged by Bacon are
of two sorts. He labels one *Experimenta lucifera* (those that shed light) and the
other *Experimenta fructifera* (those that bear fruit). Science requires both. The
first, which are utterly devoid of a theoretical bias on the experimenter's part,
are simple inquiries into elementary cause-effect sequences:

> Now experiments of this kind have one admirable property and con-
> dition; they never miss or fail. For since they are applied, not for the
> purpose of producing any particular effect, but only for discovering
> the natural cause of some effect, they answer the end equally well
> whichever way they turn out; for they settle the question.[14]

The *experimenta lucifera,* modest in aim and useless by themselves, produce
findings to be stored in the *Table of Discovery* to which other scientists can
refer. In time, the great *Table* is filled in and the searching questions become
obvious. Then, the *experimenta fructifera* are possible, and the results of these
are not only of great benefit to humanity but also unearth the basic and invio-
lable laws of nature. And if one wishes to know to how broad a range of nature
Bacon is willing to extend his model, the answer is unequivocal:

I form a history and tables of discovery for anger, fear, shame, and the like; for mattes political; and again for the mental operations of memory, composition, and division, judgment and the rest; not less than for heat, and cold, or light, or vegetation, or the like.[15]

We find throughout the work exhortations to the reader not to despair, to be mindful of how long and arduous is the path to useful knowledge, but how great the success will be once it is navigated. The message is movingly conveyed in the ninety-second aphorism of Book I:

But by far the greatest obstacle to the progress of science and to the undertaking of new tasks and provenances therein, is found in this—that men despair and think things impossible.[16]

What is the circumstance calling for these repeated encouragements? We can answer this by recalling the social climate with which Bacon and, later, Hobbes, Descartes, and Locke were contending. Once again the voices of doom were announcing from many pulpits that the end was near, that corruption and moral degradation were but pale copies of the gigantic destruction just over the earth's horizon. Once again, and with the well-worn tools of astrology, Hermetic frenzy, and biblical interpretation, the common people were convinced that the time had come, that the six thousand years had elapsed. Humanity has once more become "the dregs of Adam's race," and the metaphysical poet, John Donne (1573–1631) can only lament:

The world did in her cradle take a fall,
And turn'd her braines, and tooke a generall maime,
Wronging each joynt of th' universall frame.
The noblest part, man, felt it first; and then
Both beasts and Plants, curst in the curse of man.
So did the world from the first houre decay,
That evening was beginning of the day,
And now the Springs and Sommers which we see,
Like sonnes of women after fiftie bee.[17]

The mood had been set by the fatigue of the Reformation, by the cooling of Renaissance achievement, by the collapse of governments, by poverty and hunger, plague and civil wars. The fall of Rome had been an invitation to this sort of thing before, and the incomplete fall of the papacy served as a similar signal. The *Novum Organum* was as much a rejoinder to this mood as it was a detached exercise in scientific scholarship. Bacon was the first of a series of philosophers, principally British, to attempt to cal a halt to all this. When, just a few decades later, the Age of Newton would announce itself, the curtain was fully raised. The implications for psychology were most clearly recognized by John Locke, a "Newtonian" psychologist.

John Locke (1632–1704)

There were two conflicting intellectual movements opposing the empiricism that Locke was to advance. Each of them, skepticism and rationalism, was at odds with the other. Each, however, was even more at odds with empiricism. Since rationalism was (and has remained) an essentially Continental movement, and since its principal architect, Descartes, was hardly a leading figure in British circles, it may be safely assumed that Locke's *Essay Concerning Human Understanding* was written as much in response to his skeptical countrymen as it was to Descartes' *Meditations*. Cambridge in Locke's time was hosting a revived form of Platonism, replete with anti-empiricist arguments. Even if, during his period of self-exile in Amsterdam, Locke had not studied Descartes' rationalistic philosophy, he would have had enough to reply to at home.

The combination of Luther's Reformation, the Copernican revolution, Calvin, young Newton's universal laws, Elizabeth I, Bacon, and civil war was too much for any age to assimilate without raising some serious questions about whether anything can be known with confidence and abidingly. Thus, Locke's contemporaries included a number of persuasive skeptics such as Joseph Glanvill (1630–1680), whose *Scepsis* and *Lux Orientalis* revived an almost ancient form of contempt for what presented itself as human understanding. Pyrrhonism, named after the Greek skeptic Pyrrho, was on the verge of becoming the new Scholasticism. The Doomsday perspective of Bacon's age was evolving into Know-Nothingism, and Locke set out to challenge and replace it. In the Introduction to the essay, he sounds Bacon's theme:

> If we can find out those measures whereby a rational creature, put in that state which man is in this world, may and ought to govern his opinions and actions depending thereon, we need not be troubled that some other things escape our knowledge.[18]

He also leaves no doubt about the sort of analysis he is attempting. Unlike Hobbes and Descartes, whose psychological systems will be discussed later, Locke is completely neutral on the question of the biological or physical factors that may be responsible for the character of mental states and activities. He specifically avoids all questions regarding the physiological basis of thought:

> I shall not at present meddle with the physical consideration of the mind, or trouble myself to examine wherein its essence consist or by what motions of our spirits, or alterations of our bodies, we come to

have any sensation by our organs, or any *ideas* in our understanding; and whether those ideas do, in their formation, any or all of them, depend on matter or not.[19]

The essay, then, is not to be about brains, motions, or theological disputes. It is to be about the human understanding, which, for Locke, is no more than the ideas possessed by the mind and reflected on by it. Locke's philosophy is addressed to the origin, validity, and utility of ideas, and these are only *"whatsoever is the* object *of the understanding when a man thinks."* [20] In regard to the sources of our ideas (i.e., understanding), there are only two: sensation and reflection. The former is no more than the sensory apprehension of the particular objects of the physical world. We do not sense "universals," "species," "truths," or "principles": we sense things and only things. As the organs of sense engage the material stuff of the world, the mind also has a perceptive faculty according to which it is able to examine its own contents. This faculty is reflection, and, since the mind is furnished only through experience, reflection is comprised of that which the senses have provided.

However, it is not comprised only of these. In addition, reflection includes the mind's ability to examine its operations (i.e., not only its contents) and these operations include the effects of the passions on our ideas. What Locke proposes is a theory of knowledge based on two interacting processes. One of these is sensation, pure and simple, which leads to factual knowledge of the material world. The second is reflection, which is an internal sense able to examine the deposited sensations in the larger context of our general, emotional state. Reflection is that which removes understanding from the purely photographic process of recording items in space. Instead, through reflection, the human understanding is a psychological entity. This is not to say, however, that it is a spiritual one, since the feelings themselves are to be understood in empirical terms. Locke's empirical psychology is associationistic, even in the mechanical sense, but it never denies those mental attributes of human psychology which everyone senses oneself to possess. He does not advance the associationistic form of psychology that pavlov was to come to recommend, nor the materialistic form so prominently displayed by the early Epicurean philosophers or the Renaissance Averroists. In fact, he specifically rejects the suggestion that perception is merely the activation of the organs of sense. He speaks of the inattentive man in the *Essay* who, while inspecting an object of interest, is oblivious to a variety of sounds, despite the fact that there is no defect in his sense of hearing.[21] For perception to occur, the mind must be directed to the object. It is not a passive process but, rather, an active transaction between the observer and the knowable world.

On the question of innate ideas, Locke is uncompromisingly opposed. He argues that we enter the world as a *tabula rasa* and that all we will come to know is furnished by experience. He leaves room for the possibility that the fetus, once its sensory apparatus has taken form, will experience rudimentary sensations and, therefore, may enter the world with certain primitive ideas, but on no account can such ideas, if they exist, be considered innate.[22] Since knowledge is of particular, material things and since the infant could not possibly have had experience with such things prior to birth, there can be no sense in which the infant has "ideas." Nor, for that matter, can an adult have ideas about that with which the senses have had no commerce. In one of the most famous passages in the *Essay,* Locke applauds a proposition passed on to him by William Molineux in Ireland, who anticipated an experimental issue that the twentieth century has found irresistible:

> I shall here insert a problem of that very ingenious and studious promoter of real knowledge, the learned and worthy Mr. Molineux, which he was pleased to send me in a letter some months since: "Suppose a man born blind, and now adult, and taught by his touch to distinguish between a cube and a sphere of the same metal and nighly of the same bigness, so as to tell, when he felt one and the other, which is the cube, which the sphere. Suppose the cube and sphere placed on a table and the blind man to be made to see: *quaere,* whether by his sight, before he touched them, he could now distinguish and tell which is the globe, which the cube?" To which the acute and judicious proposer answers: "Not. For though he has obtained the experience of how a globe, how a cube, affects his touch; yet he has not yet had the experience, that what affects his touch so or so, must affect his sight so or so. . . ." I agree with this thinking gentleman. . . . The blind man, at first sight, would not be able with certainty to say which was the globe, which the cube.[23]

Locke contends that thinking and perceiving are but different words for the same process. The ideas resulting from perception are initially simple, but, through associated experiences and memory, the simple ideas are combined to form complex ones. However, no matter how complex our ideas become, they remain rooted in experience and nurtured by the reflective faculty. To say that we know is no more than to say that we have ideas. Knowledge itself is "nothing but the perception of the connection of and agreement, or disagreement and repugnancy of any of our ideas." [24] This agreement (and its converse) may be treated as a four-fold affair. First, there is the recognition that what is, is; that a thing cannot be and, at the same time, not be. After this "first act of the mind" comes our knowledge of relations, which is no more than the mind's awareness that certain ideas are related (e.g., mass and weight) and others not at all or less so (e.g., telephone numbers and annual rainfall). The third form of our knowledge of agreement is what Locke calls coexistence or what, in modern parlance, we would call a correlation. The word "gold" for example merely signifies color, malleability, weight, and so forth, such that when the word is uttered we know these properties by habitual association. Finally, we have the knowledge of real existence, such as "That is an apple." To summarize, we say that we know things in four ways: by identity (e.g., black is not white), by relation (e.g., the whole is equal to the sum of its parts), by coexistence (e.g., iron is attracted by a magnet), and by real existence (e.g., Big Ben is in London).

The foregoing modes exhaust the foundations of our knowledge. However, they do not explain the degrees of knowledge. There are some things we are more confident about than others. Locke orders the certainty attaching to human knowledge as intuitive, demonstrative, and sensitive. By intuitive he means the sudden, undeliberated awareness of a truth. For example, a circle is not a triangle, or black is not white. No other idea is necessary for the mind to perceive such truths, and anyone who seeks to discredit the certainty of this form of knowledge "has a mind to be a skeptic." [25] Without intuitive knowledge, demonstrative certainty would not be possible. Thus, for us to recognize that two things that are each equal to a third must be equal to each other (a recognition ensured by the incontrovertability of deductive proofs) we must have the capacity for intuitive knowledge of the kind just described. As intuitive knowledge forms the basis of sagacity or wisdom, demonstrative knowledge is the product of reasoning. Through this reasoning, items that cannot actually be juxtaposed for purposes of comparison are reasonably juxtaposed *by a juxtaposition of ideas*. The major distinction between intuitive and demonstrative knowledge is that although both are ultimately resistant to doubt, only the former is absolutely undoubted initially. For example, no one doubts, from the moment the sentence is uttered, that "Black is not white." Initially, however, one might doubt that "two triangles with equal bases and constructed between parallel lines are equal." The latter can be proved and, once proved, is known to be certainly true. This certainty, however, may come only after the proof.

Between this pole of certainty occupied by intuition and demonstration and the opposite pole of the merely probable or possible (i.e., mere faith or opinion), there is the sensitive knowledge of particulars. Sensitive knowledge is of two types: that which apprehends the primary qualities of substances, and that which involves secondary qualities. The distinction between these is fundamental and warrants close consideration. The primary qualities, says Locke, "are discovered by our senses, and are in [things] even when we perceive them not; such as the bulk, figure, number, situation, and motion of the parts of bodies, which are really in them, whether we take notice of them or no." As for the *secondary qualities,* it is best to let Locke state precisely what he means by the term:

> Secondly, the sensible secondary qualities which, depending on these, are nothing but the powers those substances have to produce several ideas in us by our senses; which ideas are not in the things themselves otherwise than as anything is in its cause. . . . Had we senses acute enough to discern the minute particles of bodies, and the real constitution on which their sensible qualities depend, I doubt not but that they would produce quite different ideas in us, and that which is now the yellow color of gold would then disappear, and instead of it we should see an admirable texture of parts of a certain size and figure.[26]

The Newtonian "corpuscular" reality is what gives rise to the experience of, e.g., "yellow," but the color is the effect, not the cause. It is the effect that ultimate physical properties have on the organs and processes of perception. Sensitive knowledge, then, is part objective and part invented. Physical objects have specific primary qualities that the senses are equipped to perceive. These include size, shape, number, and motion. However, matter is particular and atomic, and the senses are not able to deal with matter at its own elemental level. Instead, the elementary particles stimulate the senses, and the latter report perceived matter whose secondary qualities are products of the act of perception itself. Were there no perceiver, these qualities would not exist. As a result of this, there will always be an "incurable ignorance," for "there is no discoverable connection between any secondary quality and those primary qualities that it depends on."[27]

Those who would find a skeptical element in Locke's psychology need look no further than the theory of secondary qualities and the claim that we must remain *incurably* ignorant of the nature of the connection between primary and secondary qualities. He does not raise this specifically in the context of the mind-body problem, but he concludes that it is simply not possible to establish a necessary connection between the actual physical attributes of matter and the psychological (perceptual) effects produced by them. The problem is not merely technical. That is, the problem would not disappear were we to possess instruments by which to discern the elementary composition of matter. Even if we knew, for example, the subatomic composition of gold, we would not be able to establish the connection between these particles and the experience of "yellow" produced by them. Our senses, just as our intuition and our reason, are of such a nature as to allow us to survive and prosper in the world as we find it but are not able to yield certain knowledge of all things. Our knowledge can go so far and no further. Accordingly, our knowledge will always bear the stamp of the percipient, the mark of those special qualities that attach to experiences *because* they are experiences.

Locke does not reach skeptical conclusions, however. He does not suggest that nothing can be known for certain or that all knowledge is illusory. To those who contend that we cannot tell the difference between dreams and reality, who argue that all may be illusion, Locke replies with contemptuous indifference. Locke is concerned with the practical affairs of life and has no patience with sophistry. We have an immediate and intuitive comprehension of our existence and anyone who doubts it (Descartes?) it is not worth debating.[28]

Having established this much, Locke then defines what actually can be known. All that is knowable falls into three categories. We know ourselves by *intuition*. We know God by reason (*demonstration*). Everything else that can be said to be known and that, in fact, exists is known by *sensation*.[29] In the dispute between nominalist and realist, Locke's position is undeviatingly nominalistic. The truth of so-called universal propositions is either merely tautologous or a proposition about actual, physical entities whose essences are known. Thus, "gold is malleable" is true only to the extent that the word "gold" includes malleability in the sense of coexistence. Hence its malleability is not established by the proposition but by its actual, sensible properties: "No existence of anything without us, but only of God, can certainly be known farther than our senses inform us." [30]

It is through memory that we do not need an endless repetition of events in order to be satisfied that we have knowledge of them. Our knowledge of causes and their effects is produced in us by repeated experiences. The durability and vividness of this knowledge are enhanced when our experiences are augmented by pleasure and pain. The certain knowledge allowed by these processes is not perfect but it is "as great as our condition needs." [31] Memory fails not only because it weakens with time (i.e., because its "trace" decays) but also because we are not all gifted enough to retrieve what we want from the store of our memories.[32] However, within the limits of our sensory acuity, of our ability to focus attention, and of memory's ability to retain that with which the senses have furnished it, knowledge of the real world can become so highly probably correct as to be, for all practical purposes, certain. Since what we do depends upon what we know, at least according to the Lockean empiricist, the solution of the epistemological problems is tantamount to solving the problem of conduct: Locke's *Essay* thus addresses itself to the question of morality and briefly. His fuller development of the principles of governance and social conduct appears in his *Two Treatises on Government,* also published in 1690. In light of his historic position in the cause of civil liberty and his insistence that the legitimacy of governments is rooted in the idea of a "social contract" between the governed and their appointed leaders, it is sometimes assumed that Locke's position on moral matters was utilitarian and relativistic. Not only this, but it also is occasionally suggested that empiricism, as an epistemological system, is inescapably context-oriented on moral questions. This is not the place to examine Locke's theory of government, but we may pause to note his approach to moral questions in order to recognize that the empiricist position on such questions need not reduce moral propositions to perceptions and opinions.

Locke introduces his discussion of moral knowledge by first exploring the certainty of mathematical propositions. This requires a distinction between simple and complex ideas:

> [S]imple ideas are not fictions of our fancies, but the natural and regular productions of things without us really operating upon us. . . . Thus, the idea of whiteness or bitterness, as it is in the mind, exactly answering that power which is in any body to produce it there, has all the real conformity it can or ought to have with things without us. . . . [However], all our complex ideas except those of substances being archetypes of the mind's own making, not intended to be the copies of anything, not referred to the existence of anything, as to their originals, cannot want any conformity necessary to real knowledge. For that which is not designed to represent anything but itself, can never be capable of a wrong representation, nor mislead us from the true apprehension of anything by its dislikeness to it; and such, excepting those of substances, are all our complex ideas: which . . . are combinations of ideas which the mind by its free choice puts together without considering any connection they have in nature.[33]

Is this to say that our complex ideas have nothing to do with reality? Not at all. Our complex ideas, except for those we have of substances, are not ideas about physical entities in the external world, but they are ideas, nonetheless. And, just as mathematical ideas are "not the bare, empty vision or vain, insignificant chimeras of the brain," [34] so, also, moral ideas arise from the intuitive and demonstrative faculties of the ordered mind. Thus, "our moral ideas as well as mathematical, being archetypes themselves, and so adequate and complete ideas, all the agreement or disagreement which we shall find in them will produce real knowledge, as well as in mathematical figures." [35] Locke's thesis is not that moral propositions have the sort of validity that attaches to our perception of motion or shape, or that the canons of justice can be known the way we know, for example, that it is raining. Rather, he is proposing that most of what we know is not of the type "It is raining" or "The tram is moving." Our knowledge, from first to last, is *idea.* The idea results either from direct, sensory responses to the physical properties of actual objects (and is thus a *simple idea*) or from the mind's own compounding of simple ideas into unified constructs. By the operation of logical demonstration, these complex ideas can come to figure in moral theories and take the form of axiomatic truths not unlike those of mathematics. None of the modes of knowing—intuitive, demonstrative, perceptive—is perfect or all-inclusive. The geometer who reasons that a circle is a form containing 360 degrees, all of whose points are the same distance from its center, may never be able to find such a figure in nature. This is not to say that such a figure is "unreal" or merely a creation of the mind. It is to say that, were nature to yield such a figure, that figure would be a circle and, furthermore, that given this definition of a circle, certain other properties would necessarily follow. Moreover, given the nature of a circle, it matters not how we name it. The truths pertaining to it will be established by the demonstrative faculty of the mind and will be as certain as human knowledge can be.

> For the attaining of knowledge and certainty, it is requisite that we have determined ideas: and to make our knowledge real, it is requisite that the ideas answer their archetypes. . . . Nor will it be less true or certain because moral ideas are of our own making and naming.[36]

How are these "archetypes" to be understood? In the case of simple ideas, they are *representations*. The external physical object, through the processes of perception and the processing of perceptual information, comes to be represented in the mind as part of its contents. Experience and memory result in the melding and weaving of such representations into ideas that cannot be referred to anything *about* a physical object *qua* physical object. That is, a pointed object with a colored liquid at the tip is a "writing instrument" only to the extent that experience and memory have informed us of such uses. Whether or not such an instrument is meaningless, or is a weapon, or is used by a literate society depends on actual practices. Thus, the complex idea, "writing instrument," refers not to something physical but to a mental archetype formed from simple ideas but now going beyond them. Needless to say, the pointed object with a colored liquid at the tip *is* a pen, and is no less so simply because the idea of it is a cultural product. While the "archetypes" of physical objects refer to what resides outside the mind, the archetypes of moral ideas originate within the mind and generate moral ideas that are real and certain to the extent that our arguments and observations of the human condition are made conformable to them. It is here, of course, that Locke's "archetypes" come uncomfortably close to the rationalist's notion of an innate, moral sense. On a certain construal, the leap from these Lockean archetypes to Kant's categorical imperative is at its narrowest point in this discussion of the truth of moral propositions. It is here also that Locke's credentials as a rationalist are above reproach.

George Berkeley (1685–1783)

It is often suggested that Berkeley's major psychological work is his *New Theory of Vision*,[37] which lays down the experiential and geometric determinants of depth perception and takes Hobbes and Descartes to task for not recognizing the empirical foundation of such perceptual phenomena as the "moon illusion." [38] Were we concerned with the history of research in visual perception, it would be necessary to dissect Berkeley's theories which, for our present purposes, are able summarized by the philosopher himself:

> As we see distance, so we see magnitude. And we see both in the same way that we see shame or anger in the looks of a man. Those passions are themselves invisible; they are nevertheless let in by the eye along with colors and alterations of countenance which are the immediate object of vision, and which signify them for no other reason than barely because they have been observed to accompany them. Without which experience we should no more have taken blushing for a sign of shame than of gladness.[39]

What is most interesting in this passage is the emphasis on experience as the means by which bare sensations come to be interpreted and made meaningful. It is a clue to Berkeley's larger and striking metaphysical thesis according to which all reality requires a percipient for its very subsistence. This thesis is developed in his most influential essay, *A Treatise Concerning the Principles of Human Knowledge,* published in 1710 and, in revised form, in 1734.[40] It is this work that earned Berkeley such titles as "subjective idealist," "immaterialist," and "spiritualist." And these are the titles which, in turn, helped to make his small book one of the more misunderstood essays in philosophy.[41]

Berkeley's treatise, even more than Locke's *Essay,* is a psychological philosophy. At the time he wrote it, all three of the European "textbook" rationalists (Descartes, Spinoza, Leibniz) had contributed their major works. In fact, of the three, only Leibniz was still alive. Locke's most famous philosophical essay was already a classic, and Newton's celebrated *Principia* and *Mechanics* were joined by Galileo's achievements, together making the seventeenth century that most illustrious period of scientific creativity. We address the rationalist tradition in the next chapter, but we may note here that, by 1700, a tradition of European rationalism conspired with both Locke's empiricism and with the science of Newton and Galileo to create a decidedly materialistic perspective on the part of the intelligentsia. The downfall of Aristotle's ageless authority and the resulting disavowal of Scholasticism were already responsible for a form of religious skepticism bordering on atheism. Materialism, while still more than a century from becoming a religion in its own right, was fueling the skeptical fires. Even Descartes' dualism, which we shall study in the next chapter, was a grudging dualism according to which many, if not most, of the day-to-day affairs of the human community could be understood in essentially mechanical terms. It was this mechanical feature of social organization that Hobbes had presented as a full-blown political theory. What Berkeley set out to achieve, then, was nothing short of removing every trace of validity from materialism and to do this by refuting the latent or explicitly materialistic content in the works of Locke, Descartes, Hobbes, and the rest. His *Treatise* and his *New Theory of Vision* have this objective. It will be easier to follow the more complicated reasoning of the former by considering for a moment the more limited objectives of the latter.

The rationalist account of visual perception, particularly of depth perception, was based on the geometric theory of optics. According to this theory, developed by Descartes but having very ancient roots, we see objects at a distance or near at hand on the basis of the angle formed by these objects when the interocular distance is treated as the base of a hypothetical triangle. The further the object is moved away, the smaller the apex angle becomes, such that, by Euclid's theorems, all visual, perceptual data can be deduced from the geometric principles that govern the science of optics. Perception, by this reasoning, is the way it is necessarily. A convergence of the eyes is necessary to maintain a clear view of an approaching object; a divergence of the eyes for a receding object. The muscles of the eyes therefore report the distance between the object and the observer. By calculating the degree of convergence or divergence, the observer is able to estimate the distance. Berkeley's reply to this analysis is simply that no one sees things this way! When we examine an approaching or a receding object we do not, in fact, either compute the convergence of our eyes or the angle formed at the (invisible) apex of (invisible) triangles. Furthermore, the mathematically untutored are as accurate in judging distances as are masters of optics and so it is not likely that a knowledge of optical rays, lines, and angles ever participates in our perceptual judgments. Quite simply, distance is not visible itself and it makes no sense at all to refer to "the perception of distance." Berkeley accepted that interocular distance is the effective cue to distance, but it is the resulting experience, not a geometric calculation that provides the perception of distance. In other words, we can only sense events and objects, not "empty" distance. Thus, the idea of distance must be rooted not in geometry or optics, neither of which the average person uses in judging distance, but in that which is available to all who might ever be able to experience distance. And this is sensation. We learn to judge distance and we learn it by being exposed not to the axioms of geometry but to the affairs of sense:

> Not that there is any natural or necessary connection between the sensation we perceive by the turn of the eyes and greater or lesser distance. But—because the mind has, by constant experience, found the different sensations corresponding to the different dispositions of the eyes to be attended each with a different degree of distance in the object—there has grown a habitual or customary connection between those two sorts of ideas so that the mind no sooner perceives the sensation arising from the different turn it gives the eyes . . . but it withal

perceives the different idea of distance which was wont to be connected with that sensation.[42]

This passage, which is suggestive of the reasoning underlying the entire essay, commits Berkeley to an account of experience that is based entirely on other experiences. He specifically rejects the necessitarian argument of rationalists and mathematicians alike. He insists upon a "common sense" account which takes advantage of what all of us know to be true of the manner in which we make our judgments of distance and depth. He applies the same "common sense" standard to all other experience: touch, hearing, smell, and taste. An object, he argues, does not "look" round in the same way it "feels" round when the fingers explore its shape. We use the same words by an accident of learning, not by an identity of experience, and less by the similarity of species.

He begins his larger inquiry into the nature of human knowledge from this starting point. He begins by taking the real world as a congeries of experiences and thus as immaterial. Why? Well, on precisely the same basis as he rejects the notion of distance as something seen. We do not see "distance" and we do not see "matter." Rather, *we see*. Now, what is it that we see when we do? Berkeley answers:

> It is indeed an opinion strangely prevailing amongst men, that houses, mountains, rivers, and, in a word, all sensible objects, have an existence, natural or real, distinct from their being perceived by the understanding. But, with how great an assurance and acquiescence soever this principle may be entertained in the world, yet whoever shall find in his heart to call it into question may, if I mistake not, perceive it to involve a manifest contradiction. For, what are the forementioned objects but things we perceive by sense? and what do we perceive besides our own ideas or sensations? and is it not plainly repugnant that any one of these, or any combination of them, should exist unperceived?[43]

More than a century later, J. S. Mill would define matter as "the permanent possibility of sensation" and thereby reveal himself to be something of a reluctant Berkelean. Is this to say that Mill is a "subjective idealist"? No, and for the same reason that we are not to call Berkeley one if that term means that one regards matter as being created by the ideas we have of it. It surely was not Berkeley who argued that the trees in a wood disappear if he is not there to see them. For Berkeley to argue that matter is invented by the mind would require that he consider the separate reality of both and then offer a demonstrable proof of the existence of matter independent of mind. This is not a dualism Berkeley is likely to embrace:

> [H]uman knowlege may naturally be reduced to two heads—that of *ideas* and that of *spirits.* . . . And *first* as to ideas of unthinking things.

Our knowledge of these hath been very much obscured and con-
founded, and we have been led into very dangerous errors, by supposing a twofold existence of the objects of sense—the one *intelligible*
or in the mind, the other *real* and without the mind; whereby unthinking things are thought to have a natural subsistence of their own distinct from being perceived by spirits. This . . . is the very root of
Scepticism.[44]

This passage illustrates Berkeley's conviction that materialism leads inescapably to skepticism, and the latter is Berkeley's nemesis. The label "subjective
idealist" often is painted with the colors of skepticism. But Berkeley requires
that the skeptic's dead-end be averted. The only successful means of accomplishing this is by erecting a system of "sound and real knowledge" which
includes an understanding of what is implied by such terms as *thing, reality,*
and *existence.*[45] With respect to the first, *thing is being* and the latter is of two
radically different sorts. One is being through *idea;* the other, being through
spirit. The former impresses itself on the mind through the senses, and such
entities "*are real things, or really do exist; this we do not deny, but we deny
they can subsist without the minds which perceive them or that they are resemblances of any archetypes existing without the mind; since the very being
of a sensation or idea consists in being perceived, and an idea can be like
nothing but an idea.*" [46]

Berkeley thus rejects Locke's "archetypes." What he is saying, in effect, is
that Locke's primary and secondary qualities must finally be reducible to secondary qualities only. Motion can be nothing if not perceived motion; number,
only perceived number; form, but perceived form. Accordingly, there can be no
difference between these so-called primary qualities and the secondary ones of
color, heat, cold, and so forth. The "ideas," in Berkeley's unique use of the
term, are things in the mind with a permanence and validity that cannot continue in the absence of mind. It is because of this that the materialists must fail
to comprehend reality. It is also why the scientific theorist who abandons the
facts of his own sensations and wanders forth into the purely verbal realm of
physical laws of nature must contribute no more than nonsense in the garb of
tautology. Thus, when Mr. Newton, the author of that "*great mechanical principle now in vogue,*" postulates the mutual attraction of matter, "*I do not perceive that anything is signified besides the effect itself,*" [47] and this effect can be
nothing more than a perceived effect. It is the same Mr. Newton whose treatise
on *Mechanics* grants an existence to time, space, and place outside the mind,
and it is the same Bishop Berkeley who, in reply, must confess that he can
conceive of none of these except as relative. Thus, "*to conceive motion there
must be at least conceived two bodies. . . . [I]f there was one only body in being
it could not possibly be moved.*" [48]

Locke's theory of ideas, in Berkeley's view, was confused on several counts.

It was hopelessly dualistic in requiring both a world of matter (primary quali-ties) able to appeal directly to mental "archetypes" and an additional, private, experiential world of perception dealing with and creating secondary qualities. Worse, there is no way of deriving the latter from the former. We may list Berkeley's major epistemological contribution as the ridding of this dualism, or, at least, as a very commendable attempt to do so. Ultimately the world, the universe, and all that lives are but perceptions held eternally in God's eye. So-called material objects exist in relation to our minds in the way that all exists in God's. This is Berkeley's simultaneous rebuttal of skepticism, materialism, and atheism. He did not succeed in halting the progress of materialism, nor will convinced atheists, if such there be, find their soul in Berkeley's works. How-ever, later empiricists, and especially David Hume, did find in the *Treatise* an able attack on non-sensory approaches to the problem of knowledge. In pro-posing that perception was the first and last criterion by which reality may be known and judged, Berkeley succeeded in directing the attention of philoso-phers to the psychological dimension of all philosophical problems. In a word, he rendered epistemology a branch of psychology, and the two have never been utterly divorced since. His empiricism was of the extreme sort and, as such, served as a justification for later forms of idealism which Berkeley himself would have considered nonsensical. His theories of perception were couched in the language of experimental science and constituted a model for subsequent investigators. Carrying on in the steps of Bacon and Locke, he further clarified the distinction between words and the things signified by them; between propo-sitions that are (as Kant would call them) analytical and those that are synthetic, the former being logical and the latter empirical. In making these distinctions, he insisted that necessity does not inhere in nature but in the logical proposi-tions we invent in our attempt to understand nature. In this, he anticipated Hume but, of course, did not equal him. To the extent that psychologists are interested in perception, in the contextual determinants of perception, in the relation between things and their (verbal) significations, Berkeley's *Treatise* has been seminal. He did not write as much as Locke or Hume, nor did his scholarship reach, as theirs did, to the farthest corners of political, civic, and moral concerns. But, on the specific epistemological issue to which all empiri-cists have been addressed, and on the specific points of contact between theo-ries of knowledge and theories of psychology, many empiricist contributions following his have been footnotes and epilogues.

David Hume (1711–1776) and the Scottish Enlightenment

Twentieth-century philosophy in the English-speaking world is very nearly a creation of Hume's. It is always precarious to attempt to locate so important a figure within any context because his influence has persevered throughout such a variety of contexts. Still, as much as Hume was the founder and father of a new philosophy, and one that continues to show his influence, he was also the product of eighteenth-century thought, and of that special epoch in Western civilization, the Scottish Enlightenment. Not only had Newton revolutionized science, but such luminaries as John Milton, John Locke, Shaftesbury, Hutcheson, and Samuel Butler had provided the moral and intellectual foundations for a political liberalism that had not been witnessed since antiquity.

Apart from this were the special local conditions nurturing science and scholarship. The University of Edinburgh had become one of several major centers of legal and medical education in the world, attracting students far and wide. William Cullen's *Nosology* would be translated into French by the not yet famous Philippe Pinel, as it would be studied at Edinburgh by Benjamin Rush before he returned to America to reform the treatment of the mentally ill—and to sign the Declaration of Independence. Glasgow, Aberdeen, and St. Andrews were distinguished as well. Modern economics would be articulated at Glasgow by Adam Smith, followed there soon after by Joseph Black in chemistry, at about the time Thomas Reid was founding Scottish "common sense" philosophy at Aberdeen. Reid's philosophy would profoundly influence Kant in Prussia, as well as a small but influential cadre of founding fathers in the rebellious colonies. Reid's student, Dugald Stewart, would be judged by Thomas Jefferson to be one of the two leading metaphysicians of the entire age (the other being Destutt le Comte d'Tracy), and Reid himself would be the inspiration behind the jurisprudence espoused by James Wilson, another signer of the Declaration of Independence and the best legal mind in the new republic. In architecture, Scotland led the world in the classic revival through such luminaries as Robert Adam and William Playfair, at the same time that Sir Walter Scott captured the literary allegiance of readers on three continents. Scotland's educators would be summoned to headships of schools in Europe and America, Scottish surgeons were widely regarded as the world's best, and Scottish lawyers were without parallel. Edinburgh, the "Athens of the north," and other centers of Scottish high culture and thought sent radiating and radiant influences in many directions and with lasting consequences. Among these influences was David Hume's *A Treatise of Human Nature,* published in 1739–1740 and little noticed at the time, but in time having revolutionary effects on the thinking parts of the world.

The *Treatise* appeared a century after Britain's civil war, and a half century after Locke's *Essay.* Milton's *Areopagitica* (1644) had assaulted censorship of the press so effectively that J. S. Mill's *On Liberty* two centuries later was proud to paraphrase much of it. Intellectual freedom was still a long way off, but by the date of Hume's *Treatise* the historic sources of truly oppressive authority were quite simply outnumbered. Hume died in the year of the American Revolution, thirteen years before the French Revolution. The former is as inconceivable without Milton, Locke, and Hume as the latter is without Voltaire, Diderot, Rousseau, and Montesquieu. Nor were the scholars on either side of the Channel unaware of each other. Voltaire celebrated the English empiricists in more than one essay, and Rousseau actually lived with Hume and was left a small pension in Hume's will. Thus it does not detract from the *Treatise* to place it on a continuum of liberal arguments presented in proud defiance against all who would defend the status quo in any branch of knowledge or social practice.

In the introduction to his *Treatise* Hume singles out several of the new "natural philosophers" who, by his lights, have advanced the cause of truth. They include "Mr. Locke, my Lord Shaftesbury, Dr. Mandeville, Mr. Hutcheson, Dr. Butler, &c." [49] Hume's inclusion of Shaftesbury and Butler warrants brief comment, for both were leaders of the empiricist approach to moral philosophy. They were two of perhaps a dozen writers in the English of the early eighteenth century who labored to place ethics in that realm which, today, we would call psychology. Shaftesbury and Locke were very close, Locke being Shaftesbury's friend, secretary, and physician. Butler and Hume were contemporaries. Both Shaftesbury and Butler contended with persuasive eloquence that the moral and ethical dimensions of human life are to be understood in human terms. Both argued that our very self-interests, our need to live in society, our ability to perceive and learn from the consequences of our actions, and our capacity to articulate objectives and behave in a manner compatible with their attainment constituted the ultimate theory of values.[50] According to this view, moral "absolutes" can be so only to the extent that we are endowed with a natural faculty for perceiving wherein our interests lie; only to the extent that we can benefit from social instruction and are committed to labor to secure the praises of other human beings. Butler's emphasis on action as the standard of moral worth and on reward and punishment as the tools that forge our character would be repeated by Hume's good friend, Adam Smith, and would be installed as a systematic philosophy by Jeremy Bentham and the disciples of utilitarianism in the nineteenth century. In acknowledging the contributions of Shaftesbury and Butler, Hume was not, thereby, subscribing to the latter's instinct theory of virtue or Shaftesbury's empathy theory, about which we will say a few words, but was applauding their consistently psychological approach to the issue of values. He was applauding their rejection of rationalism and Scripture as the best means by which to unearth the nature of morality.

In *An Inquiry Concerning Virtue or Merit* (1699), Shaftesbury had located all morality in the domain of intended actions and had argued that these actions, on the part of a reflective animal such as man, were to be understood as the product of a natural appetite or disposition or affection. With Darwin's vision (but not his data), he circumscribed all living things within a great system in which survival was the common motive. For man to survive and to know happiness, he balances his self-serving affections against those directed at the public good:

> TO HAVE THE NATURAL AFFECTIONS . . . IS TO HAVE THE CHIEF MEANS
> AND POWER OF SELF ENJOYMENT: AND THAT TO WANT THEM IS CER-
> TAIN MISERY AND ILL.[51]

The goal is this self-enjoyment—*the pursuit of happiness* as it would come to be called—and we proceed to regulate the moral dimensions of our conduct in such a way as to secure it. Loyal to the teachings of Christ's church, Shaftesbury concludes nonetheless that "religious Conscience supposes moral or natural Conscience," [52] and the latter is based upon a sentiment or affection according to which "no Creature can maliciously and Intentionally do ill, without being sensible at the same time that he deserves ill." [53]

Butler carried forth the naturalistic school of virtue and made still clearer and wider the break with necessitarian doctrines:

> That which renders beings capable of moral government, is their having a moral nature, and moral faculties of perception and action . . .
> we naturally and unavoidably approve of some actions, under the peculiar view of their being virtuous . . . and disapprove others as vicious. . . . That we have this moral approving and disapproving faculty, is certain from our experiencing it in ourselves, and recognizing it in each other.[54]

The sentimentalist theories of Shaftesbury and Butler were soon to undergo revolutionary transformations at the hands of Adam Smith and Jeremy Bentham, neither of whom was satisfied that we are endowed with an abiding longing for the *summum bonum.* Smith's *Theory of the Moral Sentiments* appeared about twenty years after Hume's *Treatise* and about twenty years before Bentham's *Principles of Morals and Legislation.* The steady progression from the innate moral sentiments propounded by Shaftesbury and Butler to the bold, Benthamist assertion (in which Hume concurred) of morality-as-utility provides one of the most uncluttered chapters in the history of political theory. Now to Hume's *Treatise.*

At the very outset, in his introduction to the six-hundred-page work, Hume leaves no doubt as to his goal, his basic assumptions, and his opinion of earlier and contrary perspectives. His mission is to found a human science which ultimately must include or set the boundaries on all other sciences. Human beings are not simply those who reason but are also the worthy objects of reason and must be studied by the emerging instruments of that "experimental philosophy" so recently forged by Bacon, Newton, Locke, and others. All truly important questions finally await an understanding of the human mind whose powers can be disclosed only through careful and exact experiments and observations.

> And tho' we must endeavour to render all our principles as universal as possible, by tracing up our experiments to the utmost, and explaining all effects from the simplest and fewest causes, 'tis still certain we cannot go beyond experience; and any hypothesis, that pretends to discover the ultimate original qualities of human nature, ought at first to be rejected as presumptuous and chimerical.[55]

"Hypothesis non fingo," said Newton!

Hume's primary objective is to establish the limits of human knowledge, which objective, alas, falls short of being able to prove "our most general and most refined principles beside our experience of their reality." [56] In other words, notice is served on all who would promise extrasensory truths or irrefutable moral maxims. The body of the *Treatise* is comprised of three broad topics: the understanding (Book I), the passions (Book II), and morals (Book III). Hume's organization is not to be improved upon, and we will discuss his psychological philosophy in the order in which he presents it.

Since the contents of the mind can come into being only through experience, it follows that the human understanding, most generally, must be based on perceptions. According to Hume, these are of two sorts: impressions and ideas. Impressions include sensations, passions, and emotions that are but the reports of stimulation of one sort or another. Ideas are only the persistence, in a weaker form, of prior impressions. In Hume's words, "every simple idea has a simple impression which resembles it and every simple impression, a correspondent idea." [57] Moreover, since the simple impressions always precede the corresponding idea in experience, we can be certain that the latter is the result of the former, and not vice versa. In other words, common experience belies the idealist assertion that the mind (idea) creates the sensation. And as for the notion of innate ideas, Hume simply dismisses it as unconfirmed in experience.[58]

Our impressions are of two varieties: sensation and reflection. The origin of sensations as psychological entities is something Hume cannot explain. He notes that a sensation—the vivid, conscious awareness of a thing—is of unknown cause and, in so noting, he joins a large army of philosophers unable to discern precisely how so many atoms, angles, or grams can come to be sensed as *qualia* in that very mental sense of the term. Once we have an impression and its corresponding idea, the mind reviews and combines different impressions by the process of reflection. Reflections are impressions also, but they are further removed from the initial impressions that give rise to sensation. Through the operation of memory we are able to repeat or revive impressions and by imagination we are able to reconstruct impressions in the form of pure ideas.[59] We can make the distinction between memory and imagination clearer with an illustration. When we picture the face of a relative, we are employing memory; that is, we are simply reviving an actual sensation. When, however, we consider government to be necessary for the social good, we are not "picturing" any longer. Instead, we are abstracting from an assortment of specific impressions and ideas a general principle, and the faculty that allows this is the imagination.[60]

Simple ideas become associated and so form complex ideas, and the bases upon which such associations are formed are "RESEMBLANCE, CONTIGUITY and CAUSE and EFFECT."[61] Experience teaches that we tend to associate those events that resemble each other, those that take place together in time and space, and those that provide instances of unerring succession. The way in which resemblance, contiguity, and causation produce associations of varying durability is to be explained through the physiology of the brain.[62] (It is worth noting that Hume's physiological explanation is lifted out whole cloth from Descartes and is rife with references to "traces," etc.) On the question of actual material existence, Hume cannot refute Berkeley on logical ground and therefore is content to recognize that, " *'tis vain to ask, Whether there be a body or not? That is a point which we must take for granted in all our reasonings.*"[63] It is not reason that establishes the separate reality of things and distinguishes between them and perception. Rather, our imaginations (as heretofore defined) are responsible, and it is only on the basis of this imaginative faculty that we are given and driven to believe that our perceptions are *of* something. Thus, Hume is not attempting to prove that there is a separate, material world and an equally separate mental one. He is concerned instead with establishing the principles that force that opinion on us. As to these principles, he reduces them to "constancy" and "coherence." The mountains we looked at yesterday are found to be in the same place when we come back and look again today. These experiences convince us that objects enjoy a constant existence separate from our immediate awareness of them. He does not pause to recognize that we could only find such things "in the same place" were there a constancy and coherence of *place* itself, so to speak, and it is on matters of this kind that Kant will discover fertile grounds of criticism. In any case, even though objects do undergo some change over time, they continue to reflect a certain coherence in their relation to other objects. That is, it is only by assuming a continuity of existence on the part of objects (even when we are not there to witness them) that we are able to establish a relation between objects instead of one between perceptions alone. In summary, we believe there are objects, distinct from our perception of them, because we assume they continue to exist in our absence, and we assume this because the same objects are perceived in roughly the same form when we return to where they were first perceived. The new impressions are compared with those earlier ones now stored in our memory, and when the resemblance is close, we suppose that the present object is the continuing one.[64]

This brings us to the question of how individual impressions and their correspondent ideas can give rise to the rich notions with which every mind is concerned. How, we ask, are complex ideas produced by the rather trivial consequences of perception? Hume's answer is that complex ideas result from the association of simple ones, as we have already noted, and that all our ideas if they are more than utter fancy, can only arise from a complication of more elemental ones. The latter are inextricably tied to sensations and impressions. Our "understanding" properly called can refer to nothing but these. Given this position, we can predict how Hume will approach the remaining problems, in Books II and III, of passions and morals.

As the arch psychological philosopher, Hume accords the passions a central place in epistemology. The passions of pride, love, humility, and hatred are connected to the ideas and sensations evinced by those objects toward which these passions are felt.[65] Thus, "agreeable" images give rise to agreeable passions. Hume assembles an admirable list of passions including benevolence, anger, malice, envy, contempt, amorousness, and respect. He presents his passions as opposing pairs and explains their appearance as resulting from life-long experiences and an internal sense called sympathy:

> I have endeavored to prove that power and riches, or poverty and meanness, which give rise to love or hatred, without producing any original pleasure or uneasiness, operate upon us by means of a secondary sensation deriv'd from a sympathy with that pain or satisfaction which they produce in the person who possesses them.[66]

We seen, then, that Hume is very much in the sentimentalist (empathist) tradition of Shaftesbury and Butler. Our emotions are natural, are elicited by sensations and impressions, are directed by the interior sense of sympathy. Thus, our loftiest sentiments and our most ennobling qualities (those of love, respect, and altruism) are, at base, effectively independent of rational considerations. We do not love because logic compels it; we are not benevolent because we have intellectual grounds for rejecting selfishness; we are not malicious with reason. These sentiments or passions are produced in us by a history of associations, by the rewarding and punishing consequences of our conduct, and by those laws of thought summarized under the headings of resemblance, contiguity, and cause and effect.

The passions enter Hume's epistemology as a means of accounting for that which is central to this epistemology: belief. In quietly accepting Berkeley's "solution" to the paradox of matter—in accepting, that is, that what we know we know through our impressions and reflection—Hume is forced to the conclusion that our knowledge is *conviction*. Philosophy's task is to account for this conviction. Experience alone is not enough to explain the belief we have in our knowledge. While empiricism can account for knowledge—that is, can explain what we know of rain, dogs, and the moon—there is nothing in the experience per se that will convey the belief we have in this knowledge. Belief, then, must be understood in terms different from sensation and impression. It is, alas, a *feeling* we have *about* our knowledge:

> The imagination had the command over all its ideas, and can join and mix and vary them. . . . It may conceive fictitious objects. . . . It may set them, in a manner, before our eyes in their true colours. . . . But as it is impossible, that this faculty of imagination can ever, of itself, reach belief, it is evident that belief consists not in the peculiar nature or order of ideas, but in the *manner* of their conception, and in their *feeling* to the mind. . . . *[B]elief* is something felt by the mind, which distinguishes the ideas of the judgment from the fictions of the imagination.[67]

He develops this argument to the point of concluding that reason is ever in the service of the passions. What passes for a rationally derived system of morals is, on closer inspection, no more than a commitment to what is pleasurable. We are moral to the extent that certain experiences and actions produce a satisfying state of affairs. Moral conduct arises from "the natural sentiments of humanity,"[68] which affect our reason but do not depend upon reason for their existence. The so-called moral virtues are no more voluntary than are other "natural" abilities. That is, we are no more voluntarily virtuous than we are voluntarily beautiful or deformed. It makes no more sense to reward or punish a person for his virtues than it does to reward or punish him for his height.[69] All we end up doing is praising or blaming people for doing that which pleases or displeases us, and, while this is understandable, it cannot be said to be rational or defended by the canons of logic. In a word,

> all arguments concerning existence are founded on the relation of cause and effect; . . . our knowledge of that relation is derived entirely from experience; . . . all our experimental conclusions proceed upon the supposition that the future will be conformable to the past.[70]

If there is existence, it is *perceived* existence. That the existing object or event is caused is a supposition—an invention of the mind—as is the indication by which we assume the future will conform to the past. What we judge to be moral are perceived acts that create agreeable or disagreeable experiences within us. The "right" and "wrong" of a situation are inextricably tied to beliefs, sentiments, dispositions, and native tendencies of the mind. "Right" and "wrong" are not *in* the events or *in* the actors, except in the sense that a feeling is *in* percipient man.

We will return to Hume's position on causation in the next chapter and will discuss the effect this position has had on the philosophy of science in general and on psychology in particular.

Hume's psychological philosophy combines in a peculiar way a number of conceptions which, in modern psychology, are normally in an antagonistic relationship. Epistemologically, he was an empiricist without reservation. Only two realms of knowledge exist: the demonstrative, which is logical and purely verbal, and the factual, which is purely experiential. Since ideas can be about things only, they cannot be innate. However, while ideas are not innate, feelings are. We are, according to Hume, so constituted as to respond passionately to certain classes of action. Our appetites and the pleasures and pains attached to them give vividness to our ideas and impressions, convictions to our knowledge. Precisely *how* our constitution does this is only hinted at. Just as sensation itself is treated as a "natural" quality of the mind to be understood by "anatomists," so also the moral sentiments are somehow part of our frame and, presumably, amenable to (ultimate) biological explanation. We see, then, that materialism is an implicit feature of Hume's psychology. The theory of knowledge is empirical and associationistic; the theory of emotion, nativistic; the final and unexpressed psychological theory, materialistic. Hume was no exception to the general rule that an empirical philosophy, soon or late, becomes either solipsism or psychological materialism.

Thomas Reid (1710–1796) and the "Common-Sense" Movement

Thomas Reid is a much underrated figure in histories of psychology and philosophy, although his insights have been rediscovered several times in both disciplines. That twentieth-century school of philosophy fathered by G. E. Moore and dubbed the "common-sense" school is, in outline, Reid's system in modern dress. Even Kant's famous if tortuous "answer" to Hume, which we will review in the next chapter, was ably anticipated by Reid, as was the theoretically neutral and practical bent of modern behavioral science. It is especially surprising that modern commentators have been perfunctory in their attention, since Reid was widely read and admired by his own contemporaries in English, Scotland, and on the Continent. Perhaps he was too clear. We identify Reid with the "common-sense" movement and we are advised to pause in order to appreciate how congenial his time was to a movement of this sort.

The great figures in the line of British empiricists from Bacon to Hume were all influential outside England. Locke was adopted by the principal reformers of France, where he paid two extended visits. His *Treatise of Civil Government,* with its emphasis upon the human need for freedom and equality and its insistence on government's obligations to its citizens, perfectly complemented the spirit of the French Enlightenment that would culminate in the Revolution of 1789. The French *philosophes* may be said to derive from Montaigne (1533–1592), as the British empiricists did from Bacon, and it should be seen as more than coincidental that the French publisher who presented Locke's essay to the world of the Enlightenment was also responsible for the 1724 edition of Montaigne's collected works.[71] And it is far from coincidental that D'Alembert should write, in his introduction to Diderot's *Encyclopaedia* (1751):

> Nothing is more indisputable than the existence of our sensations. Thus, in order to prove that they are the principle of all our knowledge, it suffices to show that they can be. . . . Why suppose that we have purely intellectual notions at the outset if all we need do in order to form them is to reflect upon our sensations?[72]

The philosophers of England and France were at war with rationalism not merely on epistemological grounds but on political and social ones as well. Descartes' philosophy was, in the hands of Malebranche, a defense of religion and the perquisites of clerical office. Hobbes's *Leviathan* served others as that form of *Hobbism* that wields deductive logic in such a way that the divine rights of monarchs, the suffering of the poor, and the abuse of power, all have the legitimacy of Euclid's theorems. It is no accident, then, that the empirical epistemologists were also the empirical moralists and political theorists—no accident because their epistemologies were, in all significant respects, preliminary discourses on a *moral* treatise. That moral treatise would take many forms in several languages and nations: the Bill of Rights, the Rights of Man, Common Sense. Thomas Reid was central to this movement. He was not opposing Hume's skepticism merely to save God but, rather, to retain a philosophical foundation for a movement that extended even beyond the issue of faith.

When we recall Locke's willingness to accept the certitude of moral propositions with the same confidence displayed toward mathematical propositions, we are more willing to include Reid in the line of empiricists *despite* his assertion that the principles of common sense are innate. Locke and Reid are just two of many philosophers who can be fit into neat categories only if we are willing to look the other way. And in both cases, the willingness to take recourse to some sort of rationalist or nativist principle was prompted by the desire to avoid skepticism. For Locke, reducing the canons of morality to empirically observable factors could lead only to a form of moral relativism which, if pressed, would remove the validity of his political propositions. For Reid, Hume's philosophy leads not only to atheism but to absurdity. It was Hume who asserted that "*the errors in religion are dangerous; those in philosophy only ridiculous,*"[73] and it is Reid who sets out to ensure that the latter not be allowed to thrive just so the former might be averted.

In describing Reid as the founder of "common-sense" philosophy, we mean to say not that he rested his case on the superstitions or shifting opinions of the individual in the street but that he required philosophy to conform to that which every human being knows to be true—what philosophers themselves know to be true when freed from the pretensions of their trade. And, of the numerous pretensions endemic to the philosophical way of thinking, none is more absurd than the notion that our ideas are conditioned by the properties of things. *"We are commonly told by philosophers,"* says Reid, *"that we get the idea of extension by feeling along the extremities of a body, as if there was no manner of difficulty in the matter. I have sought with great pains, I confess, to find out how this idea can be got by feeling, but I have sought in vain."*[74] In reply to Berkeley's brand of idealism, Reid agrees to adopt this point of view and to see what it produces:

> I resolve not to believe my senses. I break my nose against a post. . . . I step into a dirty kennel; and after twenty such wise and rational actions, I am taken up and clapped into a madhouse.[75]

To the Cartesian who shuns the senses in favor of reason, Reid can only observe that, since *"they came both out of the same shop,"* he is on as firm a ground accepting his senses as he is his reason.[76] On Reid's account, Locke, Berkeley, and Hume all failed to distinguish between sensation and perception, and as a result their use of the concept of "idea" was hopelessly muddled. A *sensation* is the direct experience of that which is in the mind, whereas a perception is of that which lies outside the mind. Note the following two sentences:

> (a) I feel a pain in my leg.
> (b) I see a rose in the garden.

In (a), while there is a verb (feel) and a grammatical object (pain), the difference is only grammatical in that "feel" and "pain" are one and the same. However, in (b), there is not only a grammatical distinction between the verb (see) and the object (rose) but an actual one.[77] Our perceptions, which come from "the mint of nature," are triggered by the actual objects in nature. It is these objects that constitute the language nature uses to speak to us. We do not reason that there must be an object when we perceive one. We know immediately and instinctively that such is the case. Hardness, coldness, and color, not less than motion, number, and form, are perceived by us through the peculiarities of our constitution. Sensation is a natural sign of hardness, no more provable than is the assertion that a thing cannot be and not be at the same time:

> [S]upposing we have got the conception of hardness, how come we by the *belief* of it? Is it self-evident, from comparing the ideas, that

such a sensation could not be felt unless such a quality of bodies existed? No. Can it be proved by probable or certain arguments? No. Have we got this belief then by tradition, by education, or by experience? No. . . . Shall we then throw off this belief, as having no foundation in reason? Alas! it is not in our power; it triumphs over reason, and laughs at all the arguments of a philosopher. Even the author of the "Treatise of Human Nature," though he saw no reason for this belief . . . could hardly conquer it in his speculative and solitary moments; at other times he fairly yielded to it, and confesses that he found himself under a necessity to do so.[78]

Reid was whimsically perplexed by Hume's willingness to doubt the validity of everything except those sensations and impressions from which his ideas were supposed to emerge. Why, asks Reid, did Hume find it necessary to stop at sensations? If causation is in the mind only, if virtue is but a certain "vivacity" our ideas have received, why accept even that there are sensations or impressions or ideas or belief or anything? The answer, of course, is that sensations, including Hume's sensations, cannot even be denied without the consequence of self-contradiction. Our sensations are not chimerical: we do avoid pain; we do believe certain impressions and not others despite the fact that there is nothing in an object that can command belief other than its reality.

To understand our sensations and our perceptions and the origins of our beliefs requires, on Reid's analysis, recourse to the notion of "natura faculties" that are intrinsic to our constitution. This is not to say that the mind has innate ideas; rather, it is naturally equipped to recognize that class of natural signs which are the "primary" and "secondary" qualities" of Locke, the impressions and reflections of Hume, and Berkeley's "ideas." It makes no more sense to ask about the rational justifications for accepting the natural faculties than to ask for a logical proof that the stomach digests food or, better, *why* the stomach digests food. In anticipation of Darwin, Reid notes that our survival in the real world requires that we be psychologically conversant with nature's objects. We are, accordingly constitutionally endowed in such a way as to reconstruct the world in psychologically meaningful and useful ways. The man in the street has neither the time nor the inclination to escape his idea of rain by thinking of entering his idea of a house. He does not keep himself alive—that is, he does not retain the idea of being alive—through the idea of eating his idea of food. Neither, for that matter, does the philosopher, including the skeptical philosopher. That no argument of logic can distinguish among rain, life, food, and houses on the one hand, and our ideas of these on the other, is simply to note that we and our world are not composed of logic; that logic did not create the world; and that when logic collides with those facts universally possessed by the human mind, it is logic that will be bent.

No one had better credentials as an agnostic than David Hume. Still, there is a respect in which his skeptical philosophy was a return to the Renaissance's inability to separate nature and spirit. More than Berkeley, because he was more influential than Berkeley, Hume seemed to "psychologize" the world while at the same time praising Newton and even trying to Newtonianize the mind. It was certainly not Hume's aim to foster a dualism of any sort, but the effect of his writings was to convey to some that we live in two radically different worlds: one, the apparently factual world of causes, effects, moral truisms, and so forth; the other, the only knowable world, which is, finally, subjective and ever conjectural. In terms of their respective influence or potential influence on psychology, Hume may be said to have psychologized philosophy to the point of making science a matter of conviction, whereas Reid, in basing psychology on the firm ground of natural causes, made conviction itself a possible subject of science. Among Reid's contemporaries was David Hartley, whose work will be considered in Chapter 9. He was one of the early theorists in the "reflex-conditioning" tradition, and his *Observations on Man* (1749) was the first important and developed expression of psychological materialism to come out of the British empiricist school. Reid attacked this essay on the grounds of its excessive hypothesizing—its departure from Newton's maxims regarding the practice of science. It was not the materialism Reid objected to but the speculation. It is in his ringing denunciation of Hartley's theories that Reid's devotion to Baconian and Newtonian methodology is revealed:

> The effluvia of bodies make an impression upon the olfactory nerves; but make none upon the optic or auditory. No man has been able to give a shadow of reason for this. While this is the case, is it not better to confess our ignorance of the nature of those impressions made upon the nerves and brain in perception, than to flatter our pride with the conceit of knowledge which we have not, and to adulterate philosophy with the spurious brood of hypothesis?[79]

In this also is Reid's influence on modern psychology felt, (primarily through his influence directly on Dugald Stewart and indirectly on James Mill and, even, John Stuart Mill). He demands that philosophical explanations conform to the facts and truths possessed by every human being. He requires that all hypothetical statements be avoided unless the overwhelming weight of indirect evidence is behind them. He insists that the world is real, that the senses are affected by this world, that perception is a report of this world. He justifies his position—which is a form of direct realism—in terms of what everyone, including the skeptic, learns from life in the world. Implicitly, he goes on to defend his brand of realism with the canons of survivalism; the skeptic's course leads us to step into a dirty kennel!

As Reid's influence spread to Europe, so too did it reach America, where a "common sense" movement culminated in the American pragmatism and functionalism of Dewey, James, and Peirce. This influence is the subject of a later chapter. In addition to laying the ground for pragmatism and functionalism, he fathered what has come to be called "faculty psychology," a form of psychological inquiry devoted to uncovering the natural faculties of man (and other animals) deemed essential to knowledge and conduct. In rooting these faculties in instincts or native dispositions, Reid began a branch of British empiricism that finally yielded Galton's theories of hereditary genius and McDougall's instinct-theory of emotion. On the way, faculty psychology would be assimilated by Gall and Spurzheim and would appear in ultramaterialist garb as phrenology. These, also, are subjects to be deferred. We note them here only to acknowledge how diffuse an influence Reid has had on psychology. Even the modern behaviorist, when he defends himself against those who demand theories, when he attempts to describe the sense in which "the law of effect" is a law, when he accounts for his indifference to neurological research and mechanisms, and when he articulates the principles of his program, and ends up borrowing the common-sense arguments of Thomas Reid. A philosopher who has been successful in shaping the thought and practice of hereditarian and environmentalist alike, humanist and mechanist alike, is worthy of far more attention than Reid has received.

As Reid's epistemological position was of the common-sense sort, so also was his position on the question of the passions, emotion, free will, and moral conduct in general. Quite simply, these characteristics, most of which can be found throughout the animal kingdom, are so essential to society and to peace among the children of nature that their origins can be in none other than nature itself. On the natural appetites, for example, which even the lowly caterpillar displays in rejecting hundreds of different leaves until reaching the one that is "natural" to its diet, Reid comments:

> The ends for which our natural appetites are given, are too evident to escape the observation of any man of the least reflection. Two of those I named are intended for the preservation of the individual, and the third for the continuance of the species.[80]

The same nature (God) responsible for these natural endowments has equipped us with the faculty of deliberation so that we may keep our appetites under control, so that we may recognize our duty to others and to God. This deliberative faculty is as self-evident as Euclid's theorems, and even the skeptic must employ it in order to doubt it. It is this very same faculty that allows our actions to be voluntary. Those who would point to exceptions, such as the madman, the delirious, or the intoxicated, are more aware of the exceptions than of the fact that they *are* exceptions.[81] The rules governing those in possession of their faculties are not overthrown by instances of disease, deformity, or constitutional (native) deficiency.[82]

We will conclude our converge of Reid with a review of his approach to the problem of universals, for it is in dealing with this problem, and especially with Hume's "solution," that he presents the early form of a cognitive psychology as contrasted with a perceptual or sensationist psychology.

Berkeley had accounted for general ideas—for example, the idea of "cat"— in terms of an associationistic principle. Berkeley's theory, shared by Locke and Hume, was that all such general ideas are merely *particular* ideas melded together by words. Thus, once we have had repeated exposure to different, particular cats and have learned the word "cat," we use the word as a representation of any individual cat; that is, all cats, any one at a time. Hume not only agreed with this explanation but judged it to be one of the major advances in the recent history of philosophy. As Hume put it:

> A particular idea becomes general by being annexed to a general term . . . [through] . . . customary conjunction. . . . Abstract ideas are therefore in themselves individual although they may become general in their representation. The image in the mind is only that of a particular object.[83]

Reid, the common-sense realist, offers the following observations on this theory which, as Reid notes, is the cornerstone of the Humean theory of ideas.

First, in asserting that every idea must finally reduce to an impression of quantity and quality, Hume makes it impossible for us to mean different things when we say, "This is a line" and "This is a line of three inches." It is not the case that when someone has the idea of "line," he has it only by picturing a particular line. He pictures no such thing.

Second, Reid agrees with Hume (and the nominalists) that there can be no "abstract" triangle in the real world. However, attributes may be common to many individual objects, and to know this does not require that we make a mental picture of each individual object in order to weigh the similarities.

Third, when we have the idea of "lion" it does not come into being by having a particular lion eat a particular sheep; less by having the idea of lion eat the idea of sheep:

> If ideas differ from the object of sense only in strength and vivacity, it will follow, that the idea of a lion is a lion of less strength and vivacity.[84]

Fourth, in accounting for our assigning a term to a collection of particulars in terms of the resemblance among them, Hume either admits that we are capable of holding general ideas or, worse, he supports one hypothesis by taking recourse to the very principle he set out to reject.

Fifth, in proposing that when we use the general word it creates the idea of the individual in our minds, Hume flies in the face of the common experiences of humankind:

> I think a farmer can talk of his sheep, and his black cattle, without conceiving in his imagination one individual, with all its circumstances and proportions. If this be true, the whole of his [Hume's] theory of general ideas falls to the ground.[85]

Finally, Hume had offered as an illustration of his theory the suggestion that, for example, in a sphere of white marble, the form and the color are indistinguishable. Hume's point is that we mean no more by "form" than a particular distribution of light of a certain perceived hue. To this Reid replies:

> How foolish have mankind been to give different names, in all ages and in all languages, to things undistinguishable, and in effect the same? Henceforth, in all books of science and of entertainment, we may substitute figure for colour, and colour for figure.[86]

In these arguments Reid was not defending the reality of universals as conceived by the medieval realists. He was only pointing to the logical and practical limitations of the associationistic account endorsed by Berkeley, Locke, and Hume. He insisted that Hume's very employment of "resemblance" required an a priori faculty, or else there would be no way a general term could be applied to particulars. Further, he was reminding the sensationists that real people do not picture individual things when they think of the general words. Instead, the mind possesses the general concept which is now mentally independent of the things so represented. As the world of objects is a language nature has in order to communicate with us, our human language is a means by which the mind can free itself of particulars and of matter in general. We sense, we perceive, and we cognize. None is more "natural" than the other two; none requires logic as a proof.

Utilitarianism and Empiricism

The father of modern utilitarianism is Jeremy Bentham, but the central ideas behind it were developed by the ancient Epicureans and embraced often thereafter by a number of important thinkers. Basic to utilitarianism is the assertion that the rightness and wrongness of actions are determined exclusively by their consequences (hence the near synonym, consequentialism), and the related claim that the only consequences that need be considered are those of human happiness or suffering. What utilitarianism specifically rejects is the thesis that some actions, by their very nature—and independently of consequences—are wrong.

The two versions of utilitarianism most commonly encountered are *act* utilitarianism and *rule* utilitarianism. The former, which was the version advanced by Bentham and his immediate successors, applies the consequentialist standard to each particular act. Rule utilitarianism proposes instead that one weigh consequences on the whole resulting from actions of a given genre. The rule utilitarian, for example, might argue that, although a specific act of fraud may yield a net addition to happiness, the long-range consequences of fraud, should it become the basis on which persons deal with each other, would be a net addition to unhappiness. On this ground, he would defend the rule, "Do not perpetrate frauds."

It was Bentham's judgment that utilitarianism was a normative system of ethics in that it set forth what ought to guide everyone's conduct. In fact, however, his defenses of utilitarianism were primarily in terms of what persons actually do. But it is not possible to deduce what ought to be from a mere description of human conduct. Let us say, then, that at best Bentham offered a *descriptive* ethics grounded in certain psychological assumptions about human motivation.

Before turning to these assumptions, we should note the hypothetical nature of these assumptions, and not be distracted by their superficial agreement with daily experience. Surely much of what we do is aimed at promoting our own happiness. Thus at times our decisions and actions might best be explained according to the utilitarian principle. These times include occasions in which we endure short-term suffering in return for calculated longer-term pleasure or freedom from pain; for example, submitting to surgery so that an inflamed appendix will no longer cause poor health and great discomfort. But do such considerations guide all our actions and decisions, and should they? Suppose, for example, it could be shown that the overall unhappiness in the world is increased and that virtually no pleasure results from our accepting a certain scientific theory? The question is not whether, under such circumstances, we would or would not accept the theory. The question, rather, is whether "happiness" ought to be the criterion by which we reject or embrace theories. Note, also, that "pleasure" and "pain" are not entirely or eternally fixed in the human frame. Some of both are the result of years of nurturance by families, schools, and the culture at large. At what age, then, should utilitarian considerations become exclusive (and allowed) in a person's life? Should children be permitted to forgo all education because they are happier not going to school? This question, which might embarrass the act utilitarian, could be answered thus by the rule utilitarian: "No, children must go to school since, over the long haul, education increases the number and quality of pleasures and allows the educated person to avert the suffering of poverty, unemployment, etcetera."

But does education truly have these effects? Has not formal schooling often led to a loss of conviction, long seasons of doubt and confusion, the frustration of competition and failure, and the misery of elevated but unrequited expectation? Since through education the world has reached the point at which all humanity can be destroyed—which, one would think, is the most dis-utilitarian outcome imaginable—can any possible "pleasure" add up to more than the possible pain?

These examples illustrate the difficulties connected to utilitarianism and one of its famous offspring, behaviorism. We will consider these matters again in the final chapter, and conclude this one with a review of several of the more significant psychological aspects of the theory.

There is a nearly irresistible urge, in attempting to account for the rise of utilitarianism, to point out that Thomas Reid's best student was Dugald Stewart (1753–1828), and Stewart's admiring pupil was James Mill, and that James Mill was the principal disciple and a major expositor of Jeremy Bentham's political philosophy. The impression conveyed by such an intellectual pedigree is that the great political reform movement in England in the 1830s was born in the quiet studies of Edinburgh, Glasgow, and Aberdeen, and that its essential character was philosophical.

It is true that Dugald Stewart was mightily impressed with Reid, so much so that his *Elements of the Philosophy of the Human Mind* was devoted largely to qualifying and correcting those features of Reid's system that Stewart considered to be in error.[87] Through it all, he retained Reid's insistence on Baconian science, Reid's belief in the constitutional determinants of perception and thought, and Reid's "faculty" psychology as the proper starting point for a science of morals. He did not have much confidence, though, in appeals to common sense, nor did he share Reid's often haughty disregard for the role of reason in settling disputes. By and large, Stewart was concerned with the differences among natural science, moral science, and mathematics. (His father was professor of the last subject at Edinburgh.) His effect on James Mill, we may guess, had most to do with the force of his personality, the liberality of his political convictions, and his great personal charm.[88] Of his specific intellectual contributions to Mill's development, we may cite two: he undermined confidence in that peculiar sort of mentalism with which Reid's psychology was rife and, in the process, he salvaged Hume's mechanical associationism. Reid's "innate mental dispositions" were of a sufficiently mysterious nature to keep theology in mental science and to keep physics out of it. Stewart, like Hume, traced many of these native qualities to language and to conventional figures of speech. Another of his students, Thomas Brown (1778–1820), may have summarized Stewart's criticism of Reid's psychology most aptly when he wrote:

> To suppose the mind to exist in two different states, in the same moment, is a manifest absurdity. To the whole series of states of the mind, then, whatever the individual, momentary successive states may be, I give the name of our *consciousness*. . . . There are not sensations, thoughts, passions, *and also consciousness*, any more than there is *quadruped* or *animal*, as a separate being to be added to the wolves, tygers, elephants, and other living creatures. . . . The fallacy of conceiving consciousness to be something different from the feeling, which is said to be its *object*, has arisen, in a great measure, from the use of the personal pronoun *I*.[89]

Brown followed Stewart in restoring a stricter associationism to psychology, and Mill followed both. Still, lest the impression of intellectual continuity become firm again, we must note that Mill did not even begin his university studies until the year after the publication of Bentham's *Principles of Morals and Legislation* (1789), and Bentham was not concerned with the philosophical hairsplitting that so occupied Thomas Reid's disciples. Thus, to discover the connection—and an intimate one it is—between utilitarianism and empiricism, we must look past that superficial and almost accidental chronology linking Reid, Stewart, Brown, Mill, and Bentham.

The distinction between politics and a political movement is that the latter seeks its defense in philosophy. When people are to be encouraged to give up certain traditions, when they are to be prodded into new political spheres, when they are incited to riot and rebel, they must, simultaneously, be convinced either that those principles to which they have subscribed have been violated or that the principles are invalid. Life in a society of law, no matter how mean and confining, generally is perceived as better than life amid anarchy. Improving one's condition is, at least in principle, always possible as long as one is alive and not imprisoned, and this fact is usually sufficient to discourage the masses from taking arms against their leaders.

Neither empiricism nor rationalism, as purely philosophical doctrines, entails any given form of government. Aristotle could argue against tyranny from the perspective of a rationalist, and Hobbes could argue for it from the same perspective. To the extent that a society of laws cannot prove, either experimentally or even empirically, the validity of those first principles upon which any system of law must be based, the legal and ethical tone of a society will necessarily be set by that form of discourse and invention normally associated with philosophical rationalism. Accordingly, when government fails and reform is in the air, philosophical rationalism becomes the innocent casualty in the war on law.

It cannot be mere chance that is responsible for the frequency with which periods of social turmoil and political reform are empirical in their philosophical complexion and the frequency with which periods of national birth and regrouping are rationalistic in their philosophical complexion. Recall the ultrarationalistic philosophies following Athens' defeat by Sparta, Rome's rise to power, the creation of the European community, and the post-Reformation period in France and Germany. Contrast this with the empirical Stoicism and materialism of late Rome, the empirical pragmatism of Europe in the Reformation, and the political empiricism of England from the eighteenth century on. In these massive cultural and political movements, the empirically inclined luminaries have not so much denied that there are first principles or "self-evident" truths as they have asserted different first principles and "self-evident" truths. It is no more "self-evident" that "all men are created equal" than it is that kings have "divine rights." John Locke, whose status as an empiricist requires no defense, begins his famous *Essay on Civil Government* with a very large number of assertions, none based on the evidence of sense, and few even possibly observable. He tries to imagine man in his original "state of nature" and concludes that, in such a state, nothing is more evident

> than that creatures of the same species and rank, promiscuously born to all the same advantages of nature, and the use of the same faculties, should also be equal one amongst another without subordination or subjection, unless the lord and master of them all should, by any manifest declaration of his will, set one above another.[90]

But, empirically, what is this state of nature? And what observation or record of observations leads to the conclusion, than which none is more evident, that all individuals are similarly endowed at birth and have the same faculties in the same degree? And what of the notable exception, unless the lord and master of all (God) sets one above another? Is it not this very exception upon which all monarchies, all religious authority, all institutions of control have been based since the first pharaoh? And when Locke goes on to say that the state of nature "has a law of nature to govern it," is it not this presumed "law of nature" that dictates Caesar will rule and not follow?

Let us be sure to note, then, that the connection, between empiricism and utilitarianism is historic, not logical. Locke, Berkeley, and Hume in England and the *philosophes* of the French enlightenment had asserted the authority of experience in the affairs of state. At the grossest level, which is the only level from which political movements can draw inspiration from philosophy, the empiricists had identified truth with sense, validity with feeling, morality with sentiment. The influence of Thomas Reid, again at the grossest level, was one of making these empiricistic claims seem to rest on the natural constitution of the human frame. Jeremy Bentham's *Principles* was the coalescence of empiricism, sentimentalism, and common-sense philosophy, despite his rejection of "common-sense" schools which he described as "ipsedixitism." While rebuking the intuitionists (e.g, Reid), he could still stand behind such aphorsisms as

> pleasure is in *itself* a good . . . the only good: pain is in itself an evil; and, indeed, without exception, the only evil.[91]

This form of ipsedixitism can, of course, be derived from nothing but intuition as this term was employed by those Bentham dismissed as intuitionists. Even in the superficially empirical language of the following passage, the strains of rationalism break through:

> Nature has placed mankind under the governance of two sovereign masters, *pain* and *pleasure*. It is for them alone to point out what we ought to do as well as to determine what we shall do. . . . They govern us in all we do, in all we say, in all we think. . . . The *principle of utility* recognises this subjection, and assumes it for the foundation of that system, the object of which is to rear the fabric of felicity by the hands of reason and law. Systems which attempt to question it, deal in sounds instead of sense, in caprice instead of reason, in darkness instead of light." [92]

Here was the English version of Rousseau's "Man is born free and is everywhere in chains"—the late eighteenth-century expansion of Locke's theory of the social contract. Bentham and James Mill had their historic meeting in 1808, and thus the "Benthamites" were founded. Utilitarianism in England would soon produce the Reform Bill of 1832. In France, its continental version had already produced Napoleon.

Bentham's contribution to psychology can hardly be ignored. While the utilitarian emphasis upon pleasure and pain was, philosophically, far from original—arising as it did from more than a century of sentimentalist thinking—Bentham's attempt to quantify the doctrine created a bridge to science. His "pleasure principle" would find new expression in the psychoanalytic theories of Freud and in the research and theory of Thorndike. Indeed, behavioral science's tangled "law of effect" is but a restatement of Bentham's "two sovereigns." Less obviously, Bentham's legal prescriptions created a need for psychological inquiry. In Chapter XVI of his *Principles,* he is at some pains to establish the legal responsibility of citizens and the conditions that might confine either their claim to happiness or their ability to make such a claim. It is here that he addresses himself to the problems posed by insanity and mental deficiency and all but pleads for an Alfred Binet:

> For exhibiting the quantity of sensible heat in a human body we have a very tolerable sort of instrument, the thermometer; but for exhibiting the quantity of intelligence, we have no such instrument." [93]

In addition to these indirect effects upon psychology, effects that are very small relative to the contributions of Freud, Thorndike, and Binet themselves, Bentham directly set the tone of American psychology, and especially educational psychology, through his writings and through his major spokesman, James Mill. The bridge from utilitarianism to pragmatism is a short one. The distance from pragmatism to behaviorism is even shorter. As we shall see in later chapters, one builder of these bridges was Charles Darwin.

Many, even most, of the ideas discussed in this chapter were written in response to an alternative perspective—rationalism. A good portion of Locke's *Essay* was devoted to issues raised by Descartes, just as Hume's *Treatise* was a reply to those who would challenge Locke's empiricism. The *Treatise* went much further than a mere defense, however, and created a rationalistic streak among British and Scottish philosophers who, otherwise, were empiricists. We now will turn to this rationalism, which so boldly prefers mind to sense, logic to experiment, and certain truth to the inescapable probabilities of perception.

Notes

1. Francis Bacon, *Novum Organum,* LXIV, in *The Works of Francis Bacon,* Vol. I, Hurd and Houghton, Cambridge, 1878.

2. Francis Bacon, *Of the Proficience and Advancement of Learning Divine and Human, Works,* Vol. I, pp. 134–135.

3. Ibid., p. 128.

4. Ibid., p. 120.

5. Ibid., p. 127.

6. Ibid., p. 224.

7. Ibid., pp. 225–226.

8. Ibid., pp. 236–237.

9. Ibid., p. 254.

10. Ibid., p. 332.

11. Ibid., p. 334.

12. Ibid., p. 338.

13. Francis Bacon, *Novum Organum,* XCVIII, *Works,* Vol. I.

14. Ibid., XCIX.

15. Ibid., CXXVII.

16. Ibid., XCII.

17. John Donne, *Complete Poetry and Selected Prose,* edited by John Hayward, Random House, New York, 1936, p. 237. For an excellent analysis of this aspect of the zeitgeist, consult *All Coherence Gone,* by Victor Harris, University of Chicago Press, 1949.

18. John Locke, *An Essay Concerning Human Understanding,* Henry Regnery Co., Chicago, 1956. An especially penetrating study of the skeptical forces operating in this period is Margaret Wiley's *The Subtle Knot: Creative Scepticism in Seventeenth-Century England,* Allen and Unwin, London.

19. Locke, *Essay, Introduction.*

20. Ibid.

21. Ibid., Book II, Chap. IX, Sec. 4.

22. Ibid., Sec. 7.

23. Ibid., II, IX, 8.

24. Ibid., IV, I, 2.

25. Ibid., IV, II, 1.

26. Ibid., II, XXIII, 9 and 11.

27. Ibid., IV, III, 12.

28. Ibid., IV, II, 14, and IV, IX, 3.

29. Ibid., IV, XI, 1.

30. Ibid., IV, XI, 13.

31. Ibid., IV, XI, 8.

32. Ibid., II, X, 8.

33. Ibid., IV, IV, 5.

34. Ibid., IV, IV, 6.

35. Ibid., IV, IV, 7.

36. Ibid., IV, IV, 8.

37. George Berkeley, *An Essay Towards a New Theory of Vision* (1709), in Berkeley's *Works on Vision,* edited by Colin M. Turbayne, Library of Arts, Bobbs-Merrill, Indianapolis, 1963.

38. Ibid., Sec. 75.

39. Ibid., Sec. 65.

40. George Berkeley, *A Treatise Concerning the Principles of Human Knowledge* (1710), Open Court Edition, La Salle, Ill., 1963.

41. A. A. Luce, *Berkeley's Immaterialism,* Russell and Russell, New York, 1968.

42. Berkeley, *Essay,* Sec. 17.

43. Berkeley, *Treatise,* #4.

44. Ibid., #86.

45. Ibid., #89.

46. Ibid., #90.

47. Ibid., #103.

48. Ibid., #112.

49. David Hume, *A Treatise of Human Nature,* Introduction, edited by L. A. Selby-Bigge, Clarendon Press, Oxford, 1973.

50. The most readily available collection of essays by Shaftesbury, Butler, and other prominent sentimentalists is the Dover edition of *British Moralists,* 2 vols., edited by L. A. Selby-Bigge, Dover Books, New York, 1965.

51. Shaftesbury, *Inquiry Concerning Virtue or Merit,* in Selby-Bigge, *British Moralists.*

52. Ibid., p. 46.

53. Ibid.

54. Joseph Butler, *Of the Nature of Virtue,* Selby-Bigge, British Moralists, p. 245.

55. Hume, *Treatise,* p. XVII.

56. Ibid., p. XVIII.

57. Ibid., Book I, Pt. I, Sec. I.

58. Ibid., I, I, II.

59. Ibid., I, I, III.

60. Ibid., I, I, III.

61. Ibid., I,I, IV.

62. Ibid., I, IV, V; I, IV, I.

63. Ibid., I, IV, II, and also I, IV, IV (quotation from p. 228).

64. Ibid., I, IV, II.

65. Ibid., Ii, II, IV, and II, II, V.

66. Ibid., II, II, IX.

67. Hume, *An Enquiry Concerning Human Understanding,* Sec. V, Pt. II, in *Essential Works of David Hume,* edited by Ralph Cohen, Bantam Books, New York, 1965.

68. Hume, *Treatise,* III, II, V.

69. Ibid., III, III, IV.

70. Hume, *Enquiry,* Sec. IV, Pt. II.

71. Ira Wade, *The Intellectual Origins of the French Enlightenment,* Princeton University Press, Princeton, 1971 (p. 89).

72. Jean Le Rond D'Alembert, *Preliminary Discourse to the Encyclopedia of Diderot,* translated by Richard N. Schwab, (quotation from p. 7 of the Introduction), Bobbs-Merrill, Indianapolis, 1963.

73. Hume, *Treatise,* I, IV, VII.

74. Thomas Reid, *An Inquiry into the Human Mind on the Principles of Common Sense,* in *Between Hume and Mill: An Anthology of British Philosophy—1749–1843,* edited by Robert Brown, Random House, Modern Library, New York, 1970, p. 161.

75. Reid, *Inquiry,* p. 175.

76. Ibid., p. 174.

77. Ibid., p. 173.

78. Ibid., p. 157.

79. Ibid., p. 187.

80. Thomas Reid, *Essays on the Active Powers of the Human Mind,* Essay III, Pt. II, Ch. I, MIT Press, Cambridge, 1969.

81. Ibid., Essay II, Ch. III.

82. Thomas Reid *Essays on the Intellectual Powers of Man,* Essay II, Ch. 5 ("Of Perception"), reprinted by MIT Press, Cambridge, Mass., 1969.

83. Hume, *Treatise,* I, I, VIII.

84. Reid, *Essays on the Intellectual Powers of Man,* Essay V.

85. Ibid.

86. Ibid.

87. Dugald Stewart, *Elements of the Philosophy of the Human Mind,* in Brown, *Between Hume and Mill.*

88. Stewart's lectures seem to have been, however, a major source of the elder Mill's lifelong interest in moral philosophy. See, in this connection, Alexander Bain's *James Mill: A Biography,* London, Longmans, Green, 1882.

89. Thomas Brown, *Lectures on the Philosophy of the Human Mind,* in Robert Brown, *Between Hume and Mill,* p. 336. Note the similarity between Thomas Brown's objection and what Professor Ryle has called the "category error" in *The Concept of Mind* (1949, Hutchinson, and Co., London). Ryle describes the foreign visitor to Oxford or Cambridge who asks, after seeing the classes, playing fields, offices, etc., "But where

is the University?" Oxford University did not come into being as a result of teachers and students simply finding themselves together on what was once a large and empty lot. First there was the *idea* of a university and then there was the choice of location, the construction of buildings, the admission of qualified students, etc. Thus, the foreigner's "category-mistake" may not be an error at all. It is quite conceivable that, after examining all the buildings and interviewing all the residents of Oxford, one still may plausibly ask, "But where is the University?" That is, he may be familiar with the original charter, may recall the original intentions and commitments of the university, and may decide that what is now going on is *not* Oxford University. In terms of Brown's versions of the category-mistake, it would seem to make little difference whether one uses the word "consciousness" or the word "feeling." Even if one agrees that "I am conscious of my toothache" means no more than "I have a toothache," one still must explain the feeling.

90. John Locke: *An Essay Concerning the True Original, Extent and End of Civil Government: II: The State of Nature,* in *Social Contract,* edited by Sir Ernest Barker, Oxford University Press, New York, 1947.

91. Jeremy Bentham, *An Introduction to the Principles of Morals and Legislation,* Ch. X., Sec. XI, in *The Utilitarians,* Dolphin Books, New York, 1961.

92. Ibid., Ch. I, Sec. I.

93. Ibid., Ch. XV, Sec. XLIV.

8 Rationalism: The Geometry of the Mind

The apology for rigid classification was first offered in Chapter 1 and has been repeated several times since. It is again in order, for "rationalism" is a term that has come to suggest so many meanings that nearly any objection to its application will stand on sure ground. It is customary, for example, to refer to the eighteenth-century "Enlightenment" as "the Age of Reason," and its chief architects, needless to say, were all dedicated to reason. Yet, most of them—Voltaire, Hume, Diderot, D'Alembert, Condorcet—were empiricists, at least according to the broad definition given in the previous chapter. Even Rousseau, so passionately romantic and, in the common sense of the term, idealistic, was durably empiricist in addressing the major political questions of his time. And we have already observed that Locke's empiricism was still able to embrace intuition and the possibility of moral maxims whose certainty was as great as those of mathematics. Nonetheless, there are fundamental differences between, on the one hand, Locke, Berkeley, Hume, James Mill, and J. S. Mill and, on the other, Descartes, Spinoza, Leibniz, and Kant, Moreover, a distinguishable assortment of "psychologies" has arisen from these differences. In this chapter, we will focus on two of the more important of the differences, the two that seem to be especially at odds and that played the greatest part by far in setting the infant science of psychology off on two separate paths. One difference is epistemological and center on the issue of innate ideas. The other is methodological and concerns the relative importance of rational versus empirical modes of inquiry and explanation. Naturally enough, the epistemological bias virtually determines the choice of method. Because these two issues are central, we might best begin by examining the meaning of each.

Innate Ideas

Recall that Plato's theory of knowledge was, at the root, nativistic. He argues in several dialogues that the eternal truths are locked within our souls, even before birth, and that learning, properly conceived, is a species of recollection or reminiscence. In the *Theatetus,* Socrates makes short shrift of that marvelously Humean maxim, "Man is the measure of all things," and notes that, were the senses enough, dog-faced baboons could qualify as philosophers. The major portion of the Platonic legacy received by Aristotle, who was otherwise aloof to the theory of ideas, involved this skepticism toward the evidence of perception. Recognizing that the senses were responsive to changes in the material world, Aristotle too concluded that they were of little service in the search for the unchanging and universal principles. Accordingly, it was metaphysics that was to be the "first philosophy," and it was this ranking that Francis Bacon challenged.

But let us remain clear as to what stands behind the ranking. Aristotle was something of a "common-sense realist" in his psychological writings, not unlike Thomas Reid. He took for granted the essential accuracy and usefulness of the senses throughout the animal kingdom, for survival would be imperiled otherwise. Nature, as Aristotle said more than once, does nothing without a purpose. The senses, the perceptual and motor systems, the processes of memory and learning—all this is what serves the local as well as the long-range interests of the individual organism and the species as a whole. Thus, as fact-gathering instruments, the senses are essential and, except under peculiar conditions, faithful and accurate in their reports. However, what the evidence of experience can provide is precisely what calls for an explanation, not the explanation itself. We can observe, for example, that animal with hearts also have kidneys, but nothing in the observation will explain *why* this is so. Accordingly, scientific knowledge, which includes both facts and the explanation of these facts, must be grounded in nonempirical and irreducibly rational considerations. Ultimately, we have only understood something when we can supply the reason for its being what it is. This is where the "final causes" enter, and where the more empiricistic theorists politely (or not so politely) exit!

In the medieval period, as we have noted, contentions between adherents of each of these perspectives reappeared, this time in the form of the problem of universals. The eye can see one cat at a time, but the mind "knows" of the "universal cat." Such knowledge assumes a knowing faculty that is not sensory, and it is therefore a form of knowledge that cannot be given in experience. It is knowledge the mind must have prior to experience in order for experience to teach us anything. The words *given* and *prior* are at the heart of the problem of innate ideas. In Chapter 2, we reviewed Aristotle's objections to the strict (Platonic) nativistic account which seemed to require that infants enter the world knowing a number of things which, later in life, they will have so much trouble learning. Aristotle's jibe in this connection was the prototypic one and has been repeated in every generation by those seeking to disarm the nativist. The stock reply, of course, is that children do not have difficulty with the truth but with the language they must learn in order to articulate it in the culturally approved form. And the rebuttal to this is that such so-called "knowledge" is only linguistic in the first place. This rebuttal, in turn, prompts the reply that were such truths merely verbal, there would have been no reason for every literate society in human history to have invented terms to express them. This is not a debate likely to end in victory.

Everyone tutored in arithmetic knows that there is no number so large that one cannot be added to it. No one knows this by experience, for no one has actually conducted the experiment and determined that the prediction has succeeded. To say that we know this by inference or generalization is to impute to the fact the less-than-certain status that attaches to all other inferences. (This once prompted Jean Piaget to quip that the radical empiricist believe the series of positive integers was discovered one at a time!) It is not merely *likely* that there is no number so large that one cannot be added to it. It is absolutely and irrefutably certain. If it is an inference, it is not the sort of inference generated by sensory knowledge or by experience. Since the fact is not established by experience in the first place and since it cannot be confirmed by experience, and, finally, since it is a fact, it must be one not given in experience. Thus, it is known a priori. *That which could not possibly be given in experience* is the backbone of claimed "innate ideas." Employed this way, the term "innate idea" does not require that infants be consciously aware of the fact or facts, only that, in their sufficient maturity, they will know certain things for certain. This knowledge will be the result of maturation alone, not instruction or experience. Advocates of this thesis claim only the existence of certain innate principles or archetypes of thought that assimilate experience and determine its psychological character. This is the sense of a priori that continues to enliven psychological research and theory, It is a concept of innateness not wed to any particular period of maturation and surely not infancy. We do not have to grant the infant the ability to add in order to know that it is not experience that proves that no number is so great that is cannot be greater. The rationalists to be studied in this chapter all subscribed to the doctrine of innate ideas so defined.

Methodological Rationalism

Rationalists are further distinguished from empiricists by the method they recommend to discover nature's laws. As the term implies, rationalism is a commitment to thought, reflection, deductive rigor, an argumentative chain in which successive links are joined by the dictates of reason. The goal is a rational and intelligible account of why things are the way they are and not some other way. If the empiricist traditionally has been content to discover what is, the rationalist considers what must be. The empiricist's evidence has always been the data of experience; the rationalist's, the necessary proofs following from axioms and propositions. Integral to the rationalist tradition has been mathematics and particularly geometry. Euclid's theorems have served more than one leading rationalist as the model for epistemology, even for reality. Recall the Pythagorean theory according to which the *tetrakys* (1, 2, 3, 4) generates respectively the point, the line, the plane, and the solid, such that reality itself is spawned by number. Logic, a kind of verbal geometry, has been the rationalist's method, standing in its own defense. If a skeptical attitude toward the senses has been added to this, it has been based on the conviction that there are abiding truths, immutable and deducible, which attributes or essences exist beyond the reach of the senses. A truth that cannot be seen must be present in the mind independently of experience. Thus, the rationalist does not reject the purely local and ephemeral facts of perception, but requires that they be incorporated within a logical system the truth of which is beyond sense but accessible to reason.

In its prescientific period, psychology was largely a debate between rationalists and empiricists. The great intellectual achievement, from Copernicus to Newton, was the marriage of these two perspectives. Science could not have been developed by way of a radical empiricism; nor could it have flourished if extreme forms of rationalism were adopted at the expense of systematic observation and classification. Aristotle had brilliantly combined the best features of empiricism and rationalism, but had subsumed both under a broader metaphysical system that conferred excessive authority on "final causes." The laws of science describe nature as it is experienced, not the way it "ought" to be. To the extent that Aristotle had preconceived notions about how nature "ought" to be, and to the extent that succeeding generations of rationalists were guided by his preconceptions, then, to these extents, the modern era begins with the rejection of the authority of Aristotle. Bacon, the empiricist, led the British movement away from Aristotle and his Scholastic following. On the Continent, the leading spokesman for the new era was René Descartes.

René Descartes (1596–1650)

Descartes stands in the same relation to the Continental tradition of rationalism as Bacon does to British empiricism. Like Bacon, whom he mentions only rarely in his works and whose *Novum Organum* was probably unknown to Descartes as he began his own *Discourse,* the French philosopher saw unbridled skepticism as his most formidable adversary. In this sense, and notwithstanding the considerable differences between the Baconian and the Cartesian approaches to science, both philosophers share a common intellectual challenge: Renaissance skepticism and its tendency to breed superstitious alternatives.

We have already touched upon the skeptical strains of English scholarship in Bacon's time and we need not add much to the previous chapter in order to describe the climate in France early in the seventeenth century. Copernicus's *De Revolutionibus Orbium* (1543) developed the case for the motion of the earth against Scholastic arguments to the contrary. Recall that Aristotle was quite willing to entertain the possibility, and that, in finally coming out against it, he accepted the geocentric alternative only weakly. Still, in the politically charged climate of the late Renaissance, distinctions between Aristotle and his Scholastic disciples were crudely blurred such that any successful attack on the latter was taken as a telling refutation of the philosopher as well.

In 1609, Johannes Kepler published the first two of his laws of planetary motion which required that planetary orbits be elliptical and, further, that the sun be the immobile center of the revolutions. In the same year Galileo had observed the moons of Jupiter with the aid of a telescope fashioned by his own hands. Astrology, theology, and much of the Hermetic corpus had an abiding commitment to a "sacred numerology" according to which "7" enjoyed a privileged place: for example, the seven days for creation, the seven days of the week, the seven sisters of the Pleiades, and the seven heavenly bodies. Galileo's observations demanded a recount. There developments, attending and following the Reformation, hardened the Church's stand against heresy. In Europe the consequences were greater and bloodier than in England. Bruno had been burned as a heretic in Descartes' lifetime (1600), and Galileo would be called before the Inquisitors in 1633, the year Descartes' *De Mundo* (which subscribed to the Copernican theory) could have been published had the author not held it back. Not long before Descartes' birth, wars between the Protestant Huguonots and French Catholics had ravaged the country. St. Bartholomew's Massacre (1572) had ushered in a year of butchery grim even by the macabre standards of religious persecution. Ramus, the sedulous anti-Aristotelian at the University of Paris, was one of the first casualties of the Massacre, and this tells us that, at least in France, the line separating philosophical and Roman Catholic orthodoxy was vanishingly thin.

Here, then, was the climate of Descartes' early development: growing achievements in science, the residue of Hermetic mysticism, the sullen skepticism and cheery materialism of the followers of Montaigne, the harsh, retaliatory measures of traditional authority, the numbing conformity of unthinking "Aristotelians" whom Aristotle himself would have abhorred. Descartes' rearing in this climate was traditional and upperclass, with his first instruction received at the hands of the Jesuits and emphasizing mathematics. From them he learned to respect learning itself, to devote himself to the ultimate purposes of Christianity, and to see in science, as it was known in his time, but another version of God's eternal wisdom and power. But from his own native genius, he came to doubt, as well. The great philosophical accomplishments of his life must be viewed as the triumphant reply to his own doubts. Even his lasting contributions to science and mathematics, his researches in optics and, especially, the optics of refraction, his application of algebra to geometry and consequent founding of analytic geometry, even these are supplementary to and intended as proofs of his larger, philosophical system, his rationalism. Descartes' psychological theses are to be found in the *Discourse on Method,*[1] the second and the sixth of his *Meditations,*[2] and *The Passions of the Mind.*[3] His own summary of his broad and complex system was published in 1644, as the *Principles of Philosophy.*[4] His controversial and most influential psychobiological work, the *Treatise on Man and the Formation of the Fetus,* was published posthumously in 1664.[5]

Descartes' "method," which is summarized in Part II of his *Discourse,*[6] is really only a four-part maxim about the way the unprejudiced mind should go about its business: first, to accept nothing as true except that whose truth presents itself to the mind with such clarity and vividness as to remove the merest element of doubt; second, to divide a problem into as many discriminable elements as possible; third, to work from the solution of the smallest problem gradually up to the solution of the grandest; finally, to guarantee that the solution thus arrived at is sufficiently general as to allow no exception. Having achieved success in uniting geometry and algebra through this method, Descartes was persuaded that it might profitably be extended to all sciences. To aid him in this grander enterprise, he adopts several "moral" maxims: he will obey the laws of his country and adhere firmly to the Faith; he will adopt his provisional skepticism as if it were a proven certitude until the weight of reason requires that it be rejected; he will adopt a kind of stoic resignation toward those matters beyond the control of any individual while recognizing that his thoughts at least were within the perimeter of his powers.[7]

Applying the method epistemology, Descartes recognizes the limitations of the senses and adopts the skeptic's position that all is illusion and self-deception. Is there, then, *anything* that might justify our belief in it? Is it not just as plausible to assume that the entire fabric of apparent life is a fiction, an invention by nature assuring us of eternal ignorance?

> Whilst I thus wished to think that all was false, it was absolutely necessary that I, who thus thought, should be somewhat; and as I observed that this truth, I THINK, HENCE I AM, was so certain and of such evidence, that no ground of doubt, however extravagant, could be alleged by the Skeptics capable of shaking it, I concluded that I might, without scruple, accept it as the first principle of the Philosophy of which I was in search.[8]

The skeptic whose doubts extend as far as his very existence finally is caught in a contradiction: that which doubts must be; that which thinks must be. Even if the body is an illusion, even if all our actions and experiences are unreal, the ideas of the mind must exist, or doubt itself is impossible. It is reason, therefore, that gives indubitability to existence, not matter. The perfected existence of *triangle* is rationally provided even if there is no triangle to be found and even if a perfect one can never be formed materially. That the human mind can possess such perfected notions in an imperfect material world entails an immaterial author of perfection and this, of course, is God.

It is in the two concluding parts (V and VI) of the *Discourse* that Descartes reviews the many findings he has made in the biological and physical sciences using his method. He discusses the circulation of the blood and refers to the brilliant research "of a physician of England"[9] (William Harvey) for those who require experimental confirmation of that which reason can demonstrate. He also makes reference to unpublished treatises of his own which, because of their controversial elements, he has decided not to have published during his life. These include his essays in support of the Copernican system and, perhaps, his *Treatise on Man.* He also advises future workers to begin with reason and the immediate facts at hand before plunging forth into elaborate experiments, since the latter do not become necessary until knowledge is already well advanced. In this recommendation, Descartes conveys the essential flavor of the hypothetico-deductive method.[10]

Disputes between the ancient Greek atomists and idealists reflected an implicit disagreement over the status of mental as opposed to purely material realities. But Descartes was the first to put the mind-body problem in a form requiring the attention of scientists and philosophers alike. It may even be proper to say that the differences between modern materialism and ancient atomism—and these differences are very great—are the result of Descartes' analysis of the mind-body problem and the solution he offered to it. Psychological materialism, which in the nineteenth century became physiological psychology, is founded on the explicit rejection of both his method of analysis and his solution. Both, however, have proved to be quite durable.

It is useful to begin with Descartes' solution: mind and matter are of a qualitatively different sort, independent of each other; on no account is mind conceivably reducible to matter. Thus, Descartes is a dualist. He arrives at dualism, however, through the very skepticism to which his method commits him. He reviews his *cogito ergo sum* in the second Meditation and goes on to ask what sort of thing he is. He quickly dismisses Aristotle's definition of the human as a rational animal because this definition fails to tell us what is "rational" and what is "animal." He then reviews the various attributes he believes it has: body, hands, feet, hunger and thirst, and so forth, and a number of things it does; for example, walking, hearing, sleeping. All these, however, can be illusory. That is, the only *necessary* attribute, made necessary by the *cogito,* is thinking. Thus he must be a *thinking thing:*

> I suppose there exists an extremely powerful and . . . malignant being . . . directed toward deceiving me. Can I affirm that I possess any one of all those attributes of which I have lately spoken as belonging to the nature of body? . . . The first mentioned were the pow-

ers of nutrition and walking; but if it be true that I have no body it is true likewise that I am capable neither of walking nor of being nourished. Perception is another attribute of the soul; but perception too is impossible without the body. . . . Thinking is another attribute of the soul; and here I discover what properly belongs to myself. This alone is inseparable from me. I am, therefore, precisely speaking, only a thinking thing, that is, mind . . . or reason.[11]

Matter is extended and is, therefore, in a place. Mind is unextended. That the soul's intellection comes to affect the body indicates that it must be in some sort of intimate contact with the body. This conclusion, introduced in the sixth Meditation, is the entire topic of the *Passions of the Mind.* Before turning to it, we should note an especially subtle point raised in the sixth Meditation representing that aspect of Cartesian philosophy that declares the existence of innate ideas. He accepts the empirical (Aristotelian) assertion that we possess a passive faculty of perception by which the external world impresses the senses. He does not doubt but that our knowledge of sensible things is created this way.

But this would be useless to me, if there did not exist in me, or in some other thing, another active faculty capable of forming and producing those ideas.[12]

The truths of speculative geometry are, we may concede, truths *about* actual figures but not truths that can impress the passive faculty of perception. Only reason can analyze the figures of geometry in such a way as to discern the general truths. He is even more explicitly Platonic in the *Principles of Philosophy* when he explains the failure of some to know these general truths in terms of their being taken in by the sensations of the body.[13]

Descartes' major psychological problem and the one that has vexed dualists ever since is that of accounting for the manner in which an immaterial, unextended agent (the soul) can influence an extended, material substance (the body). How can an idea move a muscle if the idea has no mass? The seeming impossibility of this has led some to adopt either a materialistic monism, insisting that all finally reduces to matter, or a mentalistic monism, insisting that all finally reduces to mind, or a neutral monism, taking no side in the issue except to insist that, ultimately, all must reduce to one or the other or some third, presently unconceived alternative. Descartes did not resolve the dilemma in *The Passions of the Mind,* but he subscribed to a position that veritably founded the dualist tradition. His analysis is as follows.

He reserves to the body all that can be imagined as pertaining to body: sensitivity, motion, extension, growth, decay. To the mind, however, he imputes that which is inconceivable in bodies: *thought.* He divides thought into that which impels voluntary action and that which is responsible for feeling.[14] Most of our perceptions are the result of the action of external objects on the sensory nerves, and many of our feelings such as heat, cold, pain, and hunger are also referable to similar neural mechanisms.[15] Our emotional feelings, however, which can occur in the absence of any external stimulus, and our reflections on these, cannot exist in mere bodies and cannot, therefore, be attributed merely to our own bodies. These are in the province of the soul which, while not itself a body, is united as a principle to every part of the body.[16] When the body dies, the soul withdraws. But while the body lives, the will of the soul influences the body through its ability to regulate the flow of animal spirits from the brain to all the nerves associated with experience, action, and feeling. Descartes' best guess is that the site of this control is the pineal gland which, unlike the other structures of the brain, is not duplicated on each side and is located propitiously in the center of the brain.[17] Individuals differ in part because their brains differ,[18] but in all cases the passions result from the flow of the animal spirits contained in the cavities of the brain.[19] The soul's ability to direct the spirits is made possible merely by its willing to do so. Through God, this will is free.[20] He offers an illustration:

> If we see some animal approach us, the light reflected from its body depicts two images of it, one in each of our eyes. The two images, by way of the optic nerves, form two others in the interior surface of the brain. . . . The images then radiate toward the small gland which the spirits encircle. . . . The two brain images form but one image on the gland which, acting immediately on the soul, causes it to see the shape of the animal. . . . The impressions which . . . a terrifying object makes on the gland causes fear in certain men, and yet in other men can excite courage and confidence . . . all brains are not constituted in the same manner.[21]

In this passage we have Descartes' introduction of the concept of the *reflex* which would engage the mind of eighteenth-century scientists throughout Europe. His own reluctance to include the soul in this purely materialist account of sensation and behavior would not be preserved by many of his outstanding successors in a more liberal period. Still, it is Decartes who first attempted to found biology on the same mathematical foundation that Kepler had given to astronomy. It was Descartes who insisted on placing the behavior of all animals and much of human behavior in the context of natural science. Moreover, it was Descartes who advised that we turn away from the search for Aristotle's final causes,[22] that we exercise reason instead of blind faith, that we place our trust in that which about us is beyond doubt: our own thoughts.

His direct contributions to psychology were impressive. That portion of his dualism that was materialistic is the cornerstone of modern neuropsychology. His influential writings on the reflex-connections between sensation and action began a line of inquiry that culminated in the research and theory of Ivan Pavlov. His *Method,* a version of which was shared by Galileo, saved science from that vain ritual of fact-gathering that Bacon's *Novum* seemed to demand. In presenting thought and certain kinds of feeling as the distinguishing features of human beings, he prepared the way for an experimental psychology of consciousness that would be inaugurated by Wundt in the nineteenth century.

Indirectly, Descartes' influence is felt in psychology through the effect his writings had on later critics in the empiricist tradition. It was Descartes' position on innate ideas that marks the point of departure between the (British) empirical and the (Continental) rationalist schools. But even Locke, in his discussion of intuition, imagination, the axiomatic nature of moral precepts, and the reality of ideas, even Locke the empiricist in his battle with the skeptics, borrows more from rationalism than he rejects. Descartes' place in the ranks of speculative science and psychology is at the front and durably so. He was not an antiempiricist; he was just not only an empiricist. He was not an antimaterialist; he was merely not only a materialist. This is why it has been far easier to challenge one or another of his assertions than it has been to escape his presence even now, three centuries since his death.

As we shall see in the next chapter, Descartes' total effect on thought took the form of an elaborate "Cartesianism" against which the eighteenth-century *philosophes* fought so tirelessly. At a superficial level, the war on Cartesianism usually centered on its physics which, in the wake of Newton's achievement, was judged to be wrong, if not ridiculous. In comparing Newton and Descartes, the *philosophes* argued that the chief difference was one of *method.* Newton, they thought, succeeded because he had relied on the evidence of direct experience. Descartes failed because he placed his trust in the tidy rationalism of "Schoolmen." Moreover, Newton himself had said, "Hypotheses no fingo" ("I frame no hypotheses"), a phrase that made it difficult for the eighteenth-century commentator to recognize the fundamental difference between Newton and Bacon in the matter of philosophy of science. Locke, Reid, Hume, and most of the French empiricists explicitly paired Newton and Bacon when discussing the proper method of philosophizing. On this account, science and philosophy are the same: They seek the truth by confining themselves to observation and experiment, and by resisting the temptation to *rationalize* nature.

As we know, however, Newton framed a good many hypotheses, and Descartes made a good many observations, including experimental ones. At the heart of eighteenth century anti-Cartesianism was not so much a rejection of Descartes' methods as an opposition to a number of his conclusions. What, after all, follows from claims of the sort, "all brains are not constituted in the same manner," if not a direct challenge to claims of the sort, "All men are created equal"? Does not the same reasoning lead to notions of the king's "divine right" and to justifications of classes and castes? If, as Descartes' psychology holds, the human soul is indestructible and immaterial, and if the distinguishing feature of the human mind is based upon something that transcends experience, on what basis can the historic authority of Church and Crown be challenged? Note that "Cartesianism," to the eighteenth-century reformer, appeared to be something vastly more formidable than a theory of physics or a method of speculation or a system of metaphysics. It was explicitly a theory of human psychology and, as such, a theory of government and society. Although played out as a battle between "Newtonians" and "Cartesians," the intellectual collisions of the eighteenth century were between Whigs and Tories.

Descartes and the Animals

Those who have expressed a principled concern for the manner in which scientists have used and, yes, abused animals tend to place much of the blame on Descartes' psychological writings in which nonhuman animals are described as "automatons," or a kind of machine. It is true that such a position can be found in Descartes' writings, though it is also clear that he accorded nonhuman animals the full range of perceptual and (even) motivational and emotional processes. Assuming that they lacked all rationality, he was led to the conclusion that they were unable rationally to interpret stimulation and therefore were not *conscious* in the relevant respects. It goes without saying that here, too, the authority of reason—when it trumps the direct evidence of daily life—is worth resisting at every turn.

Benedict de Spinoza (1632–1677)

Spain's wars with English and the defeat of the Spanish Armada in 1588 not only loosened the hold of the Roman Church throughout Europe but also made it difficult for Spain to continue to impose orthodoxy at home. Many of the Jews who had been forced by the Spanish Inquisition to adopt Catholicism now either began to rebel openly or to find their way out of Spain and into states eager to accept non-Catholic citizens. Holland was especially tolerant and it was to there that the Spinoza family migrated toward the end of the sixteenth century. Spinoza was thus reared as a Jew in the Christian city of Amsterdam at a time when the "new philosophy" (i.e., Descartes') was all the rage in intellectual quarters. What he borrowed from this new philosophy was sufficiently unorthodox to result in his excommunication from the Jewish community. What he assimilated from his training at home and in the synagogue was enough to exclude him from the circle of important Christians. And, in finally granting to intuition a status even higher than that accorded to reason, he was unable to maintain a position among the intellectuals of the Enlightenment. He was something of a Maimonides among scholastics, a "naturalist" among mystics, a situationist among absolutists, and an absolutist among skeptics. Spinoza lived a very difficult life.

Perhaps the best approach to Spinoza's psychology is by way of one of Blaise Pascal's *Pensées:*

> I cannot forgive Descartes. In all his philosophy he would have been quite willing to dispense with God. But he could not help granting Him a flick of the forefinger to start the world in motion; beyond this, he has no further need of God.[23]

Descartes' system of philosophy took for granted the separate and independent categories of God, matter and mind. While it had not suggested that the last two could exist without the first, it was implicit in Descartes' rational psychology that, once God's forefinger flicked, the balance of nature and natural events could be studied rationally. We have already described his impatience with the search for final causes and his resulting concerns with efficient and material causes. These attitudes translate into one kind of Cartesianism, the kind that grants divine authorship in accounting for the world but restricts scientific and philosophical inquiry to methods and events with, we might say, theology held constant. The cosmology here reserves to God the creative act of bringing everything into being—and then letting it run its course according to the immutable laws of science. If divine intervention ever takes place again, it must result in the suspension of these very laws, the result being nothing less than a miracle. The *Deists* of the eighteenth century would be more transparent in adopting this perspective, but it was highly risky in the seventeenth.

Unlike Pascal, Spinoza was not influential in his own time although Leibniz, despite his politically motivated protestations to the contrary, seems to have been impressed by a number of arguments appearing in Spinoza's *Ethics.* It would be in the nineteenth century that interest in Spinoza would sharpen, chiefly among the German Romantics who often found much to condemn in Spinoza's deterministic philosophy. Lessing, Schelling, and Fichte all found occasions for discussing Spinoza. But we pause to review Spinoza's psychology briefly here because it exposes the sensed tension between reason and passion, between free will and determinism, which has been so much a part of modern psychological thought. It also reflects an awareness that science will either incorporate God completely into its deliberations or must exclude him just as completely. Spinoza is important, then, as one who saw clearly how new the new philosophy was.

Like Descartes, Leibniz, and Pascal, Spinoza was well versed and active in science. He worked as a lens-maker and had the fullest grasp of the principles of optics, which were, of course, geometric. Like Descartes and Leibniz, he was convinced that there was certain truth in a world of apparent change. Unlike Descartes, he could find no reason in logic or experience to assume that mind, matter, and God were to be relegated to distinct categories. It followed that if God is the author of all, His presence must be in all, since "things which have nothing in common cannot be one the cause of the other." [24] Thus, if God is the cause of all things, it makes no sense at all to talk about human freedom. Our very knowledge of good and evil is sufficient to prove that we were not born free, for, if we were, this constraining knowledge would not exist.[25] Our "freedom," therefore, is of a different sort. Since God is a "thinking thing," our thoughts will either share in His or be imperfect. If imperfect, our actions will be compelled by passion rather than emotion. The distinction between the two is important. For Spinoza, a passion is a feeling toward that about which we have no clear idea, whereas an emotion is a feeling shaped by a distinct idea. So-called "blind rage" is an instance of passion, while love for our fellows is an emotion.

This is a quite interesting distinction, the kernel of which can be found in parts of Aristotle's ethical writings and more fully in Hellenistic Stoicism. To respond passionately is to respond intemperately, which is to say in a manner not constrained by that knowledge and those principles that impose proportion and balance on our conduct. One of the grounds on which Aristotle withheld ascriptions of virtue from nonhuman animals was their presumed lack of a deliberative power. As their choices were assumed to be based not on deliberation but on a kind of passion, there was no controlling principle involved and, therefore, nothing of virtue in the act. To act in a virtuous way, therefore, calls for deliberation; it also calls for commitment and resolve. Thus, in Aristotle too there is room for an affective component to which we must be properly disposed. For example, we must be disposed to be angry over injustice, not justice.

Stoic philosophers, especially in the early Christian period, were more radical in their theories of the emotions. They took them to be something of a disease or malady. The actor impelled not by reason but by passion is clearly suffering from a disordered mind; *literally* disordered in that the right order of things—the rule of reason—has been upset. Spinoza's thesis is indebted to these older traditions in philosophy. If we have a clear idea of whatever it is before us, then our actions toward or in response to it will not be the result of compulsion but of commitment. Our actions will not be impelled by passion, but given resolve by emotions that are under the mind's control.

But what does it mean to have a clear idea? The clear ideas is none other than a rational awareness of the fact that whatever is the case is what it is *necessarily.* "The mind," says Spinoza, "has greater power over the emotions and is less subject thereto, in so far as it understands all things as necessary." [26] Thus, where Descartes' conception of body and soul (mind) required each to have some vague kind of relation to the other, Spinoza radically divides the psychological and the spiritual: learning, perception, memory, and emotion requires a body and end with the end of the body.[27] There is an afterlife only in the sense that God, as a thinking thing, eternally retains the idea of the essence of the given individual.[28] Since this essence is mind itself, mind is eternal.[29] As he argues in his *Proposition IV*, nothing can be destroyed except by something external to itself. To the extent that the mind is possessed of its own adequate ideas and is not totally passive (i.e., acted upon from the outside), mind as such is indestructable.

To this point Spinoza's thesis is not unlike St. Augustine's and even has Berkelean features. But when he turns to matters of direct, psychological consequence, his philosophy becomes radical. Rationalism becomes a kind of emotionalism. What we call "good" and "evil" he says are "nothing else but the emotions of pleasure and pain." [30] But pleasure and pain are, respectively, the mind's awareness of its strengths and its weaknesses.[31] It is the mind that seeks to endure eternally and it is the mind that knows that this is possible only through its own activity, which is to say through its clear ideas. Passion, as a passive state, perception as a fleeting affair, and mere imagination as the repository of purely contingent events will not permit the mind to endure. The mind equipped with these is aware of its weaknesses, fearful of its transitoriness, and thus pained. Reason, however, and, more important, intuition, bring to the mind the clear idea of necessity, whose possession is pleasure. The emotion so created is the most intense, since "an emotion toward that which we conceive as necessary is, when other conditions are equal, more intense than an emotion towards that which is possible, or contingent, or non-necessary." [32] And, as an emotion can only be affected by another emotion (*Part IV, Prop. VII*), we begin to see that the rationally achieved *adequate idea* is the only basis upon which emotions might come to be attached to the right objects of thought.

Spinoza's deterministic psychology is unreserved:

> The mind is determined to wish this or that by a cause, which has also been determined by another cause . . . and so on to infinity.[33]

The ultimate or first cause is God. To the extent that the mind is fixed on "adequate ideas," which are the ideas of necessity, it is active; otherwise, it is passive. When active, it endeavors to persist in its being, which is to say it strives for pleasure. Since this pleasure is the possession of adequate ideas and since the body is no such necessary entity and, finally, since things that have nothing in common cannot be the cause of one another, "body cannot determine mind to think." [34] However, it is in our nature to include the idea of our bodily existence in all of our ideas and, therefore, the mind is threatened by that which threatens the survival of the body. Intuitively we labor to preserve the activities of our bodies because of this sensed connection between bodily and mental survival. Since we believe that a threat to the body entails a threat to the activity of the mind, we strive to we repress those ideas of the body's injury or demise.[35]

The categories of Spinoza's psychology are passion, emotion, reason, and intuition. All are determined and, to that extent, the will is not to be described as free. What makes us unique, when we are unique, is the presence of clear or adequate ideas about the necessary causes of things, and these ideas lead back inevitably and inexorably to God. All else involves passive states of mind which, in the limiting case, would be eternal death. We are so constituted that our lives are devoted to perpetuating the existence of mental activity. The motive, at a superficial level, appears in the form of an egoistic quest for pleasure and aversion to pain. The unenlightened are moved in these respects by opinion rather than reason, and as a result of opinion and the laws of association they come to identify bodily death as the ultimate pain. Benignly, God has planted in us a higher sense, an intuitive awareness that is the active and impassable mind. But even when this mind grasps the essential nature of what is good, the emotions cannot be restrained by the truth of the knowledge, only by another and contrary emotion produced by such knowledge.[36] In this aspect of his theory, Spinoza stands in agreement with the British sentimentalists, with Hume, and with Kant: Reason alone will not produce moral conduct except to the extent that the rational exercise leads to or is correlated with the proper feeling. Our will is not the free cause of our actions but the necessary cause.[37] Given the will, action follows necessarily. Given God, the adequate idea, which is the idea that refers to Him, follows necessarily. This is worth rehearsing. What Spinoza requires is a means by which activity is initiated. Reasons for acting are not direct causes of action, for one may have a reason for doing something and still not do it. But a reason for acting may engender a certain desire of the will, and it is this that impels the action. Thus, given the will, there will be activity. But the will is not free, for it is engaged necessarily by prior events. Nor is the resulting action free, for it is impelled by the will. If the action is to be of moral worth, it must be traced back to an adequate idea, this being the necessary consequence of God.

In his uncompleted essay *On the Improvement of the Understanding,* Spinoza embellished his theory and offered a number of remarkably modern notions of learning and memory; for example, that memory is enhanced or degraded by the contextual features of the material to be memorized, that memory suffers when materials similar to those memorized are subsequently committed to memory, that memory is a brain process, and that every idea must have a correlate in the real world.[38] It is here also that he makes the traditional rationalist distinction between mere sensations or perceptions and the active assimilation of experience by the intellect. Of the various modes of knowing, only opinion is inferior to mere experience. The highest mode is that by which the essence of a thing is known. We know the essence of a circle, for example, when we know that it is the line resulting when a stylus is fixed at one end. To know the essence of a thing is to know what must follow from the fact of its existence. To know an eternal truth is to know that, if it is true at all, its contrary could not be true. Thus, the Pythagorean theorem is an eternal truth in that there could not be a right-angle triangle the square of whose hypothenuse does not equal the sum of the squared sides. To know such truths and, indeed, to know even simple facts, is simultaneously to affirm and to negate. That is, every determination entails a negation. Once we have determined that Smith is an old man, we have, in this act of determination, denied that he is young, that he is female, that he is aluminum. To affirm is to limit. To define is to confine. The infinite, then, is either undefined (which Spinoza rejects) or is self-defined. To know an eternal truth is, therefore, to know the sense in which it is self-defining, self-causing. It is to know God.

Spinoza's "God," however, is neither the Yaweh of the Old Testament nor the humanized first cause of Christian Aristotelians. Rather, it is the god of the logicians; something of a deduction stripped of theology. His predictable standing as an atheist still receives scholarly attention, though there seems little doubt but that he regarded the necessitated transactions of the cosmos to be necessitated by more than a formal argument. Spinoza did not speak to his age even if, in a peculiar way, he spoke for it. His recourse to associationistic principles of learning and memory was hardly original, and his egoistic theory of motivation, while infuriating, had been in the wind since Hobbes, and since Montaigne before him. In striving for a unique form of monism, he offered something of a pantheism. God, as a thinking thing, inspires certain substances in a way that makes them thinking things too. Such a pantheism had to be judged either as heresy or lunacy, and Spinoza's contemporaries were willing to offer both judgments. Not until the nineteenth century would his philosophy be taken seriously by a major philosopher and, in taking it seriously, Hegel would reject a good part of it and force the rest to conform to his own system.

To Spinoza, also, we owe the insistence that philosophy and theology are different enterprises, that the former must be concerned with truth, no matter where it leads, while the latter must require reverence and a degree of compliance. Further, philosophy is a search for truths pertaining to nature, and God is known in nature through the laws which are (therefore) necessary. We, as thinking things driven to create pleasure in the mind, are constrained by nature to strive toward our essence and to be pained by its denial. The goal of human psychology, then, is self-actualization. In this, Spinoza serves as a model for a number of twentieth-century humanistic psychologies.

Though not a disciple of Descartes, (openly disagreeing with him, for example, on the issues of free will, dualism, and the division of nature into compartments), Spinoza shared with Descartes a realization that Greek wisdom was not enough, that Christianity had to be more than Aristotelianism, that a rational creature, stripped of all superstition and unbiased in the face of nature, could know the truth. His rationalism was a naturalism without becoming a materialism. Spinoza did not doubt matter as did Berkeley, nor did he doubt mind as radical materialists did. He began with God and from that starting point he examined thinking matter. Implicitly, he defended materialism as applied to matter, since only materialism applies to that which is only matter; just as only idealism applies to a thinking thing when it is the thinking we wish to understand. What Spinoza was able to avoid through his complex metaphysics was the sort of trouble awaiting the usual varieties of dualism and materialism. Before Spinoza, philosophers by and large spoke in behalf of mind or matter, or mind *and* matter. Spinoza, however, was able to deduce from his major axioms the conclusion that mind and matter are the same substance viewed from different perspectives, thus introducing a quite subtle form of "double-aspect" or even "multiple-aspect" theory as a solution to the mind-body problem. At the most general perspectival level—say realism—we are prepared to claim that every existing thing triggers in us an idea of it by way of sensations, images, memories, and the like. From another perspective—call it idealism, since we are conscious only of sensations and ideas—we are obliged to accept these sensations and so forth as the only constituents of reality. From still another perspective—call it materialism, since knowledge can only be about things, and since these can enter the realm of thought only through their material effects on our material organs—reality consist only of matter.

Spinoza treats these claims as the product of certain habits of talking about truth, not as truths themselves. In their place he would offer a monistic theory according to which ideas and objects are substances of the same kind. For every object, there is a matched idea; for every idea a matched object. To have an idea, after all, is to have an idea of something. But to have an idea of what finally are only the attributes of a thing (an idea that rises no higher than the level of mere perception) is to have an incomplete and unclear idea. To know the object requires that we know the substance. The latter is what remains of a thing when its attributes are peeled away—that which allows it to be what it is eternally and immutably. Note that the idea of this is necessarily an imperishable idea, since for every object there is a corresponding idea. And it is in just this sense that the mind seeking to endure must be the mind seeking the adequate and clear idea.

Spinoza's eternal substances are of one kind, though they display a mental and a physical "aspect." They do this not because of any fundamental duality in the universe, but because we have not thought of them clearly. And by the same token, we speak of God and Nature as if there were two distinct entities involved, failing to recognize the implicit contradiction. To ascribe to God all the properties of perfection, omnipotence, and omniscience, and then to argue that such being, at a determinate time, brought the world into being, is nonsense. For God to have created the universe, it would have been necessary for God to lack something and to attempt to remedy this through an act of creation.

But if this were true, God would not have been the perfected and universally inclusive being of the Hebrews or the Christians. To make theology coherent, then, Spinoza thought it was logically necessary to take all substances—all *substantial* thing/thought items—as eternal in their substantial unity. On this construal there can be no essential difference between God and nature, mind and matter, or science and philosophy. Goethe and Coleridge, as leaders of the nineteenth-century Romantic movement, both acknowledged their debts to Spinoza, and both would give impetus to that "Religion of Nature" which is the essence of Romanticism. Others would discover in Spinoza's essays a new justification for the old-time religion; still others, for the new-fangled agnosticism. As with Plato and Aristotle, Spinoza has something for everybody, but not always what the recipient thinks is in the package!

Gottfried Wilhelm von Leibniz (1646–1716)

Leibniz was only four years old when Descartes died. He lived in an intellectual climate dominated by Decartes and the "Cartesians," a climate rapidly becoming as uniform in its reverence for Descartes as the world of letters once had been toward Aristotle. Leibniz's father was professor of philosophy at Leipzig, and, while both the father and mother were dead before he completed his university studies, Leibniz may be said to have been blessed with a stimulating early environment. He was exposed not only to the Greek and Latin classics but to the modern works of Bacon, Descartes, and Galileo. Visits to London (brief) and Paris (four years) further exposed him to the most important ideas of the time. He met with Malebranche, the great Cartesian; he studied Pascal's treatises in mathematics and invented a calculating machine even better than Pascal's; through Huygens, his interest in optics was aroused; and Hobbes' works enhanced his abiding concern with law, as the recently ended Thirty Years' War engendered a passion for peace and tolerance. By way of introduction, we should also point out his discovery of differential calculus independently of Newton, and the bitter claims of priority that sounded across the English Channel for the better part of a decade.

Louis XIV, the Sun King, had become king of France three years before Leibniz was born and died a year earlier than Leibniz. Thus, in addition to his own genius, his good head start, and the stimulating intellectual giants of his time, Leibniz passed his years in what were among the most self-consciously cultivated and achievement-oriented decades in modern history. We might underscore this by listing some of the authors whose works were published during the reign of Louis XIV or who were still alive at the time: Hobbes, Locke, Descartes, Newton, Pascal, Spinoza, Gassendi, Leibniz, Malebranche, Huygens, and Moliere. Galileo had been dead only one year before Louis XIV acquired the throne. When the king died Voltaire was already nineteen and

Hume's *Treatise* was only twenty years off. Of those on this list who were Leibniz's contemporaries or who just preceded him, Leibniz agreed with none. His role in the history of ideas is secure, his influence diffuse and recurrent. His specific contributions to psychology, while not as great, are revealing and are presented most clearly in his disagreements with Locke and with Descartes. In rejecting Locke's empirical psychology, he reasserted the cognitive, highly mentalistic, and genetic character of human knowledge and feeling. And, in opposing Descartes' dualism, he rephrased the mind-body problem in a way that would render it more interesting to the modern era. We will begin with his anti-empirical arguments.

Locke's *Essay on the Human Understanding,* published in 1690, came to Leibniz's attention in 1688. He began to draft a rebuttal straight-away, although this rebuttal, *New Essays on the Understanding,*[39] did not appear until 1765. The reason is that Locke died the year Leibniz completed the work (1704), and Leibniz was not willing to argue with his ghost. Accordingly, the *New Essays* became just one of many Leibnizian contributions not made generally available until after the author's death. Like the rest of his work, the *New Essays* continues to influence the reflective mind.

Locke had begun Book II of his *Essay* with the central claim of all empiricists before and since:

> Suppose the mind to be, as we say, white paper, void of all characters, without any ideas; how comes it to be furnished? Whence comes it by that vast store, which the busy and boundless fancy of man has painted on it with an almost endless variety? Whence has it all the materials of reason and knowledge? To this I answer, in one word, from experience.[40]

Leibniz's reply, a reply offered by all rationalists before and since, is that only some *thing* can be said to have an experience, and such a thing must be a mind somehow prepared to have experiences of a given sort. He identifies Locke's position (incorrectly) with Aristotle's, and attributes to Aristotle a statement also attributed to him by Duns Scotus, although we search in vain for this statement in any of Aristotle's works: *"Nothing is in the intellect which was not first in the senses."* To this Leibniz replies,*"Nothing except the intellect itself."* [41] He follows the introduction to his *New Essays* with a dialogue between Philalethes ("friend of sleep") and Theophilus ("friend of God"),[42] who are, respectively, the empiricist and the rationalist. After giving Philalethes the lines from Locke's *Essay* (quoted above), he has Theophilus offer the Leibnizian position on the problem of knowledge and the solution:

> This *tabula rasa,* of which so much is said, is in my opinion only a fiction which natures does not admit. . . . Uniform things and those

which contain no variety are never anything but abstractions, like time, space, and other entities of pure mathematics. There is no body whatever whose parts are at rest, and there is no substance whatever that has nothing by which to distinguish it from every other . . . those who speak so frequently of this *tabula rasa,* after having taken away the ideas, cannot say what remain. . . . Experience is necessary, I admit, in order that the soul be determined to such or such thoughts, and in order that it take notice of the ideas which are in us; but by what means can experience and the senses give ideas? Has the soul windows, does it resemble tablets, is it like wax?[43]

Experience is necessary, on Leibniz's account, in order that the soul take notice of the ideas that are in us. What experience provides is a context for our thoughts, a direction for our ideas, a means of aiming our attention, disposing ourselves to action of a certain kind. It is impossible that an experience will produce an idea for the simple reason that an experience involves the physical confrontation of matter and the organs of sense, and an idea has nothing to do with these mechanical transactions. Perception, however, which is not a mere experience and which assumes a rational, attentive mind, can lead to ideas; but perception is not mechanical and cannot be reduced to the mechanical.[44] It is precisely on this basis that Descartes' dualism is to be rejected. Leibniz is clearest on this in his *Monadology,* written two years before he died and presented as a condensation of his metaphysics:

Moreover, it must be confessed that *perception* and that which depends upon it are *inexplicable on mechanical grounds.* . . . And supposing there were a machine, so constructed as to think, feel, and have perception; it might be conceived as increased in size, while keeping the same proportions, so that one might go into it as into a mill. That being so, we should, on examining its interior, find only parts which work one upon another, and never anything by which to explain a perception.[45]

Perception is a uniquely *psychological* event. It is that of which we are conscious. It is qualitative in a way that no purely quantitative (i.e., material) phenomenon can imitate. As we walk through the great mill of the mind, observing the spinning wheels and crashing hammers, we find nothing by which the mill could have a perception; not that the mill does *not* have such perceptions or is not aware of itself but that nothing in its moving parts could convey as much. We will return to Leibniz's mill in the last chapter.

The mind-body interactionism advanced by Descartes thus seemed at once confused and meaningless to Leibniz. Mind, on the Leibnizian account, is a simple substance, a *monad,* not reducible to anything, not deriving its character from any source beyond itself, not extended. As with all simple substances, it is to be understood as a quality and not a quantity. In Leibniz's terms, its essence is *intensive* rather than extensive. The example of a point in mathematics is illustrative. A point is not a very small line or a very small fraction of a line. It is the idealized limit as extension approaches zero. Similarly, as quantity is stripped of its extensive features, there is a limit beyond which further reduction is not possible. This limit constitutes a quality of being, not a magnitude or extension. The limit of the body is also a simple substance; that is, a *monad*. Body as it is perceived is a composite whose extension arises from the assembly of simple substances. No two simple substances are alike. Each monad not only has a distinguishing quality but is the very "unit" of quality. Being dimensionless, it is not subject to modification from without. It has no window through which an external agency might enter and change it. In this respect, it makes no sense to speculate on the kind of interaction taking place between "mind" and "body" since each, properly conceived, is unique, independent, and ultimately unextended. Body and mind coexist and the relation between them is not causal but *harmonic*. If a note is loudly sounded in the presence of two resonators, we do not ask which of the resonators has established the sympathetic vibrations in the other. The two resonate in parallel as a result of being so constituted that, in the presence of the proper stimulus, each satisfies the conditions of its nature. So too with mind and body. Each, according to a preestablished harmony, exists compatibly with the other. The action of one is not caused by the other, as the mechanistic account requires; nor is the action of each reconciled to that of the other by some external "timekeeper" who occasionally steps in to make sure that all the clocks are on time—the view advanced by the philosophic occasionalists.

The universe is a collection of simple substances for which harmonic associations were established prior to their coming into being. Harmony is God's modality. If the monad is to change, it can do so only through an inner principle.[46] It is this internal principle, involving as it does a system of relations within the monad, that constitutes perception. This, however, differs from apperception or consciousness by which we not only perceive but are aware that we perceive. To the extent that every monad has an internal organization, it is perceptive. The monad whose internal principle allows both memory and perception may be called a soul.[47] It is clear from this that animals have soul. However, they do not have rational souls (i.e., minds), because, while they are able to perceive and even to retain the trace of former, consecutive perceptions, they are unaware of *necessary* truths. Human beings, too, insofar as their perceptions are simply united by memory, "*act like the lower animals, resembling the empirical physicians whose methods are those of mere practice without theory. Indeed, in three-fourths of our actions we are nothing but empirics.*"[48] It is only in our knowledge of a rule, of a necessary relationship, that we display the uniquely human quality of human life.

More than any previous philosopher, Leibniz was concerned with and wrote about an issue that would be at the core of modern psychology's problems, the unconscious. This is not to say that, in any complete sense, he anticipated Freud. Leibniz's use of the concept had little to do with the causes of motivation and nothing to do with psychopathology. He employs the concept, instead, in support of his position on the indestructibility of monads, on the distinction between perception and consciousness, and on the difference between the bare monad and the rational mind. Even in a dreamless sleep, he maintains, the monad does not perish (for it can't) and, therefore, since it also cannot exist without being affected in some way, perception, by definition, takes place. However, we are not aware of this perception because it is not accompanied by memory.[49] A number of unconscious (insensible) perceptions, stored in the mind, can add up in such a way as to break into consciousness. Indeed, there is a gradual scale separating the sleep of death from heightened awareness. We pass from one to the other in small steps, one of which constitutes a threshold. Moreover, we retain all that has happened to us even though we might not be actively aware of much of it. The representations of the past remain in the mind, creating "*an influence greater than people think. . . . The present is big with the future and laden with the past.*"[50]

As with Spinoza, it is not easy to carve out a niche for Leibniz in the history of psychology. As the dual enemy of empiricism and materialism, he cannot be located in that philosophical tradition that led to the installation of psychology as an experimental science. His writings make it clear that he believed one could deduce most of what our contemporary experimentalists are searching for, and what is not easily deduced is either trivial or readily available through common experience. We have already acknowledged his attention to the unconscious and his formal introduction of the concept of subliminal perception. That division of experimental psychology devoted to sensory thresholds has a debt to Leibniz here, but the debt is several times removed. Even his challenges to Locke were hardly more than what one can find in the *Protagoras* or the *Meno,* and Leibniz is the first to recognize that he has taken the Platonist position in the dispute.[51] His discussions of the unity of consciousness, of the role of memory in consciousness, and of the differences between consciousness on the one hand and both perception and memory on the other, would surface again and again in both the theoretical and the experimental psychology of the late nineteenth century. Perhaps his most substantial direct effect on psychology—and, we might suspect, an unforeseen one—resulted from his telling criticism of Cartesian dualism. In illustrating its defects and contradictions, he did much to overturn Descartes' authority and to liberate thinking to the point of allowing an uncomplicated physiological psychology. Leibniz would not have applauded the reduction of mentalism to materialism—he specifically opposed it—but others would dismiss his caveats and focus instead on his successful refutations of Cartesianism.

Unlike Spinoza, Leibniz did little to restore idealism to a position of philosophic significance, and, therefore, we are not even able to relate him to that tradition leading up to Hegelianism. But his emphasis upon activity and unity, the two abiding features of all and every simple substance (including mind), would reappear in the psychologies of Brentano, James, the Gestalt school, and even early behaviorism. His monism would, as we have noted, inspire confidence in those who might examine brain in order to unearth mind. His attribution of soul to animals and his insistence on the continuous evolution of various levels of organization and relation would surely not retard the development of an experimental psychology of animal intelligence. His unswerving commitment to the presence of innate characteristics and his logical arguments in the behalf of the necessity of *a priori* dispositions of the mind would constitute a starting point for one of the most influential philosophers of all time, Immanuel Kant.

Immanuel Kant (1724–1804)

The position in which philosophy, metaphysics, and science were left by Hume's *Treatise* was anything but reassuring. Rationalism was rendered deluded in its singular goal, the search for eternal truths. Epistemology was reduced to psychology, and an associationistic psychology, at that. What the *Treatise* denied was that necessity could be proved to exist in nature. It denied that logic confirms such necessity and that the senses ever perceive it. It affirmed the existence of subjective necessity as a habit of the mind. We judge B to be the effect of A when the two occur together in space and time, when A always precedes B and these perceived unions are constant. Since experience alone is responsible for our belief in causation, and since, in principle, it is "*possible for all objects to become causes or effects to each other,*" we are forced to acknowledge that "*anything may produce anything.*" [52]

Hume did not deny that events were caused. Rather, he insisted that our commitment to this view cannot rest upon any basis other than experience and, as such, the view can never acquire the added baggage of necessity. It may happen that A is always followed by B, that no one has ever recorded an exception, that the interval between the two is perfectly constant. Still, all we *know* is A and B. "*We are never sensible of any connection betwixt (them).*" [53] We have knowledge of the events and knowledge of the temporal connection between them. We have no knowledge of necessity. That is, experience will confirm only A . . . B; not A . . . necessarily . . . . B. On the same basis, moral distinctions are not derived from reason but from experience (and the feelings resulting from it). [54] And, again on the same basis, there can be no logically compelling argument against those who assert that the rational faculties themselves are but the outcome of natural material forces. For even though thought and matter seem to be different, experience suggests that "*they are constantly united; which being all the circumstances, that enter into the idea of cause and effect, when apply'd to the operations of matter, we may certainly conclude, that motion may be, and actually is, the cause of thought and perception."* [55]

In a word, the *Treatise* removed moral precepts from the domain of the rationally deducible, removed necessity from the domain of cause and effect, and removed reason itself from the domain in which we locate the determinants of knowledge, feeling, and conduct. A rational philosophy designed to unearth necessary moral prescriptions must fail. A rational philosophy seeking to penetrate what *must* occur in nature must fail as well. The *Treatise* stripped natural science of "must" and moral science of "ought." Empirical psychology was all that survived.

If we are to appreciate the lasting contribution Kant made to psychology, we begin by examining a question that has tired philosophers for nearly two centuries: What was Kant's answer to Hume? [56] The answer is to be found in his towering achievement, *Critique of Pure Reason,* [57] in his summary and clarification of this work, *Prolegomena to Any Future Metaphysics,* [58] and in his *Groundwork of the Metaphysic of Morals.* [59] An implicit theory of psychology runs through every chapter of each of these, and often the theory is explicit as well. In the following discussion, there is little chance that all of Kantian philosophy will be penetrated, but the psychology in this philosophy will be evident.

All the major philosophers of the eighteenth century were quick to note the distinction, a distinction made by Plato and by a good number of scholastics, between propositions that seek to add to our knowledge about a thing and those that only assert a semantic identity. When we say, for example, that a body is an "extended substance," the predicate-term (extended substance) is, in fact, included in our concept of the subject (body) such that the proposition does not add anything to what the subject already contains. Locke, Berkeley, and Hume all devoted sections of their epistemological works to the relationship between words and things and to the fact that, very often, the only differences between things turn out to be a difference in the words we employ in describing them. The convention appearing in the eighteenth century was to label all such propositions in which the predicate is contained in the concept of the subject as *"analytical."* Kant retained this term and employed another to represent those propositions whose predicates are not logically implied by their subjects: that is, those propositions, that expand our factual knowledge. These he called "synthetic." [60] When we say that all bodies are heavy we are making an assertion about bodies different from what is contained in the mere concept we have of a body. The same is true of statements of the sort, "The French are a people of average height," "Protein is conducive to health," and so forth. Being French does not logically entail being average in height; being protein does not logically entail good health on the part of the recipient. In noting the general position of philosophers on the matter of analytic and synthetic propositions,

Kant formalized the distinction further by observing that the principle common to all analytic propositions is the law of contradiction.[61] We cannot say, "The man at the table is not the man at the table." In an affirmative analytic judgment, a contrary predicate will produce a contradiction. This is not the case with synthetic judgments, for no such contradiction results from, "The French are a people of above-average height." This distinction is what led all empirical philosophers, and Hume particularly, to subscribe to the view that analytical propositions are (a) logically necessary, meaning that if they are true, they *must* be true, (b) certain as opposed to probable, and (c) a priori as opposed to given in experience. Since $a = a$ is necessarily true by the law of contradiction and since, on the empiricist account, nothing in experience is necessarily true, $a = a$ is said to be known a priori. By the same token, the same philosophers insisted that synthetic propositions (a) can only be contingently and never necessarily true, (b) can only be assigned a certain probability of truth and never certainty, and (c) can only be advanced or assessed a posteriori. No rational, a priori deduction will establish beyond doubt that "The French people are of average height." In the light of these terms and distinctions, we can summarize Hume's position on morals, on epistemology, and on ethics by noting that he placed all these issues in the domain of synthetical judgments. Whatever we say about knowledge or values can be true only contingently, only some of the time, and only a posteriori. Kant's task, then, is to prove that some synthetic judgments are true a priori, and Kant's answer to Hume is, in essence, that there are a priori synthetic truths. Put another way, Kant's mission is to return necessity to morals and to epistemology and thereby rescue metaphysics from mere opinion.

Before turning to Kant's analysis, we must recognize that he stands in agreement with Hume on many counts. On one of the cardinal empiricist points, he, too, insists that all judgments of experience are synthetical,[62] that objects are given to us by means of the senses, and that thought itself, directly or indirectly, finally relates back to sensibility.[63] Kant, therefore, is not to be understood as one seeking to overturn empiricism but as one striving to determine its limits. It is in this regard that one might judge the empiricist movement as *culminating* in the *Critique of Pure Reason* rather than being swept aside by it.

Kant's critical analysis of Hume's claims predictably focuses on the Humean account of the concept of causation. The account is empirical, and Kant must determine the principles according to which experience yields the concept of cause. Alas, experience cannot yield the concept; experience *assumes* the concept. The proofs of this are contained in his famous *Analogies of Experience: "The principle of the analogies is: Experience is possible only through the representation of a necessary connection of perceptions."*[64] He presents three such analogies. The first is addressed to the concept of object-permanence. For an object or event to have any real existence, it must exist in time. Time, however, is not *given* to experience by the object or event. Only through our "inner intuition," in which permanence is cognized a priori, can appearances take place *in time*.[65] There can be no relation in time unless that relation is grounded in permanence. We discover the weight of smoke, for example, by weighing the wood, burning the wood, and weighing the ashes. Matter is not conserved in the senses; that is, the wood is now gone and no smoke can any longer be seen. Still, because of the a priori category of the understanding, the category of thought we call *permanence*, we know that the weight of the smoke is just this difference between the weight of the wood and the weight of the ashes.[66] We can only call an appearance a "substance" because we are able to presuppose the existence of substance throughout time. This brings Kant to the Second Analogy: "*Everything that happens, that is, begins to be, presupposes something upon which it follows according to a rule.*"[67] It is this cryptic principle that is the kernel of Kant's answer to Hume, and it must be studied carefully.

Hume had argued that our concept of cause was to be explained in terms of contiguity (spatial), constant conjunction, and succession. Briefly, A and B take place in the same location, always together, and in an invariant order, with A always preceding B. When these conditions are met, we say that "A is the cause of B." It is in the Second Analogy that Kant raises the question of the source of succession itself. We do not "see" time. We do not "perceive" intervals. We might *imagine* that a boat floating downstream *could* float upstream, but we do not apprehend the event in any but a fixed order: the boat now is here, next is there, next is there, and so forth. But what is the empirical basis of "next"? Quite simply, unless the understanding already possessed (a priori) the category *time*, there could be no succession or constant conjunction. Conjunctions occur in time, but time is not given by the object. For us to be affected by the constant conjunction of events A and B, we must be able to experience A and B as events. What makes them events is that they stand out against a background of enduring states of affairs. Against an enduring background of silence, for example, chimes are heard. Chimes can only be events if they are separable from the enduring background. And the chimes can only be caused by the striking of the hammer if our perceptions are unfailingly ordered *in time*. Without the a priori category of understanding—in contrast to sensation alone—we would have no basis for judging the hammer to cause the chimes any better than the basis for judging the chimes to cause the hammer. Yet, we *never* make the latter mistake. The ordering is not contingent, it cannot be a posteriori, and it is hardly merely probable.

Professor L. W. Beck has summarized the argument of the Second Analogy with an elegance and simplicity seldom displayed in the many treatments of "Kant's answer to Hume":

K. Everything that happens, that is, begins to be, presupposes something upon which it follows by rule." (Kant's Second Analogy).

P. Events can be distinguished from objective enduring states of affairs, even though our apprehension of each is serial (the accomplishment of Hume's task(a)).

H. Among events, we find empirically some pairs of similar ones
which tend to be repeated, and we then make the inductive judgment: events like the first members of the pairs are causes of events
like the second (the accomplishment of Hume's task (b).

P implies K . . . H implies P, since if events cannot be distinguished,
pairs of events cannot be found, and thus P is a necessary condition
of H. Hence: H implies P and P implies K, therefore H implies K.
That is Kant's answer to Hume.[68]

What Beck has shown is that Hume's account of causation requires, and
logically requires in the sense of *necessarily* requiring, Kant's Second Analogy.
Thus, Hume is not wrong, but he can be right only if the Second Analogy is
granted and the Second Analogy confers on the understanding an a priori synthetic judgment.

The argument given in H (Hume's "succession" and "constant conjunction") can reduce the concept of causation to an inference from experience only
by granting the percipient a basis for "first" and "next," and this basis is not,
itself, grounded in experience but is assumed for experience to occur in the first
place: that is, the Second Analogy. This is what Beck means by "H implies
K." The Humean argument implies the Kantian argument.

Kant's analysis goes well beyond this question of the basis of causal inferences. The *Critique of Pure Reason* seeks to discover the foundations and principles of all knowledge, accepting at the outset that one foundation is, of
course, the empirical. But, as the Humean account is insufficient to explain
causation, Kant argues that the empirical account is insufficient to explain *anything* about human understanding except the conditions by which it becomes
furnished with objects. What the understanding achieves is judgment. Judgment is based on logical functions. The latter are imposed upon the evidence
of sense and are necessary prior to experience if experience is to have any
meaning at all. These logical functions, which we possess intuitively, are the
pure concepts of the understanding of which the following Table of Categories
is exhaustive.[69]

I. Concept of Quantity:	II. Concept of Quality:
Unity	Reality
Plurality	Negation
Totality	Limitation

III. Concept of Relation:
      Inherence and Subsistence
      Cause and Effect
      Community

IV. Concept of Modality:
     Possibility-Impossibility
     Existence-Nonexistence
     Necessity-Contingency

These are the pure concepts of synthesis[70] which the understanding possesses a priori and without which coherent experience would be impossible. These categories contain the possibility of all experience in general.[71] We confront the world of sensibility with an understanding already possessed of such pure concepts as: either a thing exists, or it does not exist, or it exists in a limited way; either A is possible or it is impossible; either it happens to follow B or it must. We can frame universal propositions (All men are mortal) only if we possess intuitively the category of totality, and there is nothing in experience that can give this. That we arrive at the proposition by induction or generalization from a large number of cases is not to deny the category but merely to cite the conditions under which it is invoked. Obviously, to think of "all men" requires that we know of "men," and we can only come to know them through experience. But we can never know "all" of anything through experience. The very process of inference assumes the concept of quantity, and the very process of generalization within a class assumes the concept of relation.

The foregoing is Kant's epistemological argument against empiricism and, more important for Kant, it is the introduction to the moral argument against the empiricist's pleasure principle. There is no doubt but that if one subscribes to an empiricist epistemology one will have little patience with a moral science based on the "truths" of reason. Locke was willing to grant an axiomatic status to moral propositions, likening them to the propositions of geometry, but this truce with the rationalists was neither convincing nor long-lived. The Hume who cannot find necessity in the conjunction of natural events is hardly going to search for it in that conjunction of behavioral events we call moral conduct. Kant agrees that if epistemology were reducible to experience morality would be likewise. But having proved to his own satisfaction that epistemology cannot be reduced to the sensual domain and having established that this very world of sense is *contained* in the world of understanding, it must follow that the laws of experience are authored by those of reason.[72] Kant's metaphysics of morals thus is the crowning achievement of that rationalist morality to which Descartes, Spinoza, and Leibniz had devoted themselves. Moral precepts have the authority of reason not because they refer to something that does not occur in the real world but because our understanding of the real world is based on a rule without which understanding would be impossible. As the pure concepts of the understanding (i.e., the categories form the logical foundations upon which all our knowledge of the natural world is based, so too there is an a priori rational principle which makes moral judgments inescapable, universal in form, and absolutely necessary to any explanation of the moral dimensions of life. It is not enough to argue that we judge "good" and "evil" on the basis of feeling unless one is able to explain why and how the given feeling attaches to the given act. The very attachment assumes a rule, and this rule is what Kant dubbed the "categorical imperative": act in such a way that the maxim of your action could serve as a universal law of nature.[73]

In its varied forms, the categorical imperative includes a reverence for law, the insistence that man is an end and never a means to some other end. The very concept of law assumes a rational animal who intends good effects.[74] The mere (empirical) listing of the observed consequences of actions will never disclose this intention, but the act itself could not have taken place had there not been the intention prior to it. To acknowledge this intention, the *fact* of intention, is to acknowledge simultaneously a freedom of the will. Its freedom is constrained in this sense: the very freedom requires that the will make law.[75] Reverence of the law is not acquired. The categorical imperative could not be acquired. The factual world of events could not be judged on a moral basis if there were not, a priori, pure moral concepts in the understanding. We are not "usually" or "contingently" ends in ourselves and not means toward some other end we have. We are necessarily ends in ourselves. We do not await the consequences of our actions in order to determine whether or not we *should* do unto others as we would have them do unto us. We understand that this is so or else we could never know guilt as long as we "got away" with what we did. There may be those who preach a situation-ethics, but they still draw the line at anarchy. And even those who might argue in favor of anarchy, if they are to argue at all, will begin with a principle and if that principle is ever to enjoy logical force, it will finally reduce to the categorical imperative—at which point, of course, it will contradict the claims of the anarchist.

Kant's influence in psychology has been far greater than is generally recognized. It is a commonplace in historical summaries to acknowledge his fame as a philosopher, to point to his nativistic emphasis and suggest its effect on some later psychologists. There are even those who have decided that Kant was really authoring a kind of anthropology and was anticipating subsequent instinct-theorists. He was decidedly not. It is, in fact, to his credit that he has influenced as many psychologists who have misunderstood him as he has those who have followed his reasoning. In subsequent chapters, we will have occasion to discuss theories of cognitive development, Gestalt psychology, genetic psychology, and moral development. We will review the ideas of Wundt, Freud, Köhler, and their disciples. In these subsequent chapters, it will become clear that, with the exceptions of behavioral and physiological psychology, there is no area of contemporary psychological concern that does not rely on the major elements of Kant's philosophy. Such topics as the innate, logical structure of thought and language, the a priori principles of perceptual organization, the stages of cognitive and moral understanding, the concept of culture-neutral or culture-free methods of psychological assessment—these and many issues of lesser importance are hardly imaginable had Hume's sensationism prevailed to the exclusion of rationalism. Kant did not set out to save rationalism. Indeed, he had more admiration for Hume than for many of his detractors. He set out to establish the limits of knowledge and the conditions by which it takes place. In the process, he saved consciousness.

There were, however, several negative contributions contained in Kantian philosophy, at least as regards the appearance of experimental psychology. It was Kant's judgment that a bona fide science of the mind was something of a contradiction in terms. The mind, unlike external nature, does not stand still as we attempt to observe it. Indeed, the very attempt to observe its contents alters them. Moreover, what is most defining about the human mind are the a priori categories or pure understanding, and, as we have seen, these are not "given" by experience and do not have empirical content. Furthermore, they are necessary (rather than contingent or accidental) features of mind and, as such, are not reducible to biological or mechanical laws. Nothing in biology or machinery can be what it is *necessarily*, whereas the a priori categories are. Still, in his very pessimism, Kant did indirectly provide impetus to a psychology of perception, a study of consciousness:

> Thus the whole of rational psychology, as a science surpassing all powers of human reason, proves abortive, and nothing is left for us but to study our soul under the guidance of experience, and to confine ourselves to those questions which do not go beyond the limits within which a content can be provided for them by possible inner experience.[76]

What he argued against in this passage was the notion that a deductive science of the human mind would extend our knowledge of the real world by unearthing those rational principles that regulate the real world. Such a view, in Kant's understanding, was taking to believing that we can increase the number of persons in a room by hanging mirrors on the wall! Instead he argues for a psychology that takes the contents of the mind as the *only* phenomena we can examine directly. The value of this psychology is of a negative sort—it permits us to criticize those productions of rationalism that collide with the facts of consciousness. On the whole, then, Kant's immense analysis imparted a conservative tone to experimental psychology. Writing in the long and broad shadow of the Kantian critical philosophy, Wundt would insist repeatedly that his psychology was nonmetaphysical, nondeductive, and concerned only with the facts and internal organization of consciousness. We shall examine further this part of the Kantian contribution in later chapters.

The Rationalist Legacy

It does not detract from Descartes, Leibniz, Spinoza, and Kant to recall the extent to which seventeenth- and eighteenth-century rationalism echoed many of the major lessons of Plato, St. Augustine, and St. Thomas. The *Theaetetus* was a reply to Protagoras's empiricist contentions, and its argument is not too different from that offered by Leibniz in the *New Essays*. The agreement on substantial psychological issues between Spinoza and St. Augustine is too marked to require further comment. Kant was unique, but in several respects his uniqueness is to be traced to an antirationalist way of treating the problems. He was, for example, uncompromising in his opposition Berkeley's brand of idealism, he argued that mathematical judgments were synthetic, and he denied the existence of innate ideas, at least as these had been described in the rationalist tradition. His "transcendental aesthetic" located psychological principles above the plane or experience. The new experimental psychology of the nineteenth century would thus have difficulty finding a place for Kant and ultimately succeeded in doing this only by ignoring the rest of his philosophical system. Those who came to play the major part in the creation of psychology as an independent discipline strived to impart to the enterprise the same rigor and objectivity enjoyed by physics or mathematics. Where the model was physics, the rationalist tradition was an encumbrance, and where the model was mathematics the enterprise failed, or seemed to fail. Even those who adopted a Kantian perspective on the mind still found it necessary to employ the methods of the empiricist. The exception, of course, was Wundt, whose works will be reviewed later. Wundt sought the best of two worlds in attempting to build an empirical science on the foundation of (rational) introspection. We may judge his success by observing that there are not many Wundtians around any longer.

Had there been no other movement in philosophy and philosophical psychology, the twentieth-century psychologist would still be actively engaged in assessing Leibniz's answer to Locke, and Kant's answer to Hume. Argument and analysis would continue to be the methods of choice. However, even as the seventeenth- and eighteenth-century philosophers carried on their disputes, there was a mammoth enterprise unfolding, and at a rate that beggared the imagination of philosopher and layman alike: the enterprise of *science,* or what Bacon and, later, Newton called "experimental philosophy." In its stubbornly and innocently pragmatic way, it drew upon both empiricism and rationalism for support and inspiration, but wed itself to neither. In time, the empiricists would claim it, although no empirical philosopher of consequence ever made a contribution to it. Its real engines were skepticism and materialism: Descartes' method of doubt seeking assurance in technology. Its materialist foundations are the subject of the next chapter; of its skeptical component, enough has been said in this and the previous chapter.

What was the rationalist bequest? As we review the problems and methods of contemporary psychology, we find little direct evidence of a conscious commitment to the rationalist's vision. Research is addressed to behavioral engineering, brain physiology, social attitudes and influence, human learning and memory, and individual differences. Only the light writing addressed to a popular audience continues to concern itself with the "mind," seldom with the soul, and never with the monads. But when we move from research to theory, the picture changes. Many would agree that by 1970 the three most influential theoretical attempts or, shall we say, the three theoretical issues commanding the greatest attention involve (a) the stage-specific cognitive capacities of man during development from infancy, (b) the a priori faculties, which must be granted if human language is to be understood, and (c) the species-specific processes, which must be assumed if we are to account for the range of emotional, intuitive, and "moral" dispositions observed throughout the animal kingdom. While (c) received its direct impetus from Darwin, (a) and (b) stem unadorned from the rationalist tradition. It is only in the methods employed to investigate these issues that these contemporary concerns can be described as "empirical." They are *rationalist* issues at the core and they are the visible signs of the rationalist legacy.

Notes

1. René Descartes, *Discourse on Method*, in *The Method, Meditations, and Philosophy of Descartes*, translated by John Veitch, Tudor, New York, 1901.

2. Descartes, *Meditations*, Veitch, *Descartes*.

3. Descartes, *Les Passions de l'Ame*, translated as "The Passions of the Soul" by E. Haldane and G. R. T. Moss, in *The Philosophical Works of Descartes*, Dover Publications, New York, 1955.

4. Descartes, *Principles of Philosophy*, Veitch, *Descartes*.

5. Descartes, *Treatise on Man*, in *Descartes—Selections*, edited by R. M. Eaton, Scribner, New York, 1927.

6. Descartes, *Discourse on Method*, Part II.

7. Ibid., Part III.

8. Ibid., Part IV, p. 171.

9. Ibid., Part V, p. 184.

10. Ibid., Part VI, p. 193.

11. Descartes, *Meditations*, II.

12. Ibid., VI.

13. Descartes, *Principles of Philosophy*, Part I, Secs. XII and XLVII.

14. Descartes, *Les Passions*, Article 17.

15. Ibid., Articles 23, 24.

16. Ibid., Article 30.

17. Ibid., Article 31, 32.

18. Ibid., Article 39.

19. Ibid., Article 37.

20. Ibid., Article 41.

21. Ibid., Articles 35, 39.

22. Descartes, *Principles of Philosophy,* Part I, Sec. XXVIII.

23. Blaise Pascal, *Pensées: Thoughts on Religion and Other Subjects,* translated by William Finlayson Trotter, Washington Square Press, New York, 1965 (*Pensée* #77).

24. Benedict de Spinoza, *Ethics,* Pt. I, Prop. III, in *The Chief Works of Benedict de Spinoza,* translated by R. H. M. Elwes, Dover Publications, New York, 1955.

25. Spinoza, *Ethics,* Pt. IV, Prop. LXVIII.

26. Ibid., Pt. V, Prop. VI.

27. Ibid., Prop. XXI.

28. Ibid., Prop. XXII.

29. Ibid., Prop. XLI.

30. Ibid., Pt. IV, Prop. VIII.

31. Ibid., Pt. III, Props. LIII, LV.

32. Ibid., Pt. VI, Prop. XI.

33. Ibid., Pt. II, Prop. XLVIII.

34. Ibid., Pt. III, Prop. II.

35. Ibid., Prop. VIII.

36. Ibid., Pt. IV, Prop. XIV.

37. Ibid., Pt. I, Prop. XXXII.

38. Benedict de Spinoza, *On the Improvement of the Understanding,* in R. H. M. Elwes, *Spinoza,* especially pp. 15–31.

39. Gottfried Wilhelm von Leibniz, *New Essays,* in *Leibniz—The Monadology and Other Philosophical Writings,* translated by Robert Latta, Oxford University Press, New York, 1898.

40. John Locke, *Essay on the Human Understanding,* II, I, 2.

41. Liebniz, *New Essays,* Book II.

42. Ibid.

43. Ibid.

44. Ibid.

45. Leibniz, *Monadology,* Latta, *Leibniz,* #17.

46. Ibid., #11.

47. Ibid., #19.

48. Ibid., #28.

49. Ibid., #20, 21.

50. Leibniz, *New Essays,* Introduction, pp. 322–323.

51. Ibid., p. 358.

52. David Hume, *A Treatise of Human Nature,* I, III, XV, edited by L. A. Selby-Bigge, Clarendon, Oxford, 1973.

53. Ibid., I, IV, V.

54. Ibid., III, I, I.

55. Ibid., I, IV, IV.

56. A remarkably brief and extraordinarily lucid summary of the Hume-Kant dispute

has been provided by Professor L. W. Beck in his *Once More unto the Breach: Kant's Answer to Hume, Again,* in *Ratio,* Vol. 9, No. 1, pp. 33–37.

57. Immanuel Kant, *Critique of Pure Reason,* translated by Norman Kemp, Smith, St. Martin's Press, New York, 1965.

58. Kant, *Prolegomena to Any Future Metaphysics,* Indianapolis, Bobbs-Merrill edition, Introduction by L. W. Beck, 1950.

59. Kant, *Groundwork of the Metaphysic of Morals,* translated by H. J. Paton, Harper and Row, Harper Torchbooks, New York, 1964.

60. Kant, *Critique of Pure Reason,* Introduction to Sec. 4.

61. Kant, *Prolegomena,* Sec. 2.

62. Ibid.

63. Kant, *Critique of Pure Reason,* A19.

64. Ibid., A176; B218.

65. Ibid., A182; B225.

66. Ibid., A185.

67. Ibid., A189.

68. L. W. Beck, *Once More Unto the Breach.*

69. Kant, *Critique of Pure Reason,* B106.

70. Ibid.

71. Ibid., B167.

72. Kant, *Fundamental Principles of the Metaphysic of Morals,* translated by Thomas K. Abbott, Library of Liberal Arts, Liberal Arts Press, New York, 1949, pp. 70–71.

73. Ibid., p. 19.

74. Kant, *Groundwork of the Metaphysic of Morals,* pp. 68–69.

75. Ibid., p. 98.

76. This passage is taken from A382 of the *Critique of Pure Reason,* which treats of the "paralogisms of pure reason." Later (B421), Kant amplifies the usefulness of a rational psychology when it is developed as a discipline rather than as a doctrine: "It . . . keeps us, on the one hand, from throwing ourselves into the arms of a soulless materialism or, on the other hand, from losing ourselves in a spiritualism which must be quite unfounded so long as we remain in the present life."

9　Materialism:
The Enlightened Machine

The Vexing Alternative

In the two preceding chapters, we have reviewed the philosophical psychologies of the English and Continental empiricists and rationalists, confining ourselves to the seventeenth and eighteenth centuries, when these become sufficiently definite to serve as alternatives. In Chapter 3, the very considerable similarities between Platonist and Aristotelian systems were noted. In subsequent chapters, equivalent similarities were found between philosophers identified with one or another school. In the Renaissance, and really not until the Renaissance, did we find signs of a clear break, the break between spiritualism and naturalism. But even in the Renaissance there were only signs, not dramatic ruptures. The great separation took place in the irresolvable conflicts between Hume's *Treatise* and Kant's *Critique*. Henceforth, empiricism and rationalism would hold out possibilities for radical departures and divisions. Unlike the differences that divided Greek atomists and Platonists, the modern antagonisms would surface as completed systems of thought, rich in implications and recommendations for social organization, law, morality, economics, and religion. By comparison, the disputes between orthodox Christians and Ockhamists in the thirteenth century or the Florentine Aristotelians and Platonists in the fifteenth would be nearly negligible.

As great as the break between rationalism and empiricism was, however, there remained ground common enough for a Kant even to attempt to answer a Hume. The philosophical disputes of the seventeenth and eighteenth centuries, to the extent that they were between rationalists and empiricists, were disputes about the same kinds of problems, and they were disputes that often began from identical starting points. Hume and Kant were both moved to explain the source of human knowledge, the nature of morality, and the character of society. Broadly, they were both engaged in the "science of *mental* life."

Neither of them subscribed to the vexing and alternative: law, morality, reason, and feeling were, in the last analysis, mere expressions of matter in motion. In Chapter 7, we noted that Locke declined to address the question of the human material nature. Berkeley openly rejected materialism and defiantly so. Hume, while tilting with the possibility, finally turned the matter over to the "anatomists." [1] Kant, too, recognized that a biological interpretation of his philosophy might be advanced, that his categories might be treated as rising from a neurological process, and he spurned the suggestion, noting that such a skeptical materialism would remove the element of necessity from the pure concepts of the understanding.[2] In short, neither Kant nor Hume nor any of the earlier figures in the two philosophical movements conceived of psychology as indistinguishable from physics. For all of them, it was and must remain a science of *mental* life and not merely science. However, in the same two centuries, a third movement was born and flourished: a movement toward physics and away from logic, a movement that came to reject the very terms of the rationalist-empiricist controversy, a movement that had more to do with the founding of psychology as a scientific discipline than did all the rationalists and empiricists combined. This was materialism, whose seventeenth- and eighteenth-century character is the subject of the present chapter.

The Metaphor of the Machine

Every age of philosophical energy is animated by notions and events of a non-philosophical complexion. Philosophers seek to understand and explain the facts of the world and they must take these facts as they find them. They are to be found, of course, outside philosophy: in the cosmos, in the world of matter, in the human mind, in the affairs of state.

Scholarship is a human enterprise and, no matter how narrow and technical it may become, or how ponderous or cultist its problems and methods may be, it seldom escapes the habits of the human mind. One of the most persistent of these habits is that which forces the mind to metaphor and simile when it seeks to comprehend an elusive phenomenon. And, among the many elusive phenomena, none is endowed with greater craft and agility than the mind itself. Thus, in their tireless attempt to comprehend the mind, philosophy and, later, psychology have taken recourse to explanations of the sort, "it is like . . ." or "it is as if . . ." or "it is no more than. . . ." Over the centuries, different metaphors and similes have gained and lost popularity. The pre-Socratics, with their special interest in hydraulics and hydrostatics and with their simplified four-element physics, were given to believe that psychological phenomena were to be understood in terms of the unique combination of earth, air, fire, and water. Platonism, which never fully extricated itself from its roots in the mystery religion of the Pythagoreans, focused on spiritual metaphors in an attempt to define that ineffable quality of mental life. Aristotle, surveying the unchallengeable truths of the syllogism, and noting the contingent nature of all non-logical realities, advanced a dualistic psychology according to which some functions of the mind were "like" biological processes and others "like" eternal, logical verities. Accordingly, while the nutritive, sensitive, and locomotor faculties perish with the flesh, being of the flesh, the intellect survives "like" the truth of logic.

From the pre-Socratic period until the Age of Faith, the metaphor was nature. All the divisions and antagonisms of competing philosophies are to be understood as different views of the nature of nature and as differences of opinion regarding the ultimate nature of nature: Is it material only or spiritual as well? Is it eternal or was there a beginning? Is it particulate and statistical or an essential and invariant form? Is it as perception records it or is experience mere illusion? The failure of philosophy to answer these questions and the failure of the very civilizations that raised such questions to survive were two of the chief causes of the rapid success of the Christian alternative. In the Western world, from the Patristic period until the seventeenth century, revelation was the ultimate authority. God was the reality and nature was the metaphor. This is not to say that everyone shared this outlook. Indeed, Chapters 5 and 6 were devoted principally to those who sensed that the question was hardly settled. But the world is peopled by more than philosophers, and for all but the handful of seekers after truth, the Truth was revealed in the gospels and in the life of Christ. Those eager to understand either at another level could study the great Thomistic synthesis—the synthesis of Aristotle and Christian belief, of reason and faith.

We have pointed out, perhaps too often, that the Renaissance did not alter the essential temper of the Western mind as it grappled with the abiding questions. The major debates from 1350 to 1600 were between those who found God in "natural magic" and those who found Him in "spirit." Ficino's Academy sought to restore Platonism to Christianity, not to challenge scripture. Pomponazzi's Aristotelian school had no quibble with the Bible, nor were the heresies of Gianfrancesco Pico of a religious sort. And this speaks only to the affairs of academics. In the world at large, there were more important matters: war, pestilence, Reformation, starvation. Nature, as the metaphor, and God, as the reality, remained in their historic places.

The most significant scientific event of the Renaissance, Copernicus' theory of the earth's revolution, was judged by its author as no more than a footnote to God's great design. The same may be said of Kepler's assessment of his theory and Newton's of his. It is equally true of Galileo and of Locke and Descartes, of Leibniz and perhaps even Spinoza. The greatest scholars in philosophy and science, even through much of the nineteenth century, labored under the light of Christian faith. Hume, of course, is the tantalizing exception, an exception even among seventeenth-century skeptics, of which there were many. The skeptics trained their doubts on man, not on God; on Aristotle, not on the Nazarene. In fact, several of the noteworthy skeptics were priests.

In light of the foregoing, it is instructive to examine a remarkable feature of contemporary psychology: no major spokesperson for the discipline, no figure identified as one responsible for its methods and concerns, none who has provided a theory of consequence to contemporary endeavors, has argued that the religious dimension of life is necessary for an understanding of human psychology. Stated another way, we recognize that in the fifteen centuries beginning in A.D. 200, there is no record of a serious psychological work devoid of religious allusion and that, since 1930, there has not been a major psychological work expressing a need for spiritual terms in an attempt to comprehend the psychological dimensions of man. This striking shift in perspective cannot be explained as a product of rationalist-empiricist tensions—witness Locke and Leibniz. Nor is it to be understood in terms of some factual discovery in the natural sciences. Moreover, it cannot even be understood as a gradual shift in perspective because, in the historical measure of time, it has been sudden. We are not able to establish all the causes, of course, but it is fair to say that we can discern its origin. The origin of this dramatic transformation in perspective, this historically unprecedented abandonment of an older and pervasive vision, is the seduction of a different metaphor: in this case, the metaphor of the machine.

The cautious reader will be ready to complain that the physicalistic outlook is hardly recent; that Zeno and Epicurus were beholden to it; that Ptolemy rendered it cosmic in scope; and that the Scholastics were as wed to the machine-like precision of celestial motion as were the pre-Socratics. While this is a correct reading of history, it fails to distinguish between the mere metaphor and the seductive metaphor: between the metaphor that confirms and conforms to the balance of one's beliefs and the metaphor that creates a new belief, and between the metaphor invented to represent reality and that which is so compelling as to be accepted as reality. We must ask, then, what there was in the mechanistic philosophy of seventeenth- and eighteenth-century thought that set the stage for the radical divorce now existing between psychology and its intellectual ancestry. What new feature was added or recognized? Who fashioned the wedge? We can only begin to answer this by discussing, all too briefly, a set of sociopolitical conditions and, coincident with those conditions, the extraordinary influence of Galileo on his contemporaries and immediate successors.

The political climates of England and France in the seventeenth and eighteenth centuries were touched upon in the preceding two chapters. The issues raised by the Reformation and the responses to these during the counter-Reformation remained hopelessly and perilously unsettled. Religious persecutions would continue in both countries until the nineteenth century, and religious toleration—always difficult to gauge—would become a political maxim only at the end of that century. Added to the death struggle of Protestant and Catholic, Protestant and Protestant, and Catholic and Catholic were the endless conflicts within the fraternity of European nations. The Roman Church, which had tied its fortunes to Aristotelianism, as understood by the Scholastics, found it necessary to take hardened stands on philosophical matters at a time when intellectual freedom was coming to rank as high as life itself in Paris. Descartes' *Method,* based on doubt and skepticism as points of departure, came closely to heresy. The writings of Montaigne, filled with Erasmian charm and insolent wit, were proscribed, making them even more popular than would otherwise have been the case. The murder of Ramus, the famous anti-Aristotelian at the University of Paris, gave the liberal-leaning philosophers of the early seventeenth century still one more reason to resist the authority of Aristotelianism. But unlike all their predecessors, they had an irrefutable, scientific illustration of the failure of the philosopher's *Metaphysics:* Galileo's laws of accelerated motion and Newton's general laws of motion.

Among the Thomistic proofs for the existence of God, the phenomenon of motion was central. Relying on the Aristotelian argument for a prime cause, the scholastic proof began with the contention that that which is in motion will seek a resting state unless impelled by a force external to itself: that is, eternal motion is impossible, and, therefore, heavenly dynamics can be understood only in terms of a Prime Mover who constantly supplies the needed power. The planets move by the will of God, a proof providing a ready metaphor for human action as the result of human will. If the action of the planets required no recourse to the will of God, the actions of men might not either. More will be said on this later. What Newton demonstrated was that bodies set in motion would continue to move, linearly and eternally, unless acted upon by a force opposed to their motion. Quite simply, once the bodies were set in motion, God's will had no further work to do for the motion to continue forever. God aside—for it was not God to whom Newton addressed his remarks—*Aristotle was wrong.* He was also wrong, and Galileo proved it, in contending that the speed with which an object fell to the ground was proportional to its mass (weight). He was also wrong about the immobility of the earth and about the motion of the sun. (Recall, however, that Aristotle was far from dogmatic on this point.)

Had Galileo's contribution been limited to the unearthing of mere factual errors in the corpus of Aristotelian science, his influence on his own age and ours would have been negligible. The health of Christian belief never depended on the notions of a wise but pagan Greek about the speed with which rocks and feathers fall to the earth. It depended even less on the number of moons orbiting Saturn since, no matter how many there are, they all came from the same shop.

However, orthodox belief did depend on the manner in which we go about *discovering* the truth, and it was this that Galileo threatened.

In his *Discourse on the Two New Sciences,*[3] Galileo virtually invented the science of mechanics, passed it on in a form that has hardly been changed, and tied each of its theorems and propositions to an *experimental* demonstration. Throughout his work, Aristotle falls on hard times, even when credited with a truth. For example: Aristotle had some idea of the principle of the lever—force bears to resistance a relationship determined by the reciprocal of the distances separating the fulcrum from the force and the resistance—derived and demonstrated by Archimedes. Galileo acknowledged Aristotle's observation on the law of the lever, but with a qualification, as usual:

> Yes, I am willing to concede him priority in point of time; but as regards rigor of demonstration the first place must be given to Archimedes.[4]

The most open attack on Aristotelianism appeared in Galileo's *Dialogue on the Two Great Systems of the World*[5] (1632). The year following its appearance found its author before the Inquisitors and called upon to renounce the heresy of Copernicanism. Copernicus was dead now for ninety years, and Galileo was not the only philosopher-scientist to subscribe to his theory. We may take it, then, that it was not merely the Copernicanism of the *Two Great Systems of the World* that troubled the Inquisitors but the ridicule heaped on the Aristotelian position in the dialogue.

Kepler (1571–1630) had advanced three laws of planetary motion remarkable in their simplicity and predictive power, and these laws assumed the earth's elliptical movement around the sun. Galileo (1564–1642) accepted the Copernican-Keplerian theory and added to it a completed body of terrestrial mechanics. William Harvey (1578–1657), who had studied with Galileo and the other great Paduans from 1598 to 1602, published his famous essay on the circulation of blood in 1628. It is to be noted that Harvey was Francis Bacon's physician—noted as a curio of history—but Harvey is to be seen as a disciple of Galileo's scientific method: one begins with a careful measurement of nature as it is found; one advances the simplest possible hypothesis sufficient to account for the measures obtained; one derives from the confirmed hypothesis those corollaries logically implied by the hypothesis; one proceeds to test the corollaries experimentally. The opinions of others, no matter how grand their genius may be, count for nothing. Truth is to be found after a test, not at the end of a dispute. The inscription adopted by the Royal Society, formed in 1662, is a testimony to Galileo's influence: *Nullius in verba.* (*"On the words of no man."*)

By far the greatest contribution to science in the seventeenth century, if not in any century, was Isaac Newton's *Principia.* Newton (1642–1727) was the saint of seventeenth-century British intellectuals, a man whose influence spread far beyond the boundaries of science. He combined the theories of Kepler and Galileo in a triumphant synthesis, the universal law of gravitation. No longer was it necessary to require, as the Aristotelians did, that a different physics be applied to the heavens and the earth. Galileo had concluded the dialogue of the *Second Day* portentously:

> We do not encounter . . . difficulty, however, if we suppose the earth to move, a body so mall, so inconsiderable in comparison with the whole universe that it could have no effect at all upon this.[6]

Newton, framing no hypotheses except those demanded by the direct observation, placed this inconsiderable ball within the realm of many other balls and kept them all suspended by a law indifferent to human vanity, belief, or hope. In Book III of his *Principia,* he offers the "rules of reasoning" in philosophy:

> Rule I:
> We are to admit no more causes of natural things than such as are both true and sufficient to explain their appearances.

> Rule II:
> Therefore to the same natural effects we must, as far as possible, assign the same causes . . .

> Rule IV:
> In experimental philosophy we are to look upon propositions inferred by general induction from phenomena as accurately or very nearly true, notwithstanding any contrary hypotheses that may be imagined, till such a time as other phenomena occur by which they may either be made more accurate or liable to exceptions.[7]

Between 1609 and 1686—that is, in a span of a lifetime—Kepler had published his laws of planetary motion (1609), Galileo had described sun spots, the moon's rough surface, the "stars" of Jupiter (1609), and had invented *Mechanics* (1632), Harvey had published his *Excertatio* (1628), Descartes his *Discourse on Method* (1637), and Robert Hooke the *Micrographia* (1665), which laid the ground for microscopic anatomy. If we wonder what the implications of this creative storm seemed to be to the enlightened bystander, we need only recall Voltaire's remark in *The Ignorant Philosopher:*

> [I]t would be very singular that all nature, all the planets, should obey eternal laws, and that there should be a little animal, five feet high, who, in contempt of these laws, could act as he pleased, solely according to his caprice.[8]

The metaphor was becoming reality.

From Dualism to Monism

Descartes' contribution to the scholarship of his day produced results he could not have anticipated. Three years before he died, he received a pamphlet published anonymously in Belgium by a former disciple, Regius. The pamphlet was in a form suitable for nailing on the door of a church and it was written in a way that mimicked Descartes' style. It was a polemic on the nature of the mind and it recommended a radical materialism to those who might inquire into the subject. It offered twenty-one propositions and closed with one of Descartes' own aphorisms: "No men more easily attain a great reputation for piety than the superstitious and the hypocrites."[9] The propositions themselves were uncompromising. The mind is no more than that which permits thought to human beings; logic does not require any distinction between mind and matter; all notions enter the mind through observation and tradition; no idea can be said to be innate; perception is a process of the brain.[10]

Regius arrived at this view through his own inclinations and, probably, from a hasty reading of *Les Passions de l'Âme,* which was completed in 1646. It was not published until 1649 and Descartes regretted its publication. Regius either read the work in an early draft or decided, from conversations with Descartes about the work, that it supported the radical materialist position. Several of the Articles in the book can be so construed. The origins of the animal spirits are traced to the brain such that the turbulence produced by stimulation of the senses becomes reflected in the actions of the muscles. In Article XVI, Descartes notes specifically that stimulation can produce orderly action without any intervention by the soul and in animals that lack souls.[11] Only thought and passion are in need of a soul for their existence. Perception, appetites, and even dreams and daydreams are quite possible as a result of bodily actions.

Descartes quickly recorded his disagreement with Regius's manifesto, rejecting its propositions point by point. It was Regius's twelfth Article that had challenged Descartes on the question of innate ideas, and Descartes' reply is instructive:

> I never wrote or concluded that the mind required innate ideas which were in some sort different from its faculty of thinking; but when I observed the existence in me of certain thoughts which proceeded, not from extraneous objects nor from the determination of my will, but solely from the faculty of thinking which is within me, then . . . I termed (these) "innate."[12]

Descartes goes on in his rebuttal to observe, yet another time, that there is nothing in an object that can be said to *contain* the idea we have of it. Thus, all ideas are innate if only in the sense that they have a quality not derivable from mere extension, which, after all, is the very substance of object:

> [N]o ideas of things, in the shape in which we envisage them by thought, are presented to us by the senses. So much so that in our ideas there is nothing which was not innate in the mind. . . . Nothing reaches our mind from external objects through the organs of sense beyond certain corporeal movements.[13]

Unlike his arguments in *Les Passions de l'Âme,* which were often of a theological complexion, Descartes, in his rejoinder to Regius, presents the argument of the philosophical dualist: there is nothing in the physical stimulus resembling the mental image, nothing we can say in describing the stimulus that will permit us to deduce its psychological consequences. This argument, as we shall see in subsequent chapters, has lost none of its persuasiveness in contemporary philosophical discourse on the mind-body problem.

The pamphlet printed in Utrecht was a relatively minor irritation. In comparison, the challenges to Cartesianism presented by Pierre Gassendi (1592–1655) were titanic and were the very foundations of that anti-Cartesianism that flowered in the eighteenth century. In his own time, Gassendi was ranked among the greatest philosophers of his age. His following was large and enthusiastic, his writings were influential, and his command of science and mathematics was expert, or at least seemingly so. He established his libertine credentials early, authoring a text of *Paradoxes against the Aristotelians* in 1624. More significantly, he became the center of a revival of interest in Epicurus and later Roman Epicureans. The thrust of this revived Epicureanism was to install observational science in the place of (Cartesian) deductive, "axiomatic" science, to accept nature (including human nature) as matter, and to oppose the authority of Aristotle at every turn. While not a skeptic, Gassendi found it easier to adopt the skeptical posture when the only alternative was dogma. The orientations were marshaled for a critical attack on Cartesianism, one lasting six years and taking the form of published rebuttals answered, in print, by Descartes. We will not examine any part of the dispute other than that addressed to Descartes' dualism. It is worth noting, however, that Gassendi's critique was all-embracing and that, as Professor Craig Brush has written, "Modern philosophy can add little or nothing significant to the objections made by Gassendi."[14]

It is in the Second and Sixth Meditations that Descartes presents his *psychophysical* dualism and it is in the rebuttals to these Meditations that Gassendi offers the monist alternative. Against the principal arguments for dualism, Gassendi registers the Epicurean complaints: (1) Why deny bodies the power to move themselves without the aid of a soul? Does not water flow and do not animals (judged to be without souls) walk?[15] (2) Since it is obvious that whatever *acts* is, why all the "beating around the bush" to establish that you (Descartes) are?[16] (3) What does it mean to say that you are "only a thinking thing"? Why exclude all other possibilities, such as you are a wind or a gas or a body? Even if it be granted that you only *know* you are a thinking thing, it does not necessarily follow that you are not other things as well.[17] (4) How can the mind reason without a brain, since even you (Descartes) have required that the brain organize and unite perceptions and actions?[18] (5) Even though the senses sometimes deceive, they often do not, and we usually have the means of determining whether a given perception is valid or of questionable validity.[19] (6) If the mind without extension, it can have no idea of extended things. The mind can be furnished by experience only if equipped, by its material nature, to respond to that which is physical. It may be said that you are composed of two bodies, one coarse and one subtle, and that only the former is immediately apparent. But it makes no sense to say that *you* or your mind is unextended.[20]

Even more than the specific objections to Descartes' dualism, Gassendi's writings reflect the impatient tone that every modernist displays toward a departing age. Gassendi was the most illustrious of the circle of visionaries that was brought and held together by Father Marsenne in Paris. They were all imbued with the spirit of Galileo, their Paduan saint, whose experimental and theoretical science constituted the final and invincible challenge to authority. Gassendi's works were known to and admired by Locke, whose empirical philosophy borrowed the spirit, if not the letter, of Gassendi's scholarship. Newton, too, acknowledged Gassendi's priority in advancing the law of inertia. While contemporary thought is not directly influenced by any of Gassendi's essays or discoveries, his position among his own contemporaries was one of considerable influence. Descartes' ambition to rest biology on a Keplerian foundation, to establish a mathematics of the life sciences while sparing the soul from materialism, was the ambition of the Gassendists but without the spiritual restrictions. Descartes' desire to combine physics and mathematics into an irrefutable body of knowledge was also an integral feature of the Gassendist program but with the rationalist element replaced by the experimentalism of Galileo. The Gassendists were the first natural monists of the modern era. When we refer to the origins of this perspective as "French Materialism," we are simultaneously acknowledging the role of Pierre Gassendi in the history of scientific psychology. He did less research than he inspired, but his inspiration was nearly without parallel.

Thomas Hobbes (1588–1679) and the Social Machine

The discussion of Hobbes could have been placed in either of the preceding two chapters as easily as in the present one. Texts devoted to the history of philosophy routinely list him as an early empiricist principally on the basis of such positions as that appearing in Part I of *Leviathan:* With the exception of prudence, which is grounded in experience,

> there is no other act of man's mind, that I can remember, naturally
> planted in him, so, as to need no other thing, to the exercise of it, but
> to be born a man, and live with the use of his five Senses.[21]

But, two chapters later, Hobbes goes on to distinguish between sense and reason, noting that the latter is not "gotten by Experience only," and that science,
which is a knowledge of consequences, is something more than the mere facts
of sense and memory.[22] Thus, while an epistemological empiricist, Hobbes is
comfortable with methodological rationalism. In fact, one of the important
sources of his scientific thinking was Galileo, whose hypothetico-deductive
method provided possibilities utterly lacking in the methodological empiricism
of Francis Bacon.

Leviathan appeared in 1651, when its author was already past sixty. Hobbes'
earliest writings and training were in classics. His translations of Thucydides
were authoritative. For five years (1621–1626) he served as a student-secretary
to Francis Bacon and, later, for a time, as a tutor to Charles II. He was involved
occasionally in the intellectual affairs of Father Marsenne's "libertines," and it
was through them that his visit to Florence and with Galileo was arranged
(1634–1637). Although late in appearing, Hobbes' influence on his contemporaries was far from negligible notwithstanding the fact that the list of luminaries active during his long life included Descartes, Locke, Newton, Gassendi,
Milton, and Galileo.

Leviathan is a long and uneven work, consuming seven hundred pages of
modern print and ranging topically from nutrition to demonology. Our present
concerns warrant attention only to those parts of it which advance and defend
the materialistic perspective on man and society. The work, as a whole, is suffused with this perspective, but it is in Part One (*Of Man*) that our present
interests are most repaid. Before turning to it, however, we might note the general aim of *Leviathan*.

The work appeared two years after the conclusion of the civil war, but before
Cromwell had been named Protector (1653) and while English intellectuals
were gravely weighing alternative possibilities for government. Hobbes had
exiled himself in France, where for a time (1646–1647) he tutored another
exile, the future Charles II. The execution of Charles I (1649) was followed by
a mixture of lingering anti-royalism and the guilt of regicide. We might say,
then, that *Leviathan* was greeted by mixed reviews. It argued for absolute monarchy, but it based the argument not on a king's "divine rights." The restoration
of the monarchy placed Charles II on the throne and so, at last, Hobbes' fortunes were secure. But the restored monarch still faced a divided country
sapped of material and moral energies, the very condition that had produced
England's long season of despair and that had encouraged Hobbes to write
Leviathan. Thus, as with Francis Brown earlier and John Locke later, Hobbes
is found attempting to oppose the voices of doom, the prophets of apocalypse:

> The *Present* onely has a being in Nature; things *Past* have a being in
> the Memory onely, but things *to come* have no being at all.[23]

His goal is to establish the principles by which, in a word, civil strife might be
averted. His method requires a determination of those aspects of human nature
that conduce to war and peace and those instruments of governance that will
work on and with this nature in such a way as to ensure national tranquility.
The metaphor of the machine is adopted (although it is to be doubted that
Hobbes considered it a metaphor), and the laws of society are to be fathomed
in the same way that Kepler and Galileo discovered laws in physics.

It was Hobbes' belief that a science of society could be established with the
same rigor and sureness enjoyed by the science of mechanics. Since humans
possess reason and thereby can come to know of cause-effect relations, there is
no excuse for their continuing to live in the perilous and uncertain way endured
by their ancestors. That they do so is simply to demonstrate that even a rational
creature is given to absurdities. Hobbes considered the cardinal absurdities to
include the following: (1) a lack of *method* such that we attempt to reason
before we have agreed on the meaning of the terms we are reasoning about;
(2) confusing the immaterial with the material such that we speak of "infusing
Faith," not recognizing that "nothing can be *powered* or *breathed* into any
thing but body; and that *extension* is *body*"[24], (3) a continuing belief in the
reality of universals; (4) the use of metaphor instead of reality; and (5) a devotion to scholastic terms such as *hypostatical, eternal-Now,* etc., which have no
basis in experience at all. He will have none of these. "Words may be called
metaphoricall; Bodies and Motions cannot."[25] Hobbes will speak of bodies and
motions only. In other words, if man is to be understood scientifically, he is to
be understood as matter in motion; if society is to be understood scientifically,
it is to be understood as men in motion. Hobbes' plan is to establish first a kind
of biophysical behaviorism and, from this, to deduce sociology.

The scientific analysis of man—that is, the biophysical behaviorism—begins in Chapter VI of *Leviathan*. The major variable—what might properly be
called the seventeenth-century variable—is *motion*. Hobbes examines animal
motion and notes that it takes two forms: *involuntary* ("vitall") and *intentional*
("animal"). The latter, which is the cause of all our grief and joy or very nearly
so, is the motion "first fancied in our minds" before being executed by our
limbs. While we cannot observe directly the source of this motion, it is none
other than motion within the interior parts of the body, motion seeking to satisfy needs of the body. Thus, the motion is caused by appetites that can be
subsumed under the label *desire*. Some appetites are native, such as those for
food and drink and the avoiding of pain. But the vast majority of our desires
"proceed from Experience."[26] The most basic desire, of course, is for the preservation of life. We judge ultimate good and ultimate evil in terms of our own

desires, and, since the chief of these is the desire for life itself, we will judge as good or as evil that which promises to preserve or that which threatens our lives. To live, we must have access to the necessities of life: nourishment and protection. Since resources are limited, we are, in the absence of government, in a state of war with our neighbors. We seek power in order to avert our own destruction. We value that which possesses power or confers power upon us, and this is true of the value we place on our fellow men. "The *Value* or WORTH of a man, is as of all other things, his Price, that is to say, so much as would be given for the use of his Power."[27] What we call *dignity* is no more than this. The commonwealth prizes a man for what he can do for them. His status, then, is neither absolute nor permanent. It lasts as long as his power does.[28] What is *honorable* is victory; dishonourable, defeat. Worthiness is fitness.[29] Our desire for power motivates us to search for causes, that is, to acquire science. When we lack this, we must content ourselves with the advice of others. It is in our search for ultimate causes that we discover God and, although we can never know the nature of God, we can and do infer His existence as a blind man infers that there *is* fire when told that fire warms us and when he then feels the warmth.[30] To the extent that religion grants power, it is judged as good and is desired. It will vary across cultures and be affected by the particular circumstances being faced by a given people.[31]

It is often surprising to the modern, Western reader who learns of Hobbes' devotion to monarchy and to the absolute powers of the state to discover that Hobbes arrived at this position from the assumption of the natural equality of all humanity. In fact, *Leviathan* is an object lesson to those who assume that a given philosophical or ethical bias logically entails a given political or social program. The argument for natural equality is presented at the beginning of Chapter XIII, and a lengthy quotation here is warranted:

> Nature hath made me so equall, in the faculties of body, and mind; as that though there bee found one man sometimes manifestly stronger in body, or of quicker mind than another; yet when all is reckoned together, the difference between man and man is not so considerable, as that one man can thereupon claim to himselfe any benefit, to which another may not pretend, as well as he. For as the strength of body, the weakest has strength enough to kill the strongest, either by secret machination, or by confederacy with others, that are in the same danger with himselfe.
>
> And as to the faculties of the mind . . . I find yet a greater equality amongst men, than that of strength. For Prudence is but Experience; which equall time, equally bestowes on all men, in those things they equally apply themselves unto. That which may perhaps make such equality incredible, is but a vain conceit of ones owne wisdome,

which amongst all men think they have in greater degree, than the Vulgar. . . . For they see their own wit at hand, and other mens at a distance. But this proveth rather that men are in that point equall, than unequall. For there is not ordinarily a greater signe of the equall distribution of any thing, than that every man is contented with his share.[32]

Because of this essential equality among men, there is no sufficient barrier to the assaults of one on another. That is, were there glaring inequalities in ability, the stronger would act with impunity; the weaker would submit without opposition. It is only because each has approximately the same chances of success as the other, each individual against each other and each group against another group, that the seeds of war are planted everywhere and for all time. Natural equality creates confidence and enmity. These, however, create the possibility of a violent death, dreaded by all. Thus, the desire for security calls upon each citizen to invest personal power in the authority of a monarch. The duty to the sovereign lasts as long and no longer than "the power lasteth, by which he is able to protect them."[33]

Leviathan, in a truly remarkable way, integrated the major developments in seventeenth-century philosophy on both sides of the English Channel. With Descartes, Hobbes identified matter with extension. Siding with the Gassendists, he insisted that only body can affect body and that only matter in motion can serve as the subject of scientific inquiry. While applauding Bacon's *Novum Organum,* he saw in the works of Galileo what true science was to become: a rational approach to nature in which theory serves always as the handmaiden of fact.

Hobbes, for reasons that must be clear, was greatly respected by the nineteenth-century utilitarians. His program was pragmatic, mechanistic, objective, and, in a subtle way, egalitarian. He could find no source of human conduct other than the desire to survive and prosper. We cannot be sure that the Epicurean revival launched by Pierre Gassendi was seminal in this respect, but the presumptive evidence is substantial. His psychology was materialistic, hedonistic, behavioristic. The mind, while not doubted, is subject to the laws of physics just as is the rest of the body! To know the causes of one's actions is to know one's desires, to know one's needs. Prime among these are the basic, biological requirements for survival. All other needs, acquired by experience and tradition, finally derive their force from these. In the introduction to *Leviathan* he wrote, "*Reward* and *Punishment* (by which fastned to the seate of the Soveraignty, every joynt and member is moved to performe his duty) are the *Nerves,* that do the same in the Body Naturall . . . *Concord, Health; Sedition, Sicknesse;* and *Civill war, Death.*"[34] Plato had examined the Republic in order to enlarge the canvas enough so that the nature of man could be discerned. Hobbes, too, saw the state as but an "artificial man." The difference, of course, is that Plato judged the ends of the state in terms of virtue; Hobbes, in terms of utility. Plato's metaphor was spirit; Hobbes', the machine.

Hobbes may be said not only to have initiated one of the first and most influential systems of materialistic philosophy in modern times but also one of the least compromised versions of *egoistic* ethics. His emphasis upon the motive for personal survival reappears continually as he addresses questions of morality. Hobbes is persuaded that what passes for benevolence, altruism, and other expressions of a principled regard for others must finally be analyzable in the terms of private gain and self-regard.

In Chapter 2, we paused to examine a classical defense of *egoism,* the one presented by Thrasymachus as an alternative to Socrates' ultrarational theory of benevolence. Thrasymachus can find no reason to believe that the holder of Gyges' ring will decline to avail himself of that which invisibility allows. Why would anyone defer the gratification of each and every desire except out of fear of reprisal? What reason has Smith to come to the rescue of Jones other than the possibility of reward or because he seeks to establish grounds for reciprocation?

The Socratic relaxation of the apparent tension between self-regard and altruism is based upon the theory that an enlightened individual can find happiness only in those actions that are good in themselves. Therefore, actions devoid of benevolence or indifferent to the needs of the state cannot produce happiness in the enlightened. Note, however, that this attempted solution plays readily into Hobbesian hands. Hobbes, even if he accepted the Socratic theory, is still able to insist that the happiness of the individual survives as the root-motive. Smith may, indeed, perform an act in the interest of the state but he does so because it brings him pleasure.

As with Hobbes' materialistic philosophy, his *egoistic* ethics would also undergo revision. Forms of it persist in all utilitarian theories and in those theories of value usually classified as "situation ethics." The eighteenth-century sentimentalist or "empathy" theorists—Bernard Mandeville and Francis Hutcheson have already been cited—could not eschew the appeal or the challenge of egoism. The individual might well be so framed as to possess an inborn concern for others, but what is this concern other than the pleasure produced by actions of a certain kind? And from the proposition that value-laden thought and conduct are inescapably selfish, it is a small step to the proposition that one must tailor for oneself that set of tastes, opinions, and dispositions conformable with one's own character, one's *ego.* Here, of course, are the seeds of individualism and liberalism. From these would grow the overarching aims of personal growth and personal success that define the commercial tone of Victorian England. It was these elements of Hobbes' ethics that would reappear in Tom Paine's leaflets, in the antiroyalist literature of revolutionary France, in the British Reform Act (1832), and nearly every day in our own time as the liberated citizen celebrates the multiplication of rights and the lessening of duties. If one were to judge from the persisting and even growing signs of disillusionment displayed by those who have received the greatest share of egoistic benefits, one might conclude that Socrates had sensed something of a deep and durable nature, something missed or misunderstood even by Hobbes. Can a human being find pleasure in acts which themselves are not good?

The Reflex

Ours is a time of scientific specialization and, as the term suggests, "professionalism." When we read a history of psychology, we expect to find the names of Freud and Wundt and the usual philosophers. We are ready to admire a Descartes who, with all his concern for philosophy, still had time to contribute to optics and geometry; a Locke whose interests in government did not prevent him from providing a theory of the mind; a Galileo whose imagination was able to embrace astronomy, mechanics, and the philosophy of science. Our admiration takes the form of orphaning these figures as "universal geniuses," exceptions to the specialist's rule. While this attitude is understandable, it is not correct historically. If Locke has been an influential figure in the history of psychology, and he certainly has, then so also was Newton. The scholars of the seventeenth century were not "professionals." In the important respects, they assumed the unity of science to an extent the modern epoch hardly approximates. Newton, for example, was driven to a theory of vision and, particularly, to a theory of color vision, because he believed that the universal law of gravitation was *universal* and that its effects should be as apparent in a biological system as in planetary motion. Borelli, a colleague of Galileo's, studied the physiology of muscles in order to apply Galileo's principles of mechanics to systems that just happened to be alive. Descartes and Hobbes both sought to base their epistemologies upon the new science that was mechanics. Descartes invented the physiological theory of reflexes with this in mind. To him and to his scientific contemporaries, it would have been inconceivable that there could be a "behavioral science" that was not, by the very fact, physics. Even Locke and Hume, disinclined to speculate on the biological basis of mind (disinclined, we may assume, because of the heavy weather suffered by Descartes at the hands of the Gassendists), both agreed that the issue was to be settled by "the anatomists." That is, while not addressing the question of materialism directly, Locke and Hume recognized it to be a *scientific* question. Hume was explicit:

> [W]hen the circulation of the blood . . . is clearly proved to have place in one creature . . . it forms a strong presumption, that the same principle has place in all. . . . [A]ny theory . . . of the understanding, or the origin and connection of the passions in man will acquire additional authority, if we find, that the same theory is required to explain the same phenomena in all other animals.[35]

Descartes' Optical Model of Perception

Descartes had said very much the same but had excluded animals from the domain of reason. Gassendi, however, had challenged this restriction, arguing that animals display memory, desire, and even a certain inductive capacity. Given the scientific climate of the second half of the seventeenth century, Descartes' dualistic path seemed to many to be too winding. When Newton's physics triumphed over Descartes', the latter's authority in all subjects fell into question, if not disrepute. His theory of the reflex had made action the consequence of stimulation and was designed to embrace a great part of all our conduct. Newton's laws, able to describe the motion of everything else, were judged by some to be immediately applicable to animal motion. (Recall Voltaire's remark, and it was Voltaire who brought Newton's work to the attention of the French.)

Descartes had envisaged the mechanism of the reflex to operate in the manner illustrated in the above figure. The arrow's image is conveyed by the optic nerves to the brain. The inverted image is set aright within the optic chiasm whence it passes to the pineal gland. Motor nerves receive "spirits" from the pineal gland in amounts determined by the force of the visual impression. Descartes considered the mechanism by which the nerves are actually energized by the spirits to be unfathomable.

The *Treatise of Man* was a work that Descartes would not allow to be published in his lifetime. The work first appeared in 1662, twelve years after his death and seven years after Gassendi's. It is his most biological work, and it is in this treatise that the lurking monist in Descartes comes close to surfacing. The *Treatise* invokes the idea of a model or "machineman" with a body like a statue. To this statue, anticipating Condillac, Descartes adds powers and functions by equipping it with sensory, motor, and reflex mechanisms. The essay includes the operation of feedback, emphasizes the muscular foundations of attention, and relates the quality of our experiences (or, the machine's experiences) to the action of the spirits flowing in and out of the pores of the brain. The spirits are rendered particulate, varying in mass and other qualities. As a result of these material differences, the spirits are able to mediate different emotions and the actions deriving from these emotions. Except for the faculty of reason, this machine simulates nearly perfectly a real man:

> I desire you to consider, I say, that these functions imitate those of a real man as perfectly as possible and that they follow naturally in this machine entirely from the disposition of the organs—no more nor less than do the movements of a clock. . . . Wherefor it is not necessary, on their account, to conceive of any vegitative or sensitive soul or any other principle of movement and life than its blood and its spirits, agitated by . . . those fires that occur in inanimate bodies.[36]

Newton's laws of motion were ideally suited to account for such a system:

1. Every body continues in its state of rest or of uniform motion in a right line unless it is compelled to change that state by forces impressed upon it.
2. The change of motion is proportional to the motive force impressed and is made in the direction of the right line in which that force is impressed.
3. To every action there is always opposed an equal reaction; or, the mutual actions of two bodies upon each other are always equal and directed to contrary parts.

Today we accept these laws as determining the manner in which objects will react to imposed forces. In the seventeenth century, these were laws of "science," potentially as applicable to the mechanics of mental life as to billiard balls. Descartes' notion of "animal spirits" was transformed into the *vis nervosa* (nervous *force*), and the laws of the reflex were relegated to the science of physics. Increasingly, English empiricism and French materialism looked alike. The Gassendists and other anti-Cartesians could assert the failures of Cartesian physics (while subscribing to his theory of reflexes), and the empirical "experimental philosophers" in England could begin the grand program of reducing philosophy to science. In both quarters, the concept of the reflex was central. British empiricism had been, since Locke's *Essay,* associationistic. By the eighteenth century it was more compelling to some to speak not of the association of ideas but of a more mechanical association in a more mechanical entity. Here and there, experiments began to address the physiology of this reflex and, hand in hand, these experiments and entire mechanistic theories of psychology marched toward the century of materialism, the nineteenth.

Perhaps the most significant discovery of the eighteenth century in regard to the evolution of psychological materialism was made by Luigi Galvani, who in 1786 reported the results of experiments on the stimulation of the muscles of frogs by the application of an electrical pulse. Until the essentially *electrical* nature of nerve-muscle interactions was discovered, the mystery in Descartes' system was preserved. David Hartley (1705–1757) had already published his *Observations on Man,*[37] which had presented psychology as the science concerned with the mechanics of associationism. Hartley's thesis was self-consciously Newtonian, and prematurely so. Thomas Reid was only one of several philosophers to dismiss Hartley's "science" as having no basis whatever in observable fact. Thus, only with Galvani's discovery was it possible to advance a mechanistic psychology equipped with an actual mechanism. Such a psychology needed the biological equivalent of the colliding billiard balls, something tangible and measurable. Galvani did not discover the neural impulse itself, but he did establish that muscles could be caused to contract by the application of an electrical charge. The effect was dismissed by a number of scientists—including the great Alessandro Volta—who insisted that the human body was not even able to conduct a charge. This complaint was ingeniously set aside by one of the more elegant demonstrations in the history of the neural sciences. A small boy was suspended by a long rope. Hung from a separate beam were leaves of gold, placed just in front of the boy's nose. Then, a glass rod was rubbed briskly with cat's fur and placed against the boy's bare toes. The gold leaves were immediately drawn to the boy's nose, and the theory of electrical conduction by the human body entered the realm of indisputable fact. Experiment again had triumphed over dogma.

At about the same time that David Hartley was insisting with Hobbes, but without evidence, that "each action results from the previous circumstances of the body and mind, in the same manner and with the same certainty as other effects do from their mechanical causes,"[38] Robert Whytt (1714–1766) was repeating an experiment that had been performed years earlier by the physiologist Stephen Hales. Hales had observed that the decapitated frog could be induced to move its limbs by pinching them. Even without a brain, the frog would avoid a painful stimulus delivered to the extremities. A surgical severing of the spinal cord, however, eliminated the response. In his essay "On the Vital and Other Involuntary Motions of Animals" (1751), Whytt argued that the spinal cord was able to mediate stimuli and responses in the absence of activity of the brain.[39] What Hartley was speculating, Whytt was demonstrating and in so compelling a way that a Newtonian biology seemed to be merely a matter of time.

The concept of "association," and its mechanical equivalent, the "reflex arc," had the same status in eighteenth- and nineteenth-century physiology as "attraction" had in the physics of the seventeenth and eighteenth. Hartley illustrates the inextricable tie joining the emerging psychology to the established physics in the opening chapter of *Observations on Man:*

> My chief design . . . is briefly to explain, establish, and apply the doctrines of *vibrations* and *association.* The first of these doctrines is taken from the hints concerning the performance of sensation and motion, which Sir Isaac Newton has given at the end of his Principia . . . the last from what Mr. Locke and other ingenious persons since his time have delivered concerning the influence of *association* over our opinions and affections.[40]

The essay then goes on to present ninety-one propositions regarding mental life, propositions embracing the simplest sensations and others designed to explain language, emotion, and dreams. All are based on the notion of a mechanical interaction between the organs of sense and the motor systems. Integrating these two systems is the brain. When we are stimulated, vibrations are established in the nerves and conveyed to the brain. After the stimulus has been removed, the vibrations persist, diminishing as a function of time. Memory, by this view, is the fading vibration induced by a bygone event. Learning is the establishment of a connection of what Newton had called "subtle forces" and Hume had described as the "gentle force": that is, a connection between *physical* events in the brain. The following theorem very nearly says it all:

> If any sensation A, idea B, or muscular motion C, be associated for a sufficient number of times with any other sensation D, idea E, or muscular motion F, it will, at last, excite d, the simple idea belonging to the sensation D, the very idea E, or the very muscular motion F.[41]

Hume had said as much. Events occurring together, in unaltered sequence, often will come to be treated as causally related. Descartes, in describing the mechanism of nonspiritual action, had already specified the anatomical arrangements by which such sensory-motor events take place. Newton's "vibratiuncles" were but the earliest form of a *vis nervosa*. Hartley even finds support in Leibniz's "pre-established harmonies," which free the philosopher from having to explain how an immaterial soul directly activates a material body. Hartley's essay is, then, one of the great syntheses of the eighteenth century. He combines the physics (and method) of Newton with the physiology of Descartes; he avoids theological dilemmas by invoking the parallelism of Leibniz; he takes for granted the empirical associationism of Locke and Hume; he assumes, as Hume and every cobbler in England did, that pleasure and pain are the glue that binds our associations. Perhaps more than any single work previously published, Hartley's *Observations on Man* is a treatise in *modern* psychology. Its influence in the nineteenth century, as we shall see in the next chapter, was considerable.

"L'Homme Machine"

The luminaries of the French Enlightenment, whether atheists or theists, whether monarchists or anarchists, all shared in their contempt for dogma. Voltaire was the center and the engine of the movement. He admired England, wrote glowingly of the English people, praised them for their egalitarianism, their opposition to the Roman Church, and their tolerance. While history reveals that Voltaire's plaudits were overly generous, there is no doubt but that the Catholic influence in eighteenth-century France was oppressive. The Enlightenment in France, which cannot be said to have produced philosophic perspectives of lasting significance, was essentially a *political* movement posing as an intellectual enterprise. It is not surprising that the most original philosophical thinker in the group was Rousseau, and that his only important contribution to philosophical thought is to be found in his additions to the "social contract" theory.

As political figures, the men and women of the Enlightenment saw in the new science and philosophy of their age a weapon that could protect them against the excesses of orthodoxy and, more, could drive a wedge between the received truths of religion and the complacent mass of believers. Politicized philosophy, like "political art," may have sudden effects, but not enduring ones. It may move people, but it does not advance the discipline in which it pretends membership. Voltaire himself introduced Newton to the French intellectuals but never fully comprehended Newtonian physics. Much of his competence in this subject was the gift of Mme. du Chatelet. This is not to say that Voltaire merely translated the *Principia* and chanted its laws. Rather, it is only to note that Voltaire's real interests in Newton's physics were extrascientific. He found in Newton another reason to oppose Cartesianism. The same may be said of Condillac (1715–1780), who translated Locke's *Essay* into French and whose *Essay on the Origin of Human Knowledge* remains one of the best analyses of Locke's empirical philosophy. It was Condillac who proposed the "sentient statue," standing in the path of a universe of stimuli, taking on psychological characteristics as a result of experience. It was Condillac's "statue" that served as the template for much of the materialistic philosophy of late eighteenth- and early nineteenth-century France. But Condillac was not particularly interested in the material basis of mind, and less in physiology. Nor was he simply a disciple of Locke's.[42] Rather, he was captivated by the anti-

metaphysical alternative of the English school, and by the philosophy of experience celebrated by Voltaire and no less so by Condillac's own cousin,
d'Alembert. Again, the pull was not simply toward England but away from
Descartes. Condillac was a priest. His enemy was clearly not the Church. His
enemy was dogma, and especially philosophical dogma disguised as religion.

We have seen that Aristotelianism was an early casualty of the French religious wars. Galileo's "new sciences" were judged by the men and women of
the Enlightenment as the definitive refutations of Peripatetic philosophy, although it is to be noted that their understanding of Galileo's physics was fuller
than their understanding of Aristotle's *Metaphysics*. Cartesianism soon loomed
as authoritarian as the *ism* so recently rejected. French intellectuals were now
unanimous in their opposition to any assertion that flew in the face of fact and
to any proposition not supported by the data of experience, excepting only
those propositions set forth softly and humbly in the language of simple faith.
No longer would they accept Thomism or Cartesianism as scientific. More significantly, they would no longer base government, social discourse, or the everyday affairs of life on foundations whose only claim to attention was custom
or authority.

It is frivolous to date an idea and reckless to date large social movements.
The eighteenth century was a century of revolution, but so was the sixteenth
and, indeed, the seventeenth. The social, political, and economic changes that
began most visibly in that period we so easily isolate as the Reformation are
changes still taking place in our own time and often with revolutionary zeal
and revolutionary consequences. Thus, it can only be apologetically that one
advances the thesis that Gassendi, Newton, Voltaire, Rousseau, and their lesser
contemporaries created a revolution in perspective that culminated in a revolution of classes. It is better and more accurate to put the thesis this way: The
Reformation was an attack on institutions and this attack came to embrace the
very idea of authority. With Hobbes, there was introduced the summoning utilitarian notion that the justification of governments is to be found in the safety
they confer on human life. Galileo and Descartes, as scientists, contributed to
the demise of that Aristotelian bastion which, for so long, had protected the
claims of institutional authority. Descartes' additional contribution was that of
publishing his major works in vulgar French, a policy embraced by every major
French philosopher to follow. Locke, Hume, and their Continental counterparts, Gassendi and Condillac, translated the social reality into enduring philosophical systems. This is not to say that they were apologists for a movement.
It is, however, to avoid the suggestion that they were the causes of the movement. They were, we prefer to say, participants and very influential ones at that.

We noted at the beginning of Chapter 7 that empirical philosophy does not
entail, in the logical sense, materialism. (Berkeley was an ungrudging empiricist and a devout immaterialist.) But for the person who believes that the mind

is furnished by experience alone, for the philosopher who can find no source of knowledge beyond the senses, the step to psychological materialism is short, logic to the contrary notwithstanding. Condillac's metaphor of the statue is a *material* metaphor. Hartley's vibrations—never witnessed by Hartley or anyone else—were not even presented as metaphor but as unseen reality. Recall that Locke and Hume specifically eschewed the temptation to theorize on the material causes of sensation and association. Hartley, a contemporary of Hume's, already displayed the attraction materialism always holds for the empiricist. Hobbes succumbed to it on the authority of Galileo; Hartley, on the authority of Newton's physics. In fact, there is a sense in which Descartes' dualistic psychology is understandable in terms of its not being an extension of an empirical philosophy.

In 1748 a book appeared bearing the title *L'Homme Machine*.[43] Its author, Julien Offray de La Mettrie (1709–1751), was a physician, a self-exiled member of the court of Frederick the Great, and the boldest psychological materialist France had produced. The book caused a commotion even greater than that produced by his *Historie Naturelle de l'Âme* (1745), whose sudden infamy had caused La Mettrie to move to Bavaria. Both books were relatively short, highly polemical, and hardly either philosophical or scientific as these adjectives are properly assigned. It is in *L'Homme Machine* that the soul is reduced to "an enlightened machine" and the human psychological faculties reduced to mere brain physiology. La Mettrie's "proofs" for these assertions are drawn from such otherwise irrelevant observations as the ability of a chicken to continue to run after its head has been removed, and the fact that the freshly removed heart of an animal will "jump" out of boiling water. His authorities include military officers who have told him tales of battle casualties but also include several of his own clinical observations. There is nothing in either book that the modern neurophysiologist or physiological psychologist would take very seriously, and less what modern philosophers with their wits about them would pause to reread. The importance of *L'Homme Machine* is historical but not because it was especially influential, beyond the influence enjoyed by any book officially condemned. It is important in its style: porous, haughty, devoid of self-critical strains. It is the sort of book one writes on a dare or as a challenge. It is defiant but seldom clear on the issue or opinion to be defied. If the enemy were those who insisted that man is endowed with an immortal and incorporeal soul, the fact that chickens run after decapitation is utterly irrelevant. If the antagonist was the Cartesian who believed that the will is directed by the soul and that the soul directs its influence from a region of the brain, the behavior of hearts dropped into water is beside the point. La Mettrie, in summarizing the respects in which man is matter, went on to insist that he is only matter, but this, as we know, cannot be established by materialist methods without committing the fallacy of *petitio principii.*

If La Mettrie had a direct influence on his successors it was on Pierre Cabanis (1757–1808), one of Napoleon's senators and a leading light in the French materialist movement. Both La Mettrie and Cabanis, however, have received more attention in histories of psychology than history warrants. It was Gassendi and the Gassendists who founded psychological materialism. Even Hobbes cannot compare to them in immediate influence. La Mettrie and Cabanis, both physicians and both lacking in that subtlety of mind that philosophy demands, contributed little to the historic debate between materialists and dualists, and nothing to the science spawned by the former. Both were products of the Gassendist tradition and both displayed the neo-Epicurean attitude of Gassendist philosophy. For both, happiness and morality, no less than sensation and action, were to be comprehended in material terms and were subjects of scientific analysis. Neither authored a bona fide materialist system. Indeed, neither had Gassendi. The system, of course, was there all along. It was the system of Galileo and Newton. But unlike Galileo and Newton, the French materialists were not physicists, were not incisive philosophers and, by and large, were not even scientists. Uncritically, they assumed psychophysical isomorphism and were thereby convinced that the laws of physics, the laws of society, and the laws of psychology were but three versions of the same material principles. Ironically these writers—d'Alembert, La Mettrie, Cabanis—are closer to a dominant contemporary perspective than were any of their more illustrious brothers. What they sensed, and what either eluded or embarrassed or failed to intrigue Locke, Hume, Leibniz, and Kant, was the growing possibility of wedding philosophy to anatomy. They were an eager group, forcing compatibilities where there was no ground other than enthusiasm for even speculating on the compatibilities. They saw into the next century.

The Scientific Context

Psychological materialism did not appear in a vacuum, nor is it to be understood simply as a consequence of the Gassendist rebuke of Cartesianism. We have devoted most of our attention to Galileo and Newton, by far the most important scientists of the seventeenth century and the men who gave the greatest impetus to eighteenth-century science. But there were many other, though smaller, contributions to the scientific spirit. La Mettrie, for example, studied with the great Dutch physician Hermann Boerhaave (1668–1738), whose treatises in chemistry had rejected vitalism and had insisted that all life processes were reducible to chemical processes. Antoine Lavoisier (1743–1794) convincingly demonstrated the conservation of matter in an experiment in which the steam from boiled water was condensed and collected. The notion of matter as being "neither created nor destroyed" is a late eighteenth-century one (discounting, as usual, the ancient Greek writers) and one with obvious religious implications. The same Lavoisier advanced the theory of chemical elements, and, in a philosophical climate, this theory held out the promise of reducing the mysterious diversity of nature to its building blocks. We might underscore the growing materialism and scientism of La Mettrie's era by noting, without discussing, the scientists who were active and prominent at the time: Joseph Black (1728–1799), Charles Coulomb (1736–1806), Benjamin Franklin (1706–1790), Luigi Galvani (already cited), Stephen Hales (1677–1761), Albrecht von Haller (1708–1777), Karl Linnaeus (1707–1778), and Joseph Priestley (1733–1804). It was the scientific achievement of this group that convinced some of the finest minds of the nineteenth century that the abiding questions could be answered at last and answered positively.

Notes

1. David Hume, *A Treatise of Human Nature,* Book I, Pt. I, Sec. II, edited by L. A. Selby-Bigge, Clarendon, Oxford, 1973.

2. Immanuel Kant, *Critique of Pure Reason,* translated by Norman Kemp Smith, St. Martin's Press, New York, 1965.

3. Galileo Galilei, *Dialogues Concerning Two New Sciences,* translated by Henry Crew and Alfonso de Salvio, Macmillan, New York, 1914.

4. Ibid., p. 110.

5. Galileo Galilei, *Dialogue Concerning the Two Great Systems of the World,* in *Classics of Modern Science,* edited by William S. Knickerbocker, Appleton-Century-Crofts, New York, 1927.

6. Ibid., *The Second Day.*

7. Isaac Newton, *Philosophae Naturalis Principia: I. The Method of Natural Philosophy,* in *Newton's Philosophy of Nature,* edited by H. S. Thayer, Hafner Publishing Co., New York, 1953.

8. Voltaire, *The Ignorant Philosopher,* cited in *A History of Modern Science,* W. C. Dampier, Cambridge University Press, Cambridge, 1966, p. 197.

9. Descartes' response to the pamphlet includes the arguments of the pamphlet itself. It appears in Vol. I of *The Philosophical Works of Descartes,* translated by Elizabeth S. Haldane and G. R. T. Ross, first published in 1911 by Cambridge University Press and republished by Dover Publications, New York, 1955.

10. Ibid., pp. 433–434.

11. Ibid., p. 339.

12. Ibid., p. 442.

13. Ibid., p. 443.

14. *The Selected Works of Pierre Gassendi,* edited and translated by Craig R. Brush, Johnson Reprint, New York, 1970. The introductory remarks by Professor Brush are especially useful in establishing the context in which Gassendi's works were written.

15. Ibid., p. 174.

16. Ibid., p. 173.

17. Ibid., pp. 189–190.

18. Ibid., pp. 194–195.

19. Ibid., pp. 266–268.

20. Ibid., pp. 269–275.

21. Thomas Hobbes, *Leviathan*, Pt. I, Ch. 3, Pelican Classics, Penguin Books, England, 1974. This is the edition edited and discussed by C. B. Macpherson.

22. Ibid., Pt. I, Ch. 5.

23. Ibid.

24. Ibid.

25. Ibid., Pt. I, Ch. 6.

26. Ibid.

27. Ibid., Pt. I, Ch. 10.

28. Ibid.

29. Ibid.

30. Ibid., Pt. I., Ch. 11.

31. Ibid., Pt. I, Ch. 12.

32. Ibid., Pt. I, Ch. 13.

33. Ibid., Pt. II, Ch. 21.

34. Ibid., Introduction, p. 81.

35. David Hume, *An Enquiry Concerning Human Understanding,* Sec. IX, in *Essential Works of David Hume,* edited by Ralph Cohen, Bantam, New York, 1965.

36. René Descartes, *Treatise of Man,* French text with translation and commentary by Thomas Steele Hall, Harvard University Press, Cambridge, 1972. This excellent and much needed edition includes important notes by Professor Hall that establish the pre-Cartesian authorities upon whom Descartes relied in constructing his psychobiological theory.

37. David Hartley, *Observations on Man,* in *Between Hume and Mill: An Anthology of British Philosophy 1749–1843,* edited by Robert Brown, Random House, Modern Library, New York, 1970.

38. Hartley, *Observations on Man,* p. 84.

39. Robert Whytt, *An Essay on the Vital and Other Involuntary Motions of Animals,* Hamilton, Balfour, and Neill, Edinburgh, 1751.

40. Hartley, pp. 5–16.

41. Ibid., p. 29.

42. Etienne Bonnot de Condillac, *An Essay on the Origin of Human Knowledge, Being a Supplement to Mr. Locke's Essay on the Human Understanding.* The most available edition is the facsimile reproduction of Thomas Nugent's translation in 1756. This facsimile, with an introduction by Robert Weyant, is published by Scholars' Facsimiles and Reprints, Gainesville, Florida, 1971. Note especially, Professor Weyant's discussion on pp. 17–18.

43. Julien Offray de La Mettrie, *L'Homme Machine,* Leiden, 1748. An English translation by M. W. Calkins is available.

3 Scientific Psychology

10 The Nineteenth Century: The Authority of Science

A Note on the Debt to the Nineteenth Century

Contemporary psychology, in its broadest features, remains a nineteenth-century enterprise. This is by no means to say that modern psychology is "old-fashioned" or behind the times. Still, it must be noted that the problems that consume the energy of the contemporary psychologist were either set forth explicitly in the nineteenth century or were introduced by those whose educational and cultural backgrounds were provided by the unique perspective of the nineteenth century. Of those who can lay claim to a significant contribution to the manner in which the contemporary psychologist advances the discipline, only B. F. Skinner (1904–1989) was born in this century. All the rest—Lashley, Piaget, Freud, Adler, Hull, Pavlov, Tolman, Köhler, Watson, Dewey, and James—are products of the century now under consideration. This fact is not a mere fact, but a suggestive and a revealing one. Modern psychology is not "modern" in the sense that modern physics and modern biology are. Recent discoveries of the molecular biology of the gene have transformed genetics into a discipline that Mendel scarcely would recognize, and the general theory of relativity has required the contemporary physicist to view the Newtonian universe through Einsteinian lenses.

In psychology, the situation is really quite different. The full range of its theoretical problems—from the study of personality and child development to inquiries into the neurophysiological basis of emotion or language to attempts to comprehend the determinants of social and national movements—can be traced directly to the thoughts and experiments of the psychologists and "natural philosophers" of the nineteenth century. Physicists no longer test the validity of Ohm's law or seek to discover whether, in fact, Maxwell's displacement current exists. They do not commit their lives to repeating those experiments that confirm the law of the conservation of energy or of angular momentum. Nor do they seek the "aether," or insist that light must be either particulate *or* wave-mechanical. Physics still has problems, very substantial problems, and some are rooted in the nineteenth century. But it is not nineteenth-century physics that is brought to bear upon these problems now. The same may be said of chemistry and the more developed branches of biology. But in contemporary psychology, not only have the problems of the nineteenth century survived, but so have many of the methods developed in that century. More important than even this fact is that the contemporary perspective is largely the one bequeathed by the scholars of that time. This will be made clearer in the next chapter. For now, we need only remain mindful, as we explore the psychological endeavors of nineteenth-century philosophers and psychologists, that our subject is only partly historical.

The Legacy of the *Philosophes*

The great British philosophers who followed in Locke's path had a decidedly scientific orientation. They were all admirers of Newton and they all either recommended the new "experimental philosophy" or actually engaged in it. Some, such as Hartley, went the route of biological psychology; others, like J. S. Mill, pursued the philosophy of science. In France, too, the disciples of Locke and Gassendi included philosophers of a seriously scientific nature, addressing their energies to specific, experimental questions. In the abiding tension between nature and spirit, they stood foursquare behind naturalism. But France, in the second half of the eighteenth century, hosted more than a group of industrious scientists and thoughtful philosophers. It hosted the *philosophes* of the Enlightenment, who had a greater effect upon the manner in which the average citizens of Paris viewed themselves and their world than did the writings of any "respectable" philosopher. The intellectual groundwork for the French Revolution was not accomplished by Descartes, less by Locke and Newton. Rather, the groundwork was done by men and women who were literary people, not philosophers or scientists. It was done by dramatists, lawyers, and, as they called themselves, *dilettantes.* The most famous members of the group were of course, Voltaire, Diderot, Rousseau, Condorcet, and D'Alembert.[1] Helvetius and Baron D'Holbach, though not members of the inner circle, derived inspiration from and voiced many of the central elements of the program of the *philosophes.* If we are to judge J. S. Mill as the nineteenth-century culmination of the empirical tradition created by Bacon, Hobbes, Locke, Newton, and Hume, then we must locate the source of Auguste Comte's *Positive Philosophy* in the French Enlightenment. Mill and Comte will occupy our attention shortly. But before turning to the specific and formalized contributions the nineteenth century made to psychology, we should have the major themes of the Enlightenment before us.

In the previous chapter, we noted that the French materialists were engaged less in science than in politics or ideology. Accordingly, the works of La Mettrie, D'Holbach, and others had little influence on science in their own time or thereafter. The same must be said of the "encyclopedists." Neither D'Alembert nor Diderot contributed a method or a set of findings that served as a starting point for any major effort in science or in psychology. But, collectively, the *philosophes* did provide the starting point for all subsequent departures from orthodoxy. They established a liberated way of thinking and did this with such wit, craft, imagination, and penetration that defenders of the status quo inescapably invited ridicule. No single work summarizes their program nor can one be said to have launched the movement, but Voltaire's *Letters Concerning the English Nation*[2] came close to accomplishing both. As with a good many books of the Enlightenment, this one, too, was ordered burned by the *parlement* (1734). The *Letters* are written in an offhand way, perfectly suited to the tastes and the time of busy and influential Frenchmen. They include praise for the pioneering efforts of Descartes but an utter dismissal of his metaphysics. Newton is offered as the master, and Bacon as his harbinger. The edition of the *Letters* condemned by the *parlement* concluded with a waspish assault on Pascal's *Pensées*. Pascal's superstitions, his historical errors, his reliance on chance, his gloominess, are all pricked by the sharpest sword in Europe.

Voltaire died in 1778. For nearly fifty years, his writings and very personality were the hub from which the scholarship of the Enlightenment radiated. He was enormously wealthy, and his influence was further increased by a close friendship with Frederick the Great of Prussia, the poet-tyrant and most liberal king in Europe. Voltaire's example gave confidence to Diderot (1713–1784), whose *Encyclopédie* suffered a painful on-again, off-again experience at the hands of the *parlement*. Indeed, many of the important works banned in France saw their way to Prussian publishers through the intercession of Voltaire. In his assertion of the superiority of the Lockean-Newtonian philosophy to that of Cartesianism, Voltaire set in motion a spirit of sensualism that led to Condillac and to Helvetius. The latter (1715–1771) published *A Treatise on Man, His Intellectual Faculties and His Education*,[3] which comes as close to twentieth-century environmentalism as any work written before 1900. It is in the *Treatise* that hereditary differences are dismissed as negligible and that the effects of training, reward, punishment, and experience are accorded first place in determining the character and accomplishments of the individual.

Voltaire also played a part, through Diderot, in the formation of Baron D'Holbach's materialistic philosophy, a materialism that soon took the form of polemical atheism. It was Holbach (1723–1789), as much as La Mettrie, who found the mechanistic portions of Descartes' psychology sufficient to account for morality, emotion, intellect, and language. It was Holbach who railed so vituperatively against established religion that the entire circle of encyclopedists were soon treated as atheists, en masse.

Voltaire is credited with writing some twenty thousand letters to more than a thousand different correspondents. His plays moved the masses, his ideas, the *philosophes.* His, more than any other single voice in the eighteenth century, spoke the cause of freedom, reason, law, and humanist ethics. He believed in God but was not religious. He believed in science but made no contribution to the literature of science. As with Diderot and Helvetius, he had received a Jesuit education and respected, all his days, the authority of reason. Many feared that he learned his lessons too well! He was no revolutionary, nor was he a "democrat" as that term is now employed. He insisted, however, that the ultimate validity of any government is rooted in its contributions to the welfare of its citizens, an idea that would become immortalized in Rousseau's *Contrat Social.* He, the encyclopedists, the growing power of the middle class, and the conflicts between king and *parlement,* between *parlement* and church, between Jesuit and Jansenist—these were the seeds of revolution and reform. All the leading figures of nineteenth-century science and natural philosophy looked back to the scholars of the Enlightenment for support and inspiration.

We may summarize what they found when they looked back. First, the idea of progress. In the works of Voltaire and, most particularly, in the rational materialism of Diderot and Condorcet, we discover repeatedly the notion of personal and cultural evolution. It is in *D'Alembert's Dream*[4] that Diderot presents the view of the whole as a collection of material parts, of the statue gaining its life from material reduction and subsequent evolution. And Condorcet (1743–1794), in his *Sketch for a Historical Picture of the Progress of the Human Mind,*[5] written as its author hid from the vengeful zealots of the Revolution—that is, from those whose new freedoms Condorcet had labored to secure—set to paper the idea that powered the entirety of the nineteenth century: the idea of progress.

Second, the idea of nature. If Voltaire, Diderot, D'Alembert, Condorcet, D'Holbach, Rousseau, and the rest can be said to have agreed on any single point—and the disagreements among the group were considerable—that point is philosophical naturalism. The world and everything in it are matter. The world is to be comprehended as matter in motion. Human reason, by which this comprehension becomes possible, must aim itself at nature and unearth nature's laws. To the Pascals who insist that we can never know everything. Voltaire's reply rings down the decades to our own day:

> Let us console ourselves for not knowing the possible connections between a spider and the rings of Saturn, and continue to examine what is within our reach.[6]

Included in the idea of nature was the idea of natural law as applying to all spheres of reality. It was in this same period that Turgot and the *physiocrats* (*physis* = nature; *krateo* = power, sovereignty) argued for a "free market" economic policy whereby the "law" of supply and demand would set the "natural" value on goods and labor.

Third, the idea of personal freedom. Rousseau's greatest essay begins with the haunting picture of man, *born free and everywhere in chains.* Here is the spirit of the Enlightenment, transported by an Englishman to America and translated as *The Rights of Man.* Tom Paine would even become an elected member of the post-Revolution Assembly, notwithstanding the fact that he scarcely knew a word of French.

The Legacy of Kant

Kant died at the beginning of the nineteenth century, but he is far more a member of the Age of Reason than he is the Age of Materialism. He set the tone of German philosophy for the entirety of the nineteenth century and, as a result, German "naturalism" would be ever barren of the strong empiricist elements of the naturalistic philosophies of France and England. German philosophy, that is, would come to accentuate the transcendentalism of Kant.[7] The combination of naturalism and transcendentalism yielded that uniquely German creation, Romantic Idealism, whose principal architect was Johann Wolfgang von Goethe (1749–1832).

It was the poetry of Goethe that animated the German-speaking world with the spirit of *Sturm und Drang.* His *Sorrows of Young Werther* (1774) resulted in an epidemic of suicides. Naturalism, in the hands of Goethe, was the panpsychism of Leibniz, Spinoza's creative forces. Man stands hopelessly alone in an immense universe, torn by the stresses of life, and finding himself only through his own activity, through his *life.* The goal of life is living of it, the activity and personal evolution, the reaching upward and outward. Life is love and passion. Kant's rejection of teleology was all Goethe needed to convince himself that the only end of human life is the activity that defines it. Nature is but a system of opposing forces: life and death, light and dark, love and hate.

Empiricism in the Nineteenth Century

We concluded Chapter 7 with a discussion of the Utilitarian philosophy of Jeremy Bentham and observed that this system would be modified and promulgated by many, particularly J. S. Mill (1806–1873). We pick up the evolution of empiricism—this time as empirical *psychology*—with Mill.

There is no need to go into Mill's "conversion" to Benthamism and philosophical radicalism at age fifteen. His *Autobiography*[8] is widely quoted, and the missionary labors of his father, James Mill, in behalf of Benthamism have already been extolled. Mill never abandoned the broadest features of utilitarianism, but he did recognize that Bentham's founding version of it was insufficient to meet the demands of the time. Unlike Bentham, who came to philosophy by way of law and economics, Mill's background was in classics, logic, and science. Where Bentham wrote in the grand style of the eighteenth-century universalists, Mill was more systematic and far less intuitive. His era was far more complex than Bentham's and he appreciated the complexities. Thus, we do not find the Mill of *On Liberty* attempting to "prove" the validity of freedom in terms of physics, or mathematics, or logic. Nor do we find the Mill of the *System of Logic* defending the inductive method in terms of economic or social considerations. Mill is a distinctively nineteenth-century gentleman, sophisticated, educated, liberal, and urbane. He is, in short, a modern as we would apply the term.

J. S. Mill's most important work in psychology was his *System of Logic* (1843), which was immediately successful and, through eight editions in his own lifetime, served as the handbook of the scientific community. The bulk of the work was devoted to a description of the principles of induction, the failure of purely rational approaches to matters of fact, the methods to be employed in establishing valid inferences, and the role of deductive processes in science. Contemporary science continues to rely on Mill's "methods," and our own easy commitment to the hypotheticodeductive method can be attributed to the immediate hold the *System of Logic* took on the scientific mind. Not many scientists in the late nineteenth century read Galileo; all of them read Mill.

The portion of the *System of Logic* most important to the appearance of experimental psychology is Book VI and especially Chapters III–VIII.[9] Having laid the foundation for all science, Mill now addresses his attention to human nature as a subject of science:

> It is a common notion, or at least it is implied in many common modes of speech, that the thoughts, feelings, and actions of sentient beings are not a subject of science. . . . This notion seems to involve some confusion of ideas, which it is necessary to begin by clearing up. Any facts are fitted, in themselves, to be a subject of science, which follow one another according to constant laws; although these laws may not have been discovered, nor even be discoverable by our existing resources.[10]

Mill offers an illustration from meteorology. We are not able, he says, to specify all the antecedent variables to produce rain, but we all agree that rainfall is the result of laws of nature. Thus, forecasting the weather is a probabilistic endeavor and may always be. That is, it may never be an *exact* science, but it is a science nonetheless.

> The science of human nature is of this description. It falls far short of the standard of exactness now realized in Astronomy; but there is no reason that it should not be as much a science as . . . Astronomy.[11]

He notes that the subject matter of psychology embraces the thoughts, feelings, and actions of human beings and that these cannot be predicted with anywhere near the accuracy as that obtaining in astronomy. This limitation, however, does not indicate that psychology cannot be a science or even that it is not a science. We cannot foresee every circumstance in which an individual might be found, and the factors that conspire to form the individual character are so diverse that even if we knew these future circumstances, we might still not be able to predict how the individual will act. Nevertheless we must still accept that actions, feelings, and thoughts are caused, and that the causes are natural and, to that extent, knowable in principle. Some of the laws are, in fact, known already, according to Mill, and central among these are the laws of association, which Mill summarizes as follows.[12]

First, there is Hume's law according to which every mental impression has a corresponding idea. Once we have experienced X, we are able to recall X without its actual presentation. We are so constituted that we are able to form an idea or mental image of that which we have perceived.

Second, there is the law of connection such that the repeated, simultaneous (or immediately successive) presentation of two stimuli leads us to think of either when the other is later presented. This law, cited by all the empiricists, is most ably described, Mill tells us, by one James Mill, whose *Analysis of the Phenomena of the Human Mind* presents the law "with a masterly hand." The third law is one that renders the intensity of a stimulus relatively interchangeable with the frequency of its presentation. According to this law, a very intense X has the same effect upon the mind as does a weaker Y that has been presented more frequently.

> These simple or elementary Laws of Mind have been ascertained by the ordinary methods of experimental inquiry; nor could they have been ascertained in any other manner. But a certain number of elementary laws having thus been obtained, it is a fair subject of scientific inquiry how far those laws can be made to go in explaining the actual phenomena. It is obvious that complex laws of thought and feeling not only may, but must be generated from these simple laws.[13]

Mill's empirical psychology was not of the radical sort. He certainly looked to education and the general cultural environment for the causes of individual differences, but he was equally persuaded that organic factors might be and probably are responsible for some of the more dramatic differences. He noted the advances being made in neurophysiology and neurology and believed that in time we would have a much better understanding of the relationship between brain physiology and the laws of the mind. However, as with Locke and Hume, he refused to take a position on the material basis of thought and noted that Hartley and his own father had displayed more confidence in the materialist account than was justified by available data.[14]

There is a second respect in which Mill's empirical psychology was conservative. He tended to doubt, notwithstanding the fact that psychology could and would develop empirical laws, that the science of psychology would ever be able to go beyond the given context in its attempts to predict human conduct, thought, and feeling. By an "empirical law," Mill meant a law of regularities: Y follows X or is coincident with it in this setting, and we may infer that Y will follow X or be coincident with it in any setting greatly resembling this one. However, in a very different context, we are unable to specify precisely what the relationship, if any, will be between them. Since the human condition is ever changing, since history ordains ever different contexts, the laws of psychology will be empirical and of restricted generality.[15] Through the empirical laws, which allow us to predict the actual facts of conduct or thought or feeling in a given and restricted context, it is possible to deduce more general laws. These will not be merely empirical but will be exact. However, the price we pay for these exact laws is that they will not apply to facts but to tendencies. That is, from the empirical laws of association, we may inexactly predict that Henry Jones will form a stronger association between intense stimuli than he will between weak ones. There may be a race of men about whom this is not true; yet, from this "situational law," we can deduce that, if there is thought at all, associations will be formed. Note that the law of associations does not predict or even attempt to predict the *exact* result of an experiment. Rather, it refers to the tendency of something to occur, other things being equal. That other things are never equal in the affairs of man is only to say that our exact laws will not be testable, not that they are not laws. Mill gave (invented) the name "ethology" to cover this science, which is deduced from the empirical laws of psychology. As conceived by him, ethology was to be the science of character, or that discipline concerned with the effects of environmental conditions on the laws of thought, feeling, and conduct. It was to be "the Exact Science of Human Nature," whose propositions are "hypothetical only, and affirm tendencies, not facts,"[16] Modern ethology, of course, bears only a slight resemblance to Mill's expectations. This is not the occasion to examine the subtleties of Mill's plan for ethology, but we must, before leaving the point, draw attention to the ambiguous distinction between that which affirms a tendency and that which affirms a fact. It is worthwhile to draw attention to Mill's choice of words here because the *philosophical* behaviorism of several twentieth-century writers (e.g., Ryle) has come to rely heavily on the notion of dispositions and tendencies as opposed to observable facts.

The very mention of behaviorism, perhaps by one of those laws of association, brings one to Mill's version of utilitarianism. Over and against Kant's categorical imperative, which Mill sees as potentially admitting "the most outrageously immoral rules of conduct,"[17] Mill adopts the "happiness theory":

> Questions of ultimate ends are not amenable to direct proof. Whatever can be proved to be good, must be so by being shown to be a means to something admitted to be good without proof.[18]

We judge medical science to be "good" because it conduces to health, but we have no way of showing that health itself is good. We accept this without proof, its ultimate sanction coming from the fact that all or very nearly all humankind will attempt to possess it. To those who reject utility as being no more than the measure of goodness employed by the beasts, Mill counters that there is nothing in utilitarianism that declares that human happiness is limited to pleasures of the flesh or bestial wants. Utilitarianism accepts the facts and the needs of the human intellect and recognizes that, for humans, the greatest happiness is not confined to mere biological gratification. Pleasures differ in quality as well as in quantity, and no utilitarian denies it:

> If I am asked what I mean by difference of quality in pleasures, or what makes one pleasure more valuable than another merely as a pleasure, except its being greater in amount, there is but one possible answer. Of two pleasures, if there be one to which all or almost all who have experience of both give a decided preference, irrespective of any feeling of moral obligation to prefer it, that is the more desirable pleasure.[19]

Students of modern behaviorism will recognize in this passage the forerunner of that notion according to which it is the organism, not the psychologist, who defines the reinforcer. Pleasure and the quality of pleasure are to be judged in terms of what is sought by individuals when they have access to a variety of possibilities. Pain is that which is avoided by individuals when they are able to reveal their judgments in their conduct. And, lest the imaginative reader attempt to reconcile this view with some version of Kantian morality, Mill makes his own position utterly transparent:

> There is, I am aware, a disposition to believe that a person who sees in moral obligation a transcendental fact, an objective reality belonging to the province of "things in themselves," is likely to be more obedient to it than one who believes it to be entirely subjective, having its seat in human consciousness only. But whatever a person's opinion may be on this point of ontology, the force he is really urged by is his own subjective feeling, and is exactly measured by its strength. . . . It is not necessary, for the present purpose, to decide whether the feeling of duty is innate or implanted. Assuming it to be innate, it is an open question to what objects it naturally attaches itself. . . . If there be anything innate in the matter, I see no reason why the feeling which is innate should not be that of regard to the pleasures and pains of others. If there is any principle of morals which is intuitively obligatory, I should say it must be that. If so, the intuitive ethics *would* coincide with the utilitarian.[20]

Thus, Mill not only doubts that the intuitivist (i.e., Kantian) can ever prove his moral system but argues that even if, somehow, he is right, the utilitarian system can easily assimilate the "native" pleasures and pains. The important point, from Mill's perspective, is to understand the source of power that moral injunctions have. That source is none other than the pleasure or pain caused by our actions or anticipated as consequences of our actions. If there is such a thing as an ultimate morality, its sanctions derive from this consideration and only from this consideration.

The connection between utilitarianism and empiricism is direct. If morality is to be a knowable subject, then it must have a factual foundation. The latter requires that the elements of any moral philosophy be observable and that its propositions be, in principle, testable. On this account, utilitarianism, according to its adherents, is the *only* scientific moral system. Human acts can be assessed in terms of their consequences. When these are judged by those affected by them to have produced happiness, the acts may be judged as moral. When the consequences, as judged by the actual or potential recipients, are painful, the act is immoral. All other appeals must reduce to the utilitarian. All other standards must reduce to the subjective consequences on actual people.

We have observed the conservative elements of Mill's empirical psychology and now we must offset these with the radical character of his empirical philosophy. Unlike his forebears Locke and Hume, who were willing to exclude at least the necessary truths of mathematics from the overall realm of empirical science, Mill argued that even these truths are finally to be understood as inferences from experience. In defining matter as the "permanent possibility of sensation," he asserted phenomenalism as the fundamental epistemology. This is not to be confused with idealism, in any of its variants, since Mill was a confirmed believer in matter. Rather, he took the tenets of epistemological empiricism to their logical limit, which indeed requires that all statements about the material world must ultimately have perceptual counterparts or referents. Thus, to the extent that mathematical propositions are propositions about the real world, they too must have their source in the data of sense. This is a philosophical and not a psychological issue and is beyond our present purpose. It is enough to indicate, however, that this feature of Mill's empiricism rendered the entire enterprise suspect as, in many quarters, it now remains.[21]

Mill was influenced not only by the British tradition but by older contemporaries in France as well. Chief among these, and one of the most important figures in the experimentalism of the nineteenth century, was Auguste Comte (1798–1857), whose six-volume *Cours de philosophie positive* established positivism as a system of philosophy. By his own admission, Comte was led to his thinking principally by Condorcet's *Sketch* (cited above), with its emphasis on cultural evolution and intellectual progress. And in the best French tradition, Comte labored to convert the idea into a movement that can only be called political. In outline, Comte's position is that cultures pass through three distinct stages: the theological, which is superstitious; the metaphysical, in which hidden physical forces or causes replace deities; and finally, the scientific, in which positive knowledge replaces the superstition and metaphysics. The three stages must occur and in the order specified. At each transition point between one and the next, the culture finds itself in a "critical period." An old perspective, which has been satisfying to the masses for an extended time, is now to be replaced by one whose virtues are only slightly sensed and only by the best minds of the period. Once the transition is complete, the culture enters an "organic" period: one of synthesis and discovery, of growth and intellectual evolution.

Much along the same lines as those provided by Condorcet in his *Sketch* of intellectual progress, Comte's positivist doctrine required each new science to emerge from the principles of an older and more established one. Each science develops methods appropriate to its problems. Illustrative is the science of sociology which Comte himself named. Its method was to be that of a comparison of cultures in various stages of evolution. And, through the discovered laws of sociology, all other sciences would be better comprehended, since all other sciences are the product of social evolution. Agreeing with Kant that the mind itself is not directly observable, and recognizing that much of what presented itself as "psychology" was no more than philosophers attempting to discover the laws of the mind introspectively, Comte dismissed psychology as "an idle fancy, and a dream, when it is not an absurdity." [22] If the mind is to be studied, only two methods are available. The first, which Comte called "phrenological psychology" and which we shall discuss later in this chapter, involves research on the relation between brain processes and mental states and functions. The other is the direct observation of the *products* of mental life, and this, to Comte, was sociology. Comte believed that if the philosophical psychologists had not attempted the impossible and conceited task of looking into their own minds, and if they had, instead, appreciated the importance of *feeling* and *emotion,* they would have looked throughout the animal kingdom and discovered bona fide psychological principles. Convinced, however, that only man was rational and intelligent, they ignored this most promising investigative terrain. [23]

Comte's positivism was extended to embrace religion, ethics, and economics. He and his disciples advanced a view of science that placed it on a footing no different from the world's historic religions. Science would solve all problems, answer all questions, remove all doubts. To deny its power was to sink into metaphysical or, worse, theological torpor. To avail one's self of its methods was to set the world aright. Once more, the promises of Bacon, Hobbes, Descartes, and the *philosophes* returned to haunt the orthodox believers. Now they were told that there are two kinds of statements a person can make. One refers to the objects of sense, and it is a scientific statement. The other is nonsense!

Mill, too, was a positivist, although he disagreed on several counts with

Comte's version. He (wisely) rejected phrenology, he accepted associationistic laws of the mind (even though the method of their proof might be "introspective"), and he refused to depart from strict empiricism even to the point of dogmatizing on what science *must* be. Still, his ethology was intended to complement Comte's sociology; his rules of inference were designed to yield positive knowledge; and his appeal to the senses as the ultimate arbiters of truth was unashamedly Comtean. The ideas of Mill and Comte, paced through the revisions that all formulations undergo, culminated in the logical positivism of the Vienna circle: Wittgenstein, Schlick, Carnap, Reichenbach, and their followers. It was this group that introduced the twentieth century to the most coherent and critical attack on rationalism that had been witnessed since William of Ockham. According to the logical positivists—and they might just as well be called radical empiricists—the facts of the world are *sensations,* and all the laws of science are ultimately reducible to empirical propositions. Once we have exhausted the data of sense, there is nothing else that can be said either of the world or ourselves. These developments will be explored further in the next chapter.

Darwin and the Idea of Progress

Evolution, as a fact of perception and as a literary metaphor, is a dateless concept in the human mind. Aristotle and Empedocles argued about its ontogenetic character, and Goethe extended it to the very processes of history. Condorcet, in his *Sketch,* analyzed the human experience as a series of stages beginning with tribal communities and proceeding, in orderly succession, through nine more levels, the last being scientific in language and orientation.[24] Comte's positivism asserted the same, more formally and in far greater detail.

If evolution has been an abiding concept, so, too, has nature-as-enemy. Every age and every person confront the challenges posed by the elements: plague, starvation, drought, illness, enfeeblement. Not only are these the ever-present features of life, but they have been perceived, since recorded history, as producing "heartier stocks," "fit specimens," and the like. The Spartan regimen recommended for Athenian youth in the *Republic* is defended on grounds such as these. The strongest metal is that tempered in the hottest flame. Even tribal initiation rites seem to respect the theory that asperity breeds future success. Darwin, then, is not to be accorded the honor of "discovering" either evolution or natural selection if by discovery we mean the same sort of process as that involved in Maxwell's discovery of the laws describing electromagnetic phenomena.

What Darwin (and, independently, Alfred Russell Wallace) discovered was that evolution, or "progress," is not only possible but inevitable and even, in a very ruthless fashion, mechanical. He and Wallace, both read and acknowledged the ideas imparted by Thomas Malthus's *Essay on the Principle of Population.*[25] Malthus demonstrated, with mathematical rigor, that the breeding potential of the human race was staggering—but never realized. That is, it is possible for human beings to increase in number *geometrically,* but such increases do not occur. Clearly there must be an opposing force tending to limit populations. In describing what he called "the struggle for existence," Malthus presented war, pestilence, and starvation as the forces tending to prevent geometric increases in population. According to his analysis, whenever the rate of increase in population exceeds the rate of increase in the production of food, people necessarily will die.

In the *Origin of Species,* Darwin applied Malthusian concepts to all living systems. He presented the forces of nature as blind agents of selection, "natural selection," which either favor or reduce the probability that a species would survive. Within any species, wide (natural) variations are apparent. The press of the environment favors certain variants and these, as a result, become more numerous. Darwin (1809–1882) published his works before Mendel's findings in genetics had become generally known and, as a result, the *Origin of Species* did not contain any modern notions about the mode of transmission of the various traits involved in survival. In fact, where Darwin was specific on this count, he was wrong. But he did recognize that mutations occur, that some of these were better suited to the conditions they faced, and that, as a result, they would come to prosper, often at the expense of the parent-type. The cheerful idea of progress entertained by the *philosophes* of the Enlightenment now took on a gloomier complexion. Progress is but the consequence of annihilation. New life appears and succeeds, while older forms perish. Supporting his theory with archaeological findings, Darwin relentlessly withstood the attacks of churchmen and skeptical scientists alike. If the Bible suggests that all living forms were created at the same time, the Bible is wrong. If Christian teaching insists that God made so many forms of life and no more, Christian teaching is wrong. If philosophers believe that man, through will and accomplishments, has removed himself from the natural contexts in which survival never exceeds probability, the philosophers are wrong. Find a species of bird now in abundance and you will have the descendant of an avian form, different from the present one, which was unable to accommodate the demands of a changing environment. Man came into being as a successful variant of a primate-type no longer with us. Quite simply, he evolved.

Darwin's most important psychological work is *The Expression of the Emotions in Man and Animals,*[26] a book that may be said to have launched comparative psychology. Since the major strains of Darwinian theory are now widely known, we need only say a few words about his impact on psychological thought in the nineteenth century. By far, the greatest effect produced by Darwinism was that which located the psychological human on a continuum of biological organization. In *The Expression of the Emotions,* Darwin examines the facial musculature of many species, *homo sapiens* included, and notes not only the anatomical similarities (already well established) but the similarities in facial expression produced by conditions giving rise to similar emotions. The angry dog and the actor feigning anger both retract their lips back over their teeth, bare their teeth, and clench them. Signs of submission, of sexual attraction, and of melancholy are of a similar nature throughout phylogeny wherever we find the anatomical equipment necessary for the expression of affect. Natural selection has favored those species able to form "serviceable habits." It has led to the evolution of nervous systems so constructed as to produce behaviors leading to mating, to the avoidance of tissue-damaging stimuli, to the ingestion of nutritious foods. Not only is the present species the survivor of a long process of natural selection, but in the behavior and the emotions of this species, we will discover elaborated forms of those behaviors and feelings that characterize simpler types. In quasi-teleological terms, Darwinian ethology explained variations in the living world in terms of the ultimate goal of all life: the continuation of life. Attainment of this goal requires adaptation to the exigencies of the environment. The species that cannot adapt will vanish. In evolving into a new and different form, it, itself, must cease to be. The required evolution includes not only the anatomy of the species but its *functional* physiology as well—its habits, reflexes, and sensitivities. Psychological evolution is, therefore, concomitant with structural, anatomical evolution. Just as we are able to discern, in the structural nuances of more advanced species, the archetypical features of earlier forms, so also can we see, in the developed psychological equipment of the advanced species, traces of more primitive dispositions and abilities.[27]

It was not long after Darwin's *Origin of Species* that his cousin, Francis Galton (1822–1911), published his studies of "hereditary genius" (1869).[28] Galton had accepted the evolutionary theory without reservation and had accepted as well the epistemological empiricism that constituted England's national philosophy. Putting the two together could produce only one outcome: a theory of mental superiority based on the capacity of the senses! Galton's studies of the children of successful students indicated that mathematical ability "ran in families," although the environmentalist interpretation of the data received only a modest hearing. Galton needed only the Darwinian theory in order to explain why there is such great variability in human intelligence. He needed only an age of experimentalism and a burgeoning commercial empire to support his belief that the measurement of this variability was important. In selecting such tasks as those involving visual acuity and depth perception, Galton merely reflected one of the more innocent features of radical empiricism. In any case, a "mental measurement" tradition was begun which, lacking Malthusian restraints, has since grown geometrically.

Darwinian biology meant different things to different people, and it still does. It was illegal to teach the subject in a number of American states until relatively recently. To religionists of a fundamentalist stripe, the whole idea was out-and-out heresy. The theory required that the earth be much older than Scripture had indicated. It argued for the continuous creation of new forms; fundamentalism insisted on a "big-bang" origin of all forms. It demanded more of man than mere virtue in the struggle for survival and it allowed the gradual elimination of man himself. At best, it traduced the Day of Judgment into a painfully drawn-out affair.

To the middle-class commercialists of Victoria's empire, Darwinism was soon interpreted as a justifying ethic. The poor were poor "by nature." Since we are all engaged in an eternal struggle against the elements, since only certain "types" are properly adapted to the demand, it is inevitable that there will be the "haves" and the "have-nots." Given Galton's findings, it is even likely that the two classes will perpetuate their respective stations hereditarily.

To the Continental scholars, so much under the influence of Comtean positivism and the "three-stage" theory of cultural evolution, the Darwinian model pulled many diverse strands of thought together. The nervous system has evolved, consciousness has evolved, structure entails function. In Freud's words, "anatomy is destiny." Simultaneously there began to appear studies of comparative culture, comparative anatomy, and, alas, comparative psychology. Everything now made so much sense: the idea of progress, the enlightened machine, utilitarianism, the positive philosophy, young Werther besieged by the *Sturm und Drang* of his own feelings. A revolution was taking place, aptly titled by one historian "heroic materialism."

Nineteenth-Century Materialism

Alexander Bain (1818–1903), a close, and admired associate of J. S. Mill, wrote the following to Mill in 1851, describing the progress he was making in his text in psychology:

> I have just finished rough drafting the first division . . . which includes the Sensations, Appetites, and Instincts. All through this portion I keep up a constant reference to the material structure of the parts concerned, it being my purpose to exhaust in this division the physiological basis of mental phenomena. . . . And although I neither can, nor at the present, desire to carry Anatomical explanation into the Intellect, I think that the state of the previous part of the subject will enable Intellect and Emotion to be treated to great advantage and in a manner altogether different from anything that has hitherto appeared. There is nothing I wish more than so to unite psychology and physiology that physiologists may be made to appreciate the true ends and drift of their researches into the nervous system.[29]

It was Bain who founded the essentially psychological journal *Mind*. It was Bain who wrote the two most influential psychology texts to appear before the twentieth century.[30] What is it that prompted so physiological a bias in this empiricist? Why, unlike Locke, Hume, and even Mill, did Bain anchor empirical associationism to the science of physiology? To answer these questions we must review three extraordinary developments in neurophysiology that occurred in the decades immediately preceding Bain's own contributions. These were (1) the Bell-Magendie law, (2) the law of specific nerve energies, and (3) the "science" of phrenology.

The Bell-Magendie law is named for Sir Charles Bell (1774–1842) and Francois Magendie (1783–1855), who independently discovered the anatomical separation of sensory and motor functions of the spinal cord. Bell, reporting the finding to associates at a dinner party (1811), almost lost recognition to Magendie, who presented his work in the more durable form of a published scientific article (1822). Both demonstrated the ability to render an animal insensitive by compressing or transecting the dorsal member of a spinal nerve. The animal so treated was still able to *move* the surgically anaesthetized part of the body but did not respond to intense stimulation delivered to that region. Similarly, by ventral-root transection it was possible to elicit cries of pain from an animal who, nevertheless, could not withdraw its limb from the pain-producing stimulus. The Bell-Mangendie law was important to psychological thinking on several counts. First, it provided clear, anatomical evidence of the sort of arrangement required by the Cartesian theory of sensory-motor functioning. Perhaps most significantly, it reinforced confidence in the experimental approach to the study of sensation and behavior. While the law did not disclose the actual mechanism of reflex-formation, it provided the structural foundation on which reflex mechanisms had to be based. The contribution thus extended a line of inquiry initiated by Descartes, receiving important experimental corroboration from the work of Stephen Hales and Robert Whytt, and theoretically rich amplification by David Hartley. Bain, the most psychological of the nineteenth-century associationists, readily perceived the importance of this law to his plan for a physiological psychology.

Bell himself had anticipated a form of the law of specific nerve energies, but most of the credit for this law is owed to one of the most prominent physiologists of the century, Johannes Müller (1801–1858), whose *Handbuch des Physiologie des Menschen* (1834–1840) was the most authoritative work of the period. The law, which has become a commonplace, asserts that the quality of experience is determined not by the features of the objective stimulus but by the particular nerves responding to it. An orthodox Kantian, Müller judged the law to be only plausible; that is, we do not sense the objective world as it is but come to know about it only in the translated form delivered by the organs of sense. The "thing in itself" remains a mystery. Our knowledge is the subjectified knowledge abstracted from the objects of sense and transformed by the organs of sense. Philosophical idealism aside, however, the law of specific nerve energies placed the qualitative and quantitative aspects of experience in the nerves, that is, in *nature*. It contributed a biological feature to epistemology in that the nature of knowledge now was inextricably tied to the characteristics of the "organs of knowledge."

Where the Bell-Magendie law and the law of specific nerve energies were important additions to an emerging physiological psychology, the concept of "localization of function," as propounded and forged into a movement by Franz Joseph Gall (1758–1828), may be said to have created the discipline. We must quickly add, however, that Gall and his phrenology accelerated the appearance of physiological psychology in the sense of presenting notions that only a physiological psychology could successfully refute. Alexander Bain was much taken by phrenology, which reached the English-speaking world through J. G. Spurzheim's *The Physiognomical System of Drs. Gall and Spurzheim*[31] and the translation of Gall's, *On the Function of the Brain and Each of Its Parts: With Observations on the Possibility of Determining the Instincts, Propensities, and Talents, or the Moral and Intellectual Dispositions of Men and Animals, by the Configuration of the Brain and Head* (1835).[32] The title is presented here because it very nearly exhausts the theory contained in the work itself. It is in this work that Gall presented the four "incontestable truths" of phrenology:

> [T]he brain alone has the great prerogative of being the organ of the mind. . . . The moral and intellectual dispositions are innate; their manifestation depends on organization; the brain is exclusively the organ of the mind; the brain is composed of as many particular and independent organs as there are fundamental powers of the mind.[33]

It is fortunate for the history of the neural sciences that the phrenology of Gall and Spurzheim was quickly challenged by a more sober group of thinkers. Yet for the better part of thirty years it was all the rage. Numerous journals devoted to the "science" sprouted in Europe, England, and the United States. It is only in our own century that the last of these disappeared. A good part of the success of the venture may be traced to Gall's deserved reputation as a neuroanatomist and to the salesmanship of Spurzheim. Moreover, the fundamental idea behind phrenology received some support from the neurological clinic, from common sense, and from the materialistic bent of nineteenth-century psychology. Even the caveman must have recognized that the head-end of his prey was uniquely effective in rendering the animal quiescent and, if the same caveman received nonlethal blows to his own skull, this recognition was further established. Greek and Egyptian medicine and its survival through the Renaissance had catalogued a number of functions mediated by the brain. Descartes, more than any previous writer, had popularized the role of the brain in experience and action. Following Decartes and the Gassendists, numberless investigators, here and there, began to disseminate their findings from neuro-surgical patients, from clumsy studies of decapitated animals, from observations of the traumatically head-injured victims of war and civil strife. La Mettrie, of course, put the whole matter on the polemical foundation needed by any movement, and several of the more medically minded *philosophes*, trailing in La Mettrie's winding path, carried both the soul and the mind confidently into the brain. D'Holbach was most prominent in this respect.

Unlike La Mettrie, Gall actually had stature as a scientist and, coming several decades after La Mettrie, had considerably more data with which to make his case. Even more important to the development of his theory than the growing number of experimental findings, however, was the firmly entrenched "faculty psychology" begun by Locke and developed by the later empirical philosophers of England and Scotland. As we discussed in Chapter 7, the idea of "faculties," as old as Aristotle's *De Anima*, received new life in the sentimentalist theories of seventeenth- and eighteenth-century England. By the time of Gall, it was common to accord to humans not only the "internal light" (of reason) but also the faculties of empathy, justice, love, morality, etc. Occasionally the list would become very long. Gall accepted these faculties, which is no more than to say that he acknowledged that people possessed such characteristics. Since they did, and since "the brain is exclusively the organ of the mind," it took only a slightly breathless exercise in deduction to conclude that each faculty enjoyed specific representation within the cerebral cortex.

Among those who took phrenology seriously, Bain was not the only scholar of substance. No less a figure than Herbert Spencer (1820–1903) more than toyed with the idea, and recall that Spencer was one of the earliest, most eloquent, and most successful defenders of Darwin's theories. Spencer, like Bain and Mill, was concerned that psychology take its place among the natural sciences and that it extricate itself from the purely speculative discipline of philosophy. Gall, understandably, held out the prospect of just such a liberation. In contending that morality, every bit as much as sensation and movement, was to be understood in terms of the organization and functional physiology of the nervous system, Gall gave psychologists precisely the foundation required for the development of an independent science. In addition, the broad features of phrenology did no violence to Darwinism. The latter, which emphasized instinctual drives, inherited habits, reflex mechanisms of survival, and naturally selected nervous processes, was actually an ethological or comparative phrenology, when understood a certain way. And, of course, this was the way the followers of Gall and Spurzheim understood it.

The polemical style of Spurzheim and the cultist flavor of this new science of phrenology invited immediate opposition from many quarters. The most

telling criticisms were advanced by Pierre Flourens (1794–1867), who actually tested certain phrenological hypotheses experimentally. Flourens performed surgery on animals, including removal of the cerebral hemispheres, and
also carefully noted pathologic changes in the brains of recently deceased patients who, prior to death, displayed a variety of neurological symptoms. It was
Flourens who insisted that the cerebral mantle functions *as a whole,* that the
magnitude of a deficit is not simply reducible to the amount of brain involved,
and that similar deficits can result from lesions in any of several brain regions.
While Flourens was struggling to defend the vitalistic and dualistic elements
of Cartesianism, he was successful, nonetheless, in drawing attention to the
purely scientific deficiencies of the phrenological theory.

Gall died before Darwin's great works were written. Spurzheim carried on,
defending the master's system till his own death four years later in 1832. Both
lived long enough to attract the attention and the admiration of Bain and Spencer, and it was the latter two who were able to see the connection between the
concept of localization of function and evolutionism. Spencer, in his very
widely read *Principles of Psychology,* divorced himself from the orthodox
phrenology of his own day, but then added this:

> Nevertheless, it seems to me that most physiologists have not suffi-
> ciently recognized the general truth of which Phrenology is an ad-
> umbration. Whoever calmly considers the question, cannot long resist
> the conviction that different parts of the cerebrum must, *in some way
> or other,* subserve different kinds of mental action. Localization of
> function is the law of all organization whatever; and it would be mar-
> vellous were there here an exception. . . . Any other hypothesis seems
> to me, on the face of it, untenable.[34]

Spencer, attempting to tie this to the theory of evolution, unfortunately accepted the Lamarckian notion of the inheritance of acquired characteristics and
so was forced to argue that the learning, memory, and habits of a species appear
in the next generation. Darwin, of course, was given to the same idea. But
eliminating this error from the Spencerian system, we are left with a not-so-
dated theory of mental function: the sensory fibers project to specific regions
of the brain; repeated stimulation *somehow* results in a greater facility of neural
transmission; chemically, prior experiences are stored within the cerebral
hemispheres; and the "subjective psychology" of association is but the other
side of the "objective" or neurophysiological psychology within which the
associations are physical. Added to this is the Darwinian contribution according to which the complexities of human neuroanatomical organization are to
be understood as the evolved forms of a more primitive organization. Thus:

> The claims of Psychology to rank as a distinct science are . . . not
> smaller but greater than those of any other science. If its phenomena

are contemplated objectively, merely as nervo-muscular adjustments by which the higher organisms from moment to moment adapt their actions to environing co-existences and sequences, its degree of specialty, even then, entitles it to a separate place. The moment the element of feeling, or consciousness, is used to interpret nervo-muscular adjustments as thus exhibited in the living beings around, objective Psychology acquires an additional, and quite exceptional, distinction.[35]

Bain and Spencer did all but found experimental psychology. They fought successfully for the independent status of the science. They presented the associationistic laws in a form never fully appreciated by their predecessors in empirical philosophy, proper. They cemented the new science to the new biology and the two have never been completely divided since. Revealingly, the first chapter of Spencer's *Principles* was titled "The Nervous System," and this established a policy followed by textbook writers for nearly a century. They failed to found experimental psychology only in that they failed actually to engage in psychological research or to create the facilities with which such research could be conducted. Thus, it is to Wilhelm Wundt (1832–1920), who conducted research and who took the pains to construct and name a laboratory devoted exclusively to psychology, that the honor of "founder" is traditionally given.

Wundt prepared for psychology's first laboratory at the University at Leipzig in 1879. By that date Darwin's revolution was beginning to be generally appreciated throughout the scientific community. Bell, Magendie, Flourens, Gall, and Spurzheim had all made their varied contributions to the neural sciences. The great Hermann von Helmholtz (1821–1894) had not only written the classic works in the physiology of vision and hearing but had also measured the velocity of the neural impulse and had advanced the most compelling version of the law of the conservation of energy. The latter two contributions are worth citing in apposition since each, in its own way, challenged the remnants of eighteenth-century vitalism. In discovering that the velocity of nervous conduction was not only measurable but even rather sluggish (maximum = 120 meters/second), Helmholtz simultaneously put to rest the Cartesian view of the soul's ubiquitous influences on the body and the historic belief that mind is beyond the reach of observation. If the mind influences the body, it does so through the brain, and the conducting pathways and the brain perform their mission in measurably physical ways. Helmholtz was the most illustrious and independent of Müller's students and the most vocal opponent of the vitalistic elements in Müller's *Handbuch*. Wundt, after receiving his degree, was Helmholtz' assistant for several years and may be said to have been intimately familiar with the major neurophysiological facts and theories of his time. It is clear that no German scientist of the period escaped the physicalistic (i.e., antivitalistic) climate created by Helmholtz, but it is equally clear that Wundt, in the history of psychology, was sui generis. Even today, one cannot read his *Principles of Physiological Psychology*[36] without being impressed by his sensitivity to the experimental, philosophical, and biophysical problems uniquely affecting the new science. Of his many works, this one was the most important. It was not only one of the first (and one of the few) of his books to be translated into English, it was also the most self-consciously programmatic.

With Bain and Spencer, Wundt saw the need to join the sciences of psychology and physiology. Unlike either Bain or Spencer, Wundt was a scientist—that is, a working scientist—and, as a result, he was far more concerned with the development of proper methods and measures. For these he looked to the recent discoveries of Gustav Fechner (1801–1876), also a Leipzig scientist. Fechner had published the landmark *Elemente der Psychophysik*[37] in 1860, the book that sought to express mathematically the relationship between mental and physical events. It was in this work that Fechner's law was set forth according to which the strength of sensation is proportional to the logarithmic value of the intensity of stimulation. In the same work Fechner had demonstrated how the properly instructed laboratory subject, paced through a series of repeated measurements, would yield reliable data reducible to lawful description. No longer was it necessary to apologize for "introspective" techniques. Comte's critique of psychology seemed to vanish under the weight of Fechner's law. The *psychophysical* methods were simple to apply and were not unlike the methods used in any quantitative science called upon to deal with variable phenomena. Wundt, seldom lavish in his praise, notes only this:

> It is Fechner's service to have found and followed the true way; to have shown us how a "mathematical psychology" may, within certain limits, be realized in practice. . . . He was the first to show how Herbart's idea of an "exact psychology" might be turned to practical account.[38]

But for Wundt, neither the rational-mathematical deductive psychology of Herbart[39] nor the dualism of Fechner could serve as the foundation of psychological science. Indeed, for Wundt, none of the earlier formulations was adequate. Of the British empiricistic psychology, he had this to say:

> In the psychological portion of their works, these writers usually adopt the theory of the "association of ideas" elaborated in the English psychology of the eighteenth century. They adopt it for the good and sufficient reason that the doctrine of association, from David Hartley (1705–1757) down to Herbert Spencer (1820–1904), has itself for the most part attempted merely a physiological interpretation of the associative process.[40]

Psychology, from Wundt's perspective, must concern itself with the "manifold of consciousness" and arrive at an understanding of far more than the laws of association. The manifold of consciousness includes the mind's commerce with more than the objects of sense, with more than external stimuli. It embraces feelings, images, dreams, memories, attention, and movement. And the psychology charged with the study of these processes is "experimental psychology," the name given by Wundt himself. The science is to be the study of mind, but by "mind," Wundt is careful to dissociate himself from that long history of metaphysical speculation:

> "Mind," "intellect," "reason," "understanding," etc., are concepts . . . that existed before the advent of any scientific psychology. The fact that the naive consciousness always and everywhere points to internal experience as a special source of knowledge, may, therefore, be accepted for the moment as sufficient testimony to the rights of psychology as science. . . . "Mind," will accordingly be the subject, to which we attribute all the separate facts of internal observation as predicates. The subject itself is determined wholly and exclusively by its predicates.[41]

Here, in a conservative and rather woolly fashion, Wundt hands down a manifesto of Humean proportion. By "mind" the psychologist will mean no more than and only that which is directly reportable as an observation of an internal event. If the mind thinks, feels, remembers, attends, and forgets, then a science of mind can be no more than experimental inquiries into the determinants of thinking, feeling, remembering, etc. When its predicates are exhausted, there is no metaphysical residue.

For Wundt, "psychology," "experimental psychology," and "physiological psychology" were three terms for the same subject. To appreciate this, it is necessary to recall how the term "physiological" was understood by Wundt, and also to recognize the place of *social* psychology in Wundt's writing. By physiological (*physiologische*) the German scientist of the late nineteenth century described an essentially scientific, law-governed approach. The term, unlike its current meaning, did not refer specifically to biological events or measures, but to the entire range of lawful operations governing the "animal economy." This, then, is the sense in which Wundt's terms "psychology" and "physiological psychology" can be taken as synonyms. But what of "experimental psychology"? In this, as will be developed further in the next chapter, a distinction must be made between Wundt's conception of "social psychology" and our own. His later works on "folk psychology" are primarily devoted to what we call cultural and historical anthropology. It was Wundt's judgment that significant events in human history were, of course, beyond the reach of experimental methods and, more importantly, were not explicable in the language of natural science. That is, one cannot comprehend human history "physiologically" because this history does not proceed according to exclusively causal laws. More will be said on this in Chapter 11. It is sufficient here to understand that Wundt did not take "social psychology" to be a branch of psychology, but a separate subject requiring its own methods, measures, and explanatory principles. Psychology proper, however, is synonymous with "physiological" and "experimental" psychology. Again, however, this did not commit Wundt to an exclusively *biological* psychology. He rejected out of hand that form of materialism that insisted that mind was no more than matter or the mentalist assertion that mind was irreducible to matter. He rejected, that is, the grounding of his science in metaphysical disputes. Psychology henceforth was not to be a branch of philosophy, less of biology. It was to be the experimental science devoted to an analysis of the contents of consciousness or, as he put it, the manifold of conscious experience. Ultimately this analysis would converge upon equivalent analyses of the structure-function relationships of the nervous system. Wundt took a disinterested position in the Flourens-Gall controversy, well aware that Gall spoke without having the necessary facts and that Flourens, in opposing phrenology, dismissed more of the localization theory than the facts demanded. He reviewed in his *Principles* the recently discovered technique of direct, electrical stimulation of the living animal's cerebral cortex. Beginning in 1870, Fritsch and Hitzig published the results of their studies of the effects of direct electrical stimulation of the cerebral cortex of dogs. They demonstrated the topographic organization of sensation and movement on the cortical surface and further demonstrated the relatively separate cortical "strips" associated with sensory and motor capacities. Referring to their work, Wundt would remark on the "simplicity of the structural plan," [42] though the simplicity of 1875 has been seriously questioned in the intervening century. Wundt also considered the more established techniques of ablative surgery and the various methods of clinical assay. He recognized early the severe limitations of each of these methods and also foresaw that none of them was a substitute for the introspective data easily secured from a conscious, healthy human being. He had a sophisticated awareness of the fact that the mind-body problem was not to evaporate as a result of improved technology. He had, as well, a nearly modern disdain for the suggestion that it was psychology's responsibility to settle the matter. The topics, methods, and theories that define contemporary psychology have changed considerably since Wundt's time, but the essential attitude of the experimental psychologist today may be said to have originated in the Leipzig laboratory.

The Reflex Revisited

Even as the larger agenda of psychology was altered by research and theory drawn from neurology, more basic research in the functions of the brain and spinal cord was reinforcing an essentially mechanistic perspective. Marshall Hall (1790–1857) was one of the leaders in these developments. He was also one of the founders of the British Medical Association and a most outspoken abolitionist, addressing this cause in his travels in the United States.

In the tradition of Whytt, Hall undertook extensive research on spinal reflexes, establishing that coordinated and integrated behavior could be elicited in the absence of any regulation or participation by higher centers. But research and clinical findings also led him to conclude that such behavior is not accompanied by consciousness. To support this conclusion he suggested the following:

> [T]he cerebrum itself may be viewed as the organ of the mind on which the *psyche* sits, as it were, enthroned . . .[43]

> How different . . . are the functions which belong to the true spinal marrow! In these there is no sensation, no volition, no consciousness, nothing psychical.[44]

> The true spinal system never sleeps.[45]

> What is the hidden agent in [spinal] function? Is it *Galvanism?*[46]

Hall's work was widely disseminated through Johannes Müller's influential *Handbuch*. He was also an active spokesman in behalf of his own accomplishments, claiming as much for himself as his work would allow—and perhaps a little bit more. In the process, he helped to put the concept of the reflex more securely on the map of an emerging theoretical psychology, laying the foundations for a biologically oriented behavioral science.

The "Idealist" Alternative

The Kantian legacy was briefly touched on earlier in this chapter, but the most important influences were neglected. These were the interpretations of Kantian philosophy by his immediate successors, Fichte (1762–1814), Schelling (1775–1854), and Hegel (1770–1831). Together, they forged a unique form of idealistic psychology, which still affects the manner in which Continental psychology proceeds. We neglected this facet of the Kantian legacy in part because Kant himself may not have been eager to claim it and also because the movement was fueled by those empiristic and materialistic perspectives we have just explored—fueled in the sense that this new idealism was a conscientious antagonist of empirical and physiological psychologies. Predictably, Wundt dismissed the system as no more than a "rational psychology" emerging from Kantian "nature philosophy" and having only negative effects on scientific psychology specifically and on natural science in general.

Kant's role in the appearance of German idealism is central, if miscast. Although rejecting the subjective idealism of Berkeley, he described his own metaphysical position as "transcendental idealism," by which he meant to distinguish between the actual physical world of material objects and the sorts of knowledge we can ever have of those objects. The objects, for Kant, were "things in themselves" that could never be known empirically as "things in themselves." Rather, our empirical knowledge constitutes a translation of the real world, a translation performed by the pure categories of the understanding operating in conjunction with imperfect senses. To "know" is to interpret, not merely to sense, but in the very act of interpretation we suffuse objective nature with the categories. The mind that does this is not an "object" and, therefore, can never be known in the sense in which we know the natural world. This leads to the insistence, of course, that a *science* of psychology is scarcely imaginable. Wundt summarized the Kantian position this way:

> Kant once declared that psychology was incapable of ever raising itself to the rank of an exact natural science. The reasons that he gives . . . have often been repeated in later times. In the first place, Kant says, psychology cannot become an exact science because mathematics is inapplicable to the phenomena of the internal sense; the pure internal perception, in which mental phenomena must be constructed,—time,—has but one dimension. In the second place, however, it cannot even become an experimental science, because in it the manifold of internal observation cannot be arbitrarily varied,— still less, another thinking subject be submitted to one's experiments, comformably to the end in view; moreover, the very fact of observation means alteration of the observed object.[47]

We have seen that these objections did not prevent Wundt from committing his life to experimental psychology. Fechner had shown, to Wundt's satisfaction, that mathematics could, indeed, be applied to the phenomena of the "internal sense." Moreover, every branch of natural science must alter its objects in the process of observing them.

If Wundt did not perceive the Kantian objections to be telling, Fichte, Schelling, and Hegel accepted them as axiomatic. For Fichte, again in the Kantian tradition, the very freedom of the human will, in contrast to the deterministic character of purely physical processes, settled once and for all the question of a scientific psychology: there could be none. If there is to be a psychology, it must be a deductive, philosophical discipline that accepts as its subject the will and intentions of the self (*ego*). The ego affirms itself through itself and not through recourse to external objects. In affirming itself, it forces nature to conform to its will and, indeed, may even be said to animate nature through its will. While the will is free, the spiritual ideals of human life impose upon it the constraint of duty, the neglect of which is the essence of evil. In these various elements—the self imposing its own character on nature, the transcendental nature of duty, the freedom of the will—Kant's metaphysics appears in the form of a psychology of personality.[48] Schelling's departures from Fichte's philosophical psychology require no mention here. In the significant respects, he and Fichte agreed; freedom of the will in an otherwise determined world of matter entails an inescapable dualism that no scientific psychology can eliminate. If we are to comprehend the nature of mind, our only method is that of the mind reflecting on itself and deducing the terms of its unity. Wundt, while rejecting the prescriptions of these idealists in the matter of method, could not extricate his psychology from their prescriptions regarding topic: the manifold of conscious experience.

The capstone of nineteenth-century German idealism is to be found in the philosophical and logical works of George Friedrich Hegel, whose influence on European thought is equaled, if at all, only by Descartes and Kant. Even if our present subject were the history of philosophy, it would hardly be possible to summarize "Hegelianism" in less than a very substantial chapter. Indeed, unless one is a professional philosopher, it is hazardous to attempt a summary of any sort. As with most productive scholars, Hegel underwent a change of mind from time to time. His writing suffers from the fatal combination of genius and literary awkwardness. The subjects of greatest interest to him—the Absolute, the Ineffable, Soul, Art, and Religion—tend to frustrate authors of the greatest literary skill, and Hegel is not one of these. One of the clearer expositions of his system in the English language is the old but not dated study provided by Professor W. T. Stace,[49] but even this work is so punctuated with Hegelian locutions as to leave the uninitiated . . . uninitiated. Russell surely has many advocates of his contention that Hegel is "the hardest to understand of all the great philosophers," [50] and this from a man who believes that "almost all of Hegel's doctrines are false."[51] Be that as it may, Hegel invented phenomenology, was acclaimed by some otherwise moderate Englishmen as the new Aristotle, and caused Karl Marx to describe himself as "the pupil of that mighty thinker." [52] Hegel is not to be ignored.

Hegelian "doctrines," as Russell called them, are all derived from rational first principles. Completely rejecting Kant's caveat about the subjective nature of reason, Hegel declares that the truths of reason are necessary, nonarbitrary, and final. Through reason, the mind can dissect the apparent world such as to lay bare *its* reason. At this point, we must insert the distinction, the Hegelian distinction, between "reason and cause." Hegel, while totally opposed to the thrust of empirical philosophy, readily appreciated Hume's arguments against *necessary* causal sequences. He agreed that no logical bridge could ever be constructed whereby one could proceed from an effect back to physical causes of a necessary nature. However, while causes do not logically entail effects, reasons do. Professor Stace explains the emphasis this way:

[E]xplanation involves the idea of logical necessity. It is just the apparent absence of necessity in the world which makes us complain that it is incomprehensible. Cold produces ice. This is a fact which simply *is*. We cannot see why it *must* be. . . . If, instead of being a mere fact, we could see that it is logical necessity; if we could see the *reason* of it, and that it follows from the reason as necessarily as a logical consequent from its antecedent, then we should understand it. . . . Thus, a philosophy which would genuinely explain the world will take as its first principle, not a cause but a reason. . . . This is the fundamental Hegelian idea of explanation . . . it was for this that the Greeks, especially Aristotle, were groping, when they said that the first principle of the world is not prior to the world in time, i.e., as a cause is prior to its effect, but is *logically* prior to the world, i.e., as a logical antecedent is prior to its consequent.[53]

Reason, then, is the first principle and it explains itself. It determines itself and the world in that, unlike causal sequences, the sequence from reason to statements about the world is directed by logical necessity. To say that John Smith is dead "because a trigger was pulled" is not to explain *why* he is dead—only to note one of the antecedent causes of his demise. We explain *why* Smith is dead when we say that "Jones wanted him dead and it was Jones who pulled the trigger." The reason is prior to the cause. Where the effect is only contingently related to the cause, it is necessarily related to the reason. Harkening back to Kant's *Categories,* we can say that the characteristic of temporal succession is *logically* tied to the a priori concept of time.

Hegel's theory of psychology is presented most clearly in his *Encyclopaedia*[54] and his *Phenomenology of Mind*.[55] There has been something of a contest among historians to come up with as many people as possible who may be said to have anticipated the theories of Sigmund Freud. We do not seek this elusive prize in recognizing Hegel's part in the development of Freud's thinking. Hegel's philosophy of mind is rife with the concepts of stages of development, ego and anti-ego conflicts, and intimations of a death wish. One could not be educated in the Austria of the 1860s and 1870s and not be influenced by the thoughts of Hegel.

The Hegelian philosophy of mind begins with the theory that the mind is a stage in the evolution of soul. Initially (and Platonically) the soul shares the realities of nature; as Stace describes this "natural soul," it is the "beginning of spirit"[56] and exists as mere being. It cannot reflect upon itself nor can it assimilate to itself the objective elements of the physical world. From this state of pure egoism, the soul, through historical development, gains "sensibility" by which it is able to distinguish between itself and its contents. While the soul still cannot comprehend *external* objects, it has nonetheless an awareness of the difference between the feelings within it and itself. The "natural soul" is now the "feeling soul." This stage is followed by one in which the soul actually can receive external objects, can distinguish itself from the perceptual contents resulting from sensory experience, and can catalogue the elements of the physical world according to universal categories. The soul, now the "actual soul," recognizes itself "as its contents"; it is at one with its sensations, ideas, and feelings. Once the soul has expressed the ability to distinguish between itself and the objects of the external world, it may be said to have "consciousness" and to be "mind." It is the study of this consciousness—as Wundt would later call it, the manifold of consciousness—that is phenomenology. Consciousness too passes through stages: "sensuous consciousness," "sense perception," and "intellect." The first of these stages allows the mind to receive the "raw data" of experience, impressions devoid of cognitive features. The sensation is of an event—immediate and psychologically neutral. Perception, however, is another matter. Here, the observer adds to or brings to bear upon the merely sensuous the concept of the universal. Where the sensuous produces a "this" or a "that," the sense perception is of the form "What is *this?*":

> Whatever we say in answer to this question invests the "this" with a universal character. . . . To say that it is "here" or "now" is at once to apply concepts, or universals, to it; for "here" and "now" are both universals. . . . Everything belongs to the class of objects which are called "this." Hence, "this" is a *class*-name and imports a universal.[57]

Sense perception, in applying the universal concept to each particular, establishes a contradiction—the contradiction between a particular object and a *class.* Perception, alone, cannot resolve the contradiction. This is the task of the intellect which, through its inventions of scientific laws and principles, recognizes the ultimate reality to be universals, and particular objects to be mere appearances or instances. Once this developed form of consciousness has succeeded in abstracting the universal principle from each of the particular sense perceptions, it recognizes that its knowledge is *idea,* for universals are, by their nature, ideas. At this point, consciousness is led to self-consciousness or the awareness of the idea of self.[58]

It is hardly necessary to remind the modern reader of the "Hegelian" tone of much contemporary discourse not only within the community of professional psychologists but within the relatively nontechnical spheres of daily life. "Self-awareness," "self-actualization," "consciousness-raising," and related expressions of self-concern are directly attributable to Hegel and the neo-Hegelian idealists. The dialectical triad of thesis, antithesis, and synthesis, which Hegel advanced as laws of thought expressed in logic, has now become a permanent fixture in the lexicon of undergraduate students and news commentators alike. In a subtler way, the triad found its way into Freudian theory:

the "id" asserting that most fundamental of all theses; the "superego" standing in antithetical regard to it; and the "ego" emerging, synthetically, from the reconciliation of these counterpoised forces. Hegelianism, in its triumphant form, emerged in the middle years of the nineteenth century as Romanticism. In his *Reason in History,* Hegel proclaimed that "nothing great in the world has been accomplished without passion,"[59] and he argued that the essential nature of man—not one at a time but as a collective of consciousness—is freedom, the irrepressible freedom of spirit. Renewing the ageless dichotomy, he dismissed matter as the passive victim of natural laws and asserted spirit as the only free force in the universe. He and Beethoven were born in the same year. One set Goethe to music, the other to philosophy. In all, we find the romantic expression of Condorcet's idea of progress. Escalated to the level of social action, we find a revolutionary spirit that has hardly begun to exhaust itself.

Karl Marx (1818–1883)

The historian of ideas is sorely tempted to discuss the roots of Marxist philosophy in a way that tends to trivialize the content of the philosophy. The temptation does not betoken or at least need not betoken enmity toward the philosophy or its author. Rather, from a purely philosophical perspective, so much of Marx's thought is derivative that a mere enumeration of Marxist principles obscures the brilliant originality of his work taken as a whole.

Offsetting this tendency toward underestimation is an error in the opposite direction. Marx has proved to be such a powerful figure in the social and political affairs of our own century that it becomes tempting to read too much depth and genius into the exclusively philosophical aspects of his writings.

For our present purposes, we seek to do no violence to the stature of the man in social and political history by observing the very marginal influence of his speculations on the evolution of modern psychology. Marx did not have a great effect either on contemporaries who figured centrally in psychology or on later psychologists who, in fact, shared his materialistic orientation. It is true that in the highly politicized decades of Soviet science there would be eager attempts to establish the connection between Marxist theory and Pavlovian psychology but these attempts seldom rose higher than the level of mere propaganda and were never taken seriously by scientists of enduring consequence.

Marx's failure to influence the course of psychological scholarship can be understood in a variety of ways. The nineteenth-century developments in psychology were of a largely experimental nature. Marx's approach was historiographical and, in the loose sense, logical, and this was the very approach that the founders of experimental psychology were rejecting. Moreover, while Marx accorded "consciousness" a central role in his theory, it was a role that, on first inspection, was indistinguishable from Hegelianism and therefore not likely to be serviceable. But even more than these dissonances was the un-breachable separation between psychology's commitment to the study of the individual and Marx's undivided attention to broad social processes. In short, Marx, was a sociologist at a time when psychology was being founded along biological lines. He discovered the psychology of alienation a century before social psychology would be prepared to study it. The same may be said of his recognition of those problems we now locate in such fields as "urban psychology," "industrial relations," and "community psychology." He saw the effects of industrialization on the family, on the worker, and on the relations among states. In these effects he perceived economic forces as the engine of all social and cultural and intellectual evolution.

We have noted Marx's acknowledged debt to Hegel. It is also noteworthy that his doctoral dissertation analyzed the systems of Democritus and Epicurus, a dissertation that could only have steeped him in the Enlightenment scholarship of Diderot and Condorcet. Indeed, in *The German Ideology* (1845–1846), where he discusses the successive stages of economic evolution in terms of tribal, state, feudal, and private ownership, we find almost a paraphrasing of Condorcet's *Sketch*. But unlike Condorcet or, for that matter, Epicurus, Marx was unwilling to place rational forces at the core of such evolution. For Marx, the materialism of consequence was *historic* materialism. He had no doubts but that man is so constituted biologically as to require and conform to historic materialism, but Marx was not to digress into anatomical or physiological reflections about "Man" while a world of *men* were suffering under the yolk of industrialism. Nevertheless, he was aware of the direct connection between the psychological materialism of the Enlightenment and the philosophical justifications of communism. This awareness is almost glibly rendered in *The Holy Family:*

> As *Cartesian* materialism merges into *natural science proper,* the other branch of French materialism leads direct to *socialism* and *communism*. There is no need of any great penetration to see from the teaching of materialism on the original goodness and equal intellectual endowment of men, the omnipotence of experience, habit and education, and the influence of environment on man, the great significance of industry, the justification of enjoyment, etc., how necessarily materialism is connected with communism and socialism.[60]

It is in the same work that Marx carries the analysis to its logical terminus. Since man is formed completely by his environment, by the social forces imposed upon him throughout his development, he cannot reasonably be held responsible for his crimes. These are the result of social evils and society is to blame. Consciousness itself is the creation of society.

Perhaps it is too trite to observe that a fair fraction of the common-sense psychologies of history can be partitioned in terms of attention to similarities among men versus attention to differences. With a broad brush, we can color with the same hues egalitarianism, behaviorism, Marxism, socialism; with another hue, elitism, idealism, capitalism. Marx, like the empiricists he so admired, was persuaded that the similarities among men were far greater than their diverse stations would suggest. Noting how the steam engine alone had transformed the very character of the English nation, he was convinced that economic systems of production had imposed artificial class-differences upon the human community and that these differences must be eradicated. In retrospect, we tend to dismiss much of this as a kind of "folk" psychology, realizing that individual differences among people are not trivial and judging also that the psychological character of our race appears to survive a remarkable range of social and economic systems. Although we have hardly exhausted the set of important cross-cultural studies of cognitive and perceptual processes, there would seem to be enough data now to cast doubt upon such cornerstones of Marxist psychology as, "The nature of individuals thus depends on the material conditions determining their production." [61] But the issue here is not whether Marx was "right" or "wrong" any more than it is whether his century was "right" or "wrong." Instead, we are to discover in Marx*ism* that peculiar and fascinating theme uniting all the *ism*s of the nineteenth century: positivism, materialism, utilitarianism, Hegelianism, pragmatism, and experimentalism. And what unites these otherwise immiscible movements is the confident belief that the world or the human enterprise or the heavens or everything can finally be encompassed by a grand vision validated by a faultless method. Scholarship, neither before nor since, is displayed with such certainty and finality. In the major works of physics, political theory, psychology, biology, sociology, and even more philosophy we discover an intellectual stridency and sureness that only amazes the twentieth century witness. In yet another way, this legacy of confidence also frustrates the modern citizen who is unable to locate the year when things started to go wrong; the time when certain physics became uncertain, when the knowable mind sunk once more to its historically unreachable depths, when the ping-pong rhythm of social organization gave way to cacophony.

There is yet another element in Marxist writing that warrants comment, although it, too, has no direct bearing on the evolution of modern psychology proper. The element is that of enmity and contempt, an element last present during the late Renaissance and Reformation. The wit and charm of the Enlightenment are absent. The canons of civility are ignored. Logic and compassion lose their struggle with impatience.

Nietzsche and Irrationalism

Little is served by attempts to analyze the "psychology" of intellectual leaders. Their importance is grounded in their ideas, not in their motives or personal idiosyncracies. However, in Marx and in Nietzsche we confront an impatience and even a contempt never to be seen in the works of a Spencer or Bain or Mill. It will not do to propose that British philosophers were of a uniformly even disposition. Rather, what is at work is the passionate romanticism of Hegel: the Continental acceptance of emotion as a proper corollary of analysis. In observing this we call attention to the essential conservatism of nineteenth-century scholarship in Britain and France and the radicalism of thought flourishing in the German-speaking world. Now, it could only be a cliché to let the matter rest here. Of course, Marx, Nietzsche, and their disciples were radicals, but the utilitarians were also radicals in their ambitions. The difference is subtler. Philosophy in Germany was speaking directly to the people. Philosophers were applying the ideas and the recognized forms of philosophical discourse to ends that were clearly political. In their success, they introduced to scholarship a relevance that would have made even the *philosophes* uncomfortable.

Toward the conclusion of Part I of his *Beyond Good and Evil* (1886), Friedrich Nietzsche (1844–1900) proclaimed that once we have sailed over and past the conventional morality of the philosophers, "psychology shall again be recognized as the queen of the sciences, to serve and prepare for which the other sciences exist. For psychology is now once again the road to the fundamental problems." [62] Walter Kaufmann would come to call Nietzsche the first great "depth psychologist" owing to Nietzsche's identification of unconscious processes as the source of much of what we take to be the affairs of daily life and of culture itself.[63] It was Nietzsche who traced the common features of all traditional philosophies to "the common philosophy of grammar," and who anticipated much of today's deconstructionist projects and Wittgensteinian analyses. In Nietzsche is to be found the most developed recognition of the cultural and linguistic sources of science, philosophy, and morality. Noting Boscovich's pioneering insights in the eighteenth century, which reduced the putatively material atoms of the Newtonian world to nonmaterial centers of force, Nietzsche celebrated the elimination of "substances," "intuitions," and other spiritous features of the traditional ontology. Pressing on to levels that are pre-linguistic and primitive, he found as the veritable wellspring of culture that will to power previously analyzed by Schopenhauer. For Nietzsche, the will to power is more basic than Darwinian survival instincts. But, as with instincts, it operates beyond consciousness. It uses consciousness, with the latter being a merely superficial aspect of the mental. If the fundamental drive in life is not survival per se but the expression of power, then there are things that must always count more than life itself, not to mention more than the simple pleasures of the hedonist. The power in question is the strength that is found not in the brute or bully, but in the capacity to endure and triumph over life's menial obstacles. His hero was Goethe, and the power in Goethe is at once the power of clarity

and of courage. Goethe knew there was more to life than the material composition of the Newtonian heavens or the Lockean mind; that life required a kind
of devotion and sincerity ultimately expressive of levels of reality beyond mere
sense and prosaic thought. It was Goethe, whether in his *Elective Affinities* or
his theory of color, who relentlessly brought thought back into the thinker,
value back into the world that is known. It was Goethe whose magisterial
flights of artistry left in the distance the rubble of failed theories, each of them
a prison of the mind. Goethe fulfilled the universal impulse: *He exercised his
power.*

Then, too, there was the notorious friendship and falling out with Richard
Wagner. Nietzsche was captured by the power of Wagner's music and its delib-
erate break with tradition; its contempt toward sentimentality; its deep roots in
the most primal traditions of German culture; its purity of purpose. The letters
between them disclose an unedited passion for social and moral theories, for
theories of art and religion, and for theories of "types." As his own mind became a casualty of madness, Nietzsche's prose moved from the biting and perceptive to the dismal and merely peevish. The break with Wagner was occa-
sioned in part by what Nietzsche took to be a concession to the faithful many,
a toying with Christian superstitions. Like Marx, and unlike William James, he
saw in religion a reflection of weakness and self-deception. Had he really been
"the first great [depth] psychologist," he might have found in the universality
of religious inclinations a vein of interests running deeper than the will to
power.

In his celebrated (though rash and unconvincing) critique of Kant's metaphysics, Nietzsche found Kant charmed by the discovery of the pure categories
and the possibility of synthetic knowledge a priori. As he understood Kant, all
this is possible owing to a "faculty" of reason: So then we are able to do this
because we have something that enables us to do this—the old *virtus dorma-
tiva* of Moliere again. It is doubtful that any serious examination of Kantian
metaphysics would result in so lighthearted a dismissal, but what is worth noting here is the conclusion Nietzsche reached, having satisfied himself of Kant's
philosophical innocence (!). The conclusion reached is not that we do *not* have
such fixed features of thought and judgment, but that our having them does not
establish their validity, only their utility. We think and reason as we do in the
furtherance of desires that are all too human. The discoveries of philosophy are
but inventions, not the illumination of what is in the natural world or the world
of independently existing entities, but the crafting of things expressive of cultural or personal value.

Marx and Nietzsche promised more than they delivered. The first was striving for a scientific understanding of events too utterly under-determined to admit of strict analysis. Nietzsche would put an end to the search of any certainties beyond the ambit of a given culture and its prejudices. He is not at all to

blame for what would take place in Germany a few decades after he declared that,

> It is important that as few people as possible should think about morality—consequently it is *very* important that morality should not only one day become interesting![64]

Résumé

In this hurried survey of nineteenth-century psychological thought, the broad themes and special methods that would come to define modern psychology are all quite apparent. To this extent, the present chapter acknowledges the place of intellectual history in the evolution of modern psychology, and it is in this respect that the latter is a footnote, it were, to the recent past. But modern psychology, if not more than this recent past, is certainly different from it. Indeed, there is very little in our contemporary texts or journals that explicitly acknowledges a debt to the nineteenth century, and even less that sets out to settle the problems that so arrested past thinkers. What we need, then, is yet another bridge, one that might take us from the larger concerns and general counsel of the past to the narrower issues and more developed methods of the present. One bridge brought us from a philosophical to an allegedly scientific psychology, but not to genuinely modern psychology. In the next chapter, therefore, we will review more closely some of the developments already discussed and still other developments that succeeded in creating those perspectives, methods, and special schools that give color and contour to today's psychology.

Notes

1. No complete list can ignore Pierre Bayle (1647–1706), whose *Dictionnaire historique et critique* animated the scholarship of the Enlightenment with skeptical wit and an irreverent commitment to expose the pronouncements of authority to the twin lights of reason and evidence. Bayle's ten-volume *Dictionary* and related writings made him a figure of great consequence to such illustrious contemporaries as Locke and Leibniz. Later, he would be hailed by Voltaire and would, as well, influence the thinking of scholars as different in outlook as Berkeley and Hume. Except for his summary dismissal of distinctions between "primary" and "secondary" (Lockean) qualities, however, Bayle provides little material for psychological analyses. His immediate followers in philosophy were inspired by his courageous and telling assaults on the chained mind, but the bulk of his writings remain only peripheral to matters of central concern to psychology.

2. Voltaire, *Philosophical Letters,* translated by Ernest Dilworth, Bobbs-Merrill, Indianapolis, 1961.

3. Claude-Adrien Helvetius, *A Treatise on Man: His Intellectual Faculties and His Education,* translated by William Hooper, London, 1777.

4. Denis Diderot, *D'Alembert's Dream,* in *Diderot's Selected Writings,* edited by Lester G. Crocker and translated by Derek Coltman, Macmillan, New York, 1966, pp. 179–222.

5. Antoine-Nicolas De Condorcet, *Sketch for a Historical Picture of the Progress of the Human Mind* (1795), translated by June Barraclough, win an introduction by Stuart Hampshire, Noonday Press, New York, 1955.

6. Voltaire, *On the Pensées of M. Pascal,* in *Philosophical Letters,* p. 144.

7. In his final years, Kant had undertaken to oppose this transcendentalism. He died before completing what was to be a stern rebuke of those who would defend Romantic Idealism on Kantian grounds.

8. J. S. Mill, *Autobiography.* The edited volume by F. E. Mineka (Toronto, 1963) is especially useful.

9. J. S. Mill, *A System of Logic, Ratiocinative and Inductive: Being a Connected View of the Principles of Evidence and the Methods of Scientific Investigation,* Longmans, Green, London, 1900.

10. Ibid., Book VI, Ch. III, Sec. I.

11. Ibid., Sec. 2.

12. Ibid., Ch. IV, Sec. 3.

13. Ibid.

14. Ibid., Sec. 4.

15. Ibid., Ch. 5.

16. Ibid., Ch. V, Sec. 4.

17. J. S. Mill, *Utilitarianism,* in *The Utilitarians,* Dolphin Books, Doubleday, Garden City, N.Y., 1961, p. 404.

18. Ibid.

19. Ibid., pp. 408–209.

20. Ibid., pp. 432–433.

21. J. S. Mill spoke not only with the most eloquent voice of his age but, in major respects, with one most accurately reflecting the tone of that age. His contemporary, the essayist John Morley, observed in his eulogy for Mill, "Much will one day have to be said as to the precise value of Mr. Mill's philosophical principles. . . . However this trial may go, we shall at any rate be sure that with his reputation will stand or fall the intellectual repute of a whole generation of his countrymen." (*On the Death of Mr. Mill,* in *Nineteenth Century Essays,* University of Chicago Press, 1970.)

22. L. Levy-Bruhl, *The Philosophy of Auguste Comte,* translated by Frederic Harrison and published in English by Swan Sonnenschein, London, 1903. This was a most important edition, bringing the ideas of Comte to the English-speaking world. In America, in the 1850s, Henry Edger did much to advance the cause of positivism, but it was not really until Levy-Bruhl's careful study of the positivist program that a large number of philosophers in England and America seriously approached the system. On the earliest American forms, see *Positivism in the United States—1853–1861,* by Richmond, Laurin Hawkins, Harvard University Press, Cambridge, 1938.

23. Levy-Bruhl, pp. 191–193.

24. Antoine-Nicolas De Condorcet, *Sketch.*

25. Thomas Malthus, *An Essay on the Principle of Population as It Affects the Future Improvement of Society, with Remarks on the Speculations of Mr. Godwin, M. Condor-*

cet, and Other Writers. The essay has been reproduced in a paperback edition by the University of Michigan Press, Ann Arbor, 1959, with an introduction by Kenneth Boulding. The original appeared in 1798, in London.

26. Charles Darwin, *The Expression of the Emotions in Man and Animals,* Appleton-Century-Crofts, New York, 1896.

27. It was Ernst Haeckel (1834–1919) who formalized this Darwinian proposition into the "biogenetic law" according to which ontogeny recapitulates phylogeny. But Henri Bergson (1858–1941), almost single-handedly, began to lead an intellectual movement against Darwin's materialistic theory of evolution. His *L'Évolution créatrice* ("Creative Evolution"), which was published in 1907, challenged evolutionary biology in terms very much like those used by the Gestalt psychologists in their arguments with associationists. Bergson found little plausibility in the notion that random processes could produce, in piecemeal fashion, such enormously complicated and functionally interdependent systems as those underlying vision, for example. Only a *creative* evolution, on Bergson's account, could lead to the degree of biological organization of which the higher species are evidence. This creative evolution is powered by an *élan original,* a divine agency that has authored free will as well. That Bergson's influence on contemporary psychology and especially contemporary American psychology has been less than commanding is to be understood in terms of the failure of such notions to be reduced to experimental modes of verification. Thus, we find Bergsonian ideas most prominently and perhaps uncomfortably located in that wide-ranging literature described as "existential psychology," "humanistic psychology," and "Gestalt," psychotherapy. Perhaps their most cogent integration is to be found in the hypotheses advanced by Professor Jean Piaget, whose discussions of "cognitive development" are richly shaded by notions akin to *creative evolution.*

28. Francis Galton, *Hereditary Genius,* Macmillan, London, 1869.

29. The letter by Alexander Bain appears in R. M. Young's recent and excellent study, *Mind, Brain, and Adaptation in the Nineteenth Century,* Oxford University Press, Clarendon, 1970, pp. 102–103.

30. As Professor Young notes, Bain's *The Senses and Intellect* (1855) and *The Emotions and the Will* (1859) comprised the two-volume set that served as the standard British psychology for nearly fifty years. Both were published in London by the Parker Publishing Co. These volumes are discussed and reprinted without abridgement in Series A, Vols. IV and V, of *Significant Contributions to the History of Psychology,* D. N. Robinson, University Publications of America, Washington, D.C., 1978.

31. Johann, G. Spurzheim, *The Physiognomical System of Drs. Gall and Spurzheim,* Baldwin, Cradock, and Joy, London, 1815.

32. François Joseph Gall, *On the Functions of the Brain and of Each of Its Parts,* etc., 6 vols. translated by Winslow Lewis, Marsh, Capen, and Lyon, Boston, 1835.

33. Gall, *Functions.*

34. Herbert Spencer, *The Principles of Psychology,* Appleton-Century-Crofts, New York, 1896, p. 573.

35. Ibid., p. 141.

36. Wilhelm Wundt, *Principles of Physiological Psychology,* Vol. I, translated by E. B. Titchener (from the fifth German edition of 1902), Macmillan, New York, 1904.

37. Gustav Fechner, *Elements of Psychophysics,* translated by Helmut Adler, edited

by David H. Howes and Edwin G. Boring, Holt, Rinehart and Winston, New York, 1966. The *Elements* was originally published in 1860.

38. Wundt, *Principles,* pp. 6–7.

39. Johann Friedrich Herbart (1776–1841) was a student of Fichte's. He held the chair once occupied by Kant at Königsberg. Unlike Fichte, he envisaged a scientific psychology based on mathematical analyses, a view he advanced in his *Psychologie als Wissenschaft* (1824–1825). Herbart's scientific psychology provided no room for innate ideas nor for a priori concepts. Rather, the elements of unconsciousness, including feelings, are to be understood in terms of the dynamic, mechanical laws of physics and are, therefore, mathematically expressible and deducible. He thus anticipated Fechner's psychophysical science and, in the Preface to the *Elemente,* Fechner acknowledges the debt: "To Herbart will always belong the credit not only of having been the first to point out the possibility of a mathematical treatment . . . but also of having made the first ingenious attempt to carry out such an enterprise; and everyone since Herbart will in this respect have to be second" (Fechner, *Elements of Psychophysics*). Davis H. Howes, and Edwin G. Boring,

40. Wundt, *Principles,* pp. 9–10.

41. Ibid, p. 17.

42. Wilhelm Wundt, *Principles,* p. 193.

43. Marshall Hall, *Memoirs on the Nervous System* (1837; p. 70), in *Significant Contributions to the History of Psychology,* Series E, Vol. I, edited by D. N. Robinson, Greenwood Publishing, Connecticut, 1978.

44. Ibid., pp. 70–71.

45. Ibid., p. 74.

46. Ibid., p. 110.

47. Wundt, *Principles,* pp. 8–9.

48. To appreciate in a historical way the approach of Fichte, it is useful to consult expositions of his works written at the time he began to enjoy wide attention outside Germany. Particularly penetrating in this respect is C. C. Everett's *Fichte's Science of Knowledge* (Chicago, S. C. Griggs and Company, 1892). In the same regard, consult E. B. Talbot's *The Fundamental Principles of Fichte's Philosophy,* New York, 1906.

49. W. T. Stace, *The Philosophy of Hegel: A Systematic Exposition,* originally published by Macmillan, 1924. This authoritative analysis of Hegel's *Encyclopaedia* and his *Phenomenology of Mind* is now available from Dover Publications, New York, 1955.

50. Bertrand Russell, *A History of Western Philosophy,* Simon and Schuster, Clarion paperback edition, New York, p. 730.

51. Ibid.

52. This remark appears in Karl Marx's Preface to the second edition of *Capital.* Marx moved away from orthodox Hegelianism in later years, but for a time he was as much a product of the Hegelian view of history and philosophy as anyone in the history of the movement. Only later did he and Engels "stand Hegel on his head."

53. Stace, *Hegel,* Sec. 75.

54. English editions of Hegel's *Encyclopaedia* are available. The William Wallace translation (Oxford University Press, 1873) can be found in later editions.

55. The authoritative translation of *Phenomenology of Mind* remains that by J. B. Baillie, London, 1910; 1931.

56. Stace, *Hegel,* p. 328.

57. Ibid., p. 343.

58. The contemporary system of psychology bearing resemblance to Hegelianism is, of course, the cognitive psychology of Prof. Jean Piaget. The Piagetian stages of cognitive development begin with *egocentrism* and culminate with the ability to identify the connection between particular instances and universal propositions. Indeed, the six-stage evolution of cognition described by Piaget fits neatly into the Hegelian three stages of sensuous consciousness, sense-perception, and intellect. Piaget's psychology, often defined as an *evolutionary epistemology,* has strived to document by experimental demonstration that logical structure of thought which Hegel deduced. It is not surprising then, that contemporary challenges directed against Piagetian psychology are of a form almost indistinguishable from the British empiricists' attacks on Hegelianism.

59. G. W. F. Hegel, *Reason in History: A General Introduction to the Philosophy of History,* translated by Robert S. Hartman, Bobbs-Merrill, Indianapolis, 1953, p. 29.

60. *Karl Marx: The Essential Writings,* edited by F. L. Bender, Harper and Row, New York, 1972, pp. 145–146.

61. Karl Marx and Friedrich Engels, *The German Ideology,* Part 2, edited by C. J. Arthur, International Publishers, New York, 1976, p. 42.

62. Friedrich Nietzsche, *Beyond Good and Evil: A Prelude to a Philosophy of the Future,* translated by R. J. Hollingdale, Penguin Books, Hammondsworth, 1973, p. 36.

63. Walter Kaufmann, "Nietzsche as the First Great (Depth) Psychologist," in *A Century of Psychology as Science,* edited by Sigmund Koch and David Leary, McGraw-Hill, New York, 1985, pp. 911ff.

64. Nietzsche, *Beyond Good and Evil,* p. 138.

11 From Systems to Specialties: The Crucial Half Century (1870–1920)

If we quickly thumb through the preceding chapters, one characteristic amid the diversity of topics and claims stands out and serves to divide all of the past from what is now taken to be the proper study of psychology. That characteristic is easier to detect than to define. Perhaps the term "system" serves most economically. If we ask, for example, what is common to the psychologies of Plato, Aristotle, Aquinas, Hobbes, Descartes, Locke, and the rest, it turns out to be their uniform commitment to develop a *system* of psychology able to embrace the fullest range of psychological phenomena: thought, emotion, memory, conduct, morality, government, aesthetics, and so forth. When we compare this commitment with the activities that most clearly identify contemporary psychology, we find a quite striking difference. To a certain degree, Freudian theory has been applied to a variety of personal and cultural matters and, to a lesser degree, modern behaviorism has attempted to address various aspects of social life and social organization. But neither Freud nor the modern behaviorist would claim that his approach to psychology is designed to cover as broad a spectrum of phenomena as the approach that was routinely considered just a century ago.

The more obvious explanations of this difference are uninforming. It could be argued, for example, that since earlier attempts failed, modern psychologists have learned to be more modest in their goals, more conservative in their speculations. Yet, if we look at the history of any significant intellectual undertaking, the same failures are common, but the surrender is not. Since the time of the pre-Socratics, there have been numerous attempts to account for all physical phenomena through a small set of universal laws, and none of these attempts succeeded. Nonetheless, today's theoretical physicist is just as committed as the ancients to an all-encompassing theory of matter. Similarly, the pages of intellectual history are filled with attempts to provide a faultless theory of politics, but none has surfaced and won the unopposed approval of the human race. Still, every year thoughtful scholars begin to sketch out a new theory, a new set of basic premises, a new "solution." Thus, we cannot explain the difference between the older and the new psychologists simply in terms of a resignation produced by prior failures.

A more subtle explanation begins with a denial of the claim. Modern psychology, on this account, pursues the same objectives that prompted all earlier endeavors, but it begins with more elemental processes and does not advance to complex phenomena until these processes have been understood. What is unsatisfying about this sort of claim is that it generally is moot on the question of which processes must be understood before the larger issues can be addressed. Is it true that contemporary studies of visual perception are undertaken as the preliminary stage of what is intended to be a developed psychology of aesthetics? Does today's psychologist study the principles of learning and memory *so that* the ageless issues of epistemology might be settled? These questions are not posed to depreciate modern efforts, but to weigh the validity of the claim that these efforts have been deliberately chosen in the interest of larger objectives.

Finally, we may take as an illustration of obvious but unenlightening explanations the one that draws a sharp line between science and every other form of inquiry. Thus, modern psychology is the way it is because it is "scientific," whereas earlier psychologies were not. What is troublesome about this account is a two-fold dilemma: First, as Chapter 1 attempted to show, it is far from clear that psychology is "scientific" or is in any demanding respect somehow "more scientific" than the psychological contributions of the past. It is clearly more *experimental,* but this is a rather different matter. Secondly, it is odd to use the adjective "scientific" as a way of explaining why a subject or range of issues has been constricted. That is, from the mere fact or allegation that X is a science, it does not follow that X is addressed to or interested in a narrower range of issues than Y.

There are, of course, other kinds of explanations too frivolous to consider. For example: Today's psychologists are simply not as bright or clever as the geniuses of old; today's psychologists are more practical and less speculative than their ancestors; today's psychologists have simply chosen their problems with more precision and have accepted as problems only those that might be settled experimentally. The first of these explanations is offensive without being documented. The second and third beg the question, rather than answer it.

By way of introducing the present chapter, it is useful to test an explanation less obvious than the foregoing but truer to the history of the trends. We might best begin with the central fact: Today's psychology is dominated by a relatively small number of highly specialized fields of inquiry and practice, and, in large measure, each of these fields has developed and continues to develop independently of the others and is a manner that is often indifferent to the others. Thus, we have (a) the psychology of personality, (b) animal learning and memory, (c) human psychophysics, (d) psychotherapy, (e) social psychology, and (f) genetic psychology. This is scarcely an exhaustive list, but it does illustrate the variety of separate issues occupying today's specialists. Within this list we recognize the relative independence/indifference just noted. The social psychologist can undertake experimental and theoretical programs without ever consulting the facts or methods of psychophysics. The psychophysicist concerned with auditory discrimination need not explore the nuances of psychotherapy. The psychotherapist proceeds without any or at least much help from the literature on comparative psychology. This fact is neither "good" nor "bad," a cause for joy or sorrow, a subject for concern or aloofness. But it is a fact of history, and one to be understood at least partly in historical terms. As previous chapters have shown, it is not a fact that appeared suddenly, but one that was taking shape for centuries. Every one of the old *isms* contributed to it: empiricism, rationalism, idealism, materialism, and nativism. So too did advances in the physical and biological sciences; and so especially did the general divorce between science and philosophy, which was decreed toward the middle of the nineteenth century. At about this time and for an additional fifty years or so, a number of scientists and scholars combined the various *isms* and scientific advances within the context of the great divorce and thereupon created special fields of psychological inquiry and practice. These individuals are the subject of the present chapter, the bridge-builders allowing us to go from the promise of a "scientific psychology" to the realities of contemporary psychology.

Comparative Psychology—From Anthropomorphism to Behaviorism

Throughout the Enlightenment the attacks on Cartesianism often included special criticism of Descartes "automaton" theory of animal psychology. As early as Pierre Bayle's *Dictionary* there were vigorous defenses of the view that all human faculties can be found in one form or another among the other advanced species and that man's claimed uniqueness was little more than vanity.[1] But Franz Joseph Gall framed the issue squarely in terms of evolutionary biology fifty years before Darwin's *Origin of Species*.[2] Gall's accomplishments would become blurred by the subsequent history of his system of phrenology, but his actual place in the history of physiological psychology remains secure. By the standards of his time, he was a painstaking anatomist and a brilliant theorist. Even before the end of the eighteenth century, he had begun research on the fetal and post-natal development of the nervous systems of a number of species, including man. He offered evidence and strong arguments favoring the conclusion that the degree of moral and intellectual prowess displayed by any animal was tied completely to the degree of cerebral development attained by the animal. He called for a new kind of taxonomy, one based upon cerebral evolution and functional capacities rather than merely general anatomy.

As has been noted in earlier chapters, the idea of progress was the preoccupation of the eighteenth century. Philosophical expressions of the evolutionary point of view were common, and even scientific versions were not rare. By the early nineteenth century, this same point of view had been fortified by the veritable "religion of nature" whose priests and prophets included Schiller and Goethe in Germany, Wordsworth and Coleridge in England, and Thoreau and the new "transcendentalists" in America. Naturalism in this version was less an addition to the idea of progress than a deduction from it. Nor, in this version, was it incompatible with orthodox Christian teaching. What it took for granted was that every living thing had its place in Creation; that each more advanced form of life carried its primitive past with it; that in the dark struggles of the living world, beauty and order arise out of chaos; that man is not removed from these natural laws and forces; and that every production of nature is but a stage in the endless march of progress.

The naturalistic point of view is evident in a wide range of early nineteenth-century activities—in art, politics, religion, and science. In all of these endeavors, the leading figures employed what was called "the natural history method," a phrase found often in the books and essays of the period. What was understood by this was the firm connection between what a thing is and how it got to be that way. To understand civilization, one studies "savages." To understand government, one examines historically earlier forms of it. And, alas, to understand human nature, one consults the rest of nature. Note that one of Darwin's engaging works was his essay on the development of his own son. Not to make too much of this, it is enough to say that Darwin's studies of fossils and his careful observations of an infant son proceeded from the same naturalistic perspective that animated the artistic, historical, and philosophical creations of the nineteenth century.

The influence of this perspective is abundantly clear in the psychological thought of the period, and it is as clear before Darwin's great contribution as it is thereafter. Throughout the early decades of the nineteenth century, the general periodicals of the day were filled with accounts of life and habits of "the lower orders"—birds, bees, ants (especially), and fish. In this same period, the combination of philanthropic sentiment and the naturalistic perspective was responsible for many essays devoted to the elimination of cruelty in the treatment of domesticated animals. The arguments advanced in these essays were strengthened by constant and exaggerated reference to the *human* qualities possessed by cats, dogs, horses, and the rest.

The theory of evolution, as set forth by Darwin, was in this respect the crowning achievement of naturalism, not its earliest bible. It put in order and on a firm foundation of fact views that had already come to dominate the thinking of the age. Indeed, when we consult critical reviews of *Origin of Species* (1859) written by Darwin's contemporaries and published in the most influential journals of the day, we are impressed by the praise and general agreement displayed toward the work.[3] It should be recalled that the famous wave of anti-Darwinian rhetoric did not begin with *Origin of Species* but with *Descent of Man* (1871). The negative response was to Darwin's "metaphysics," not his naturalism.

The first writer of note to apply evolutionary theory to a broad spectrum of psychological issues was George Romanes (1848–1894) whose first large-scale work on the subject was *Animal Intelligence*[4] (1882). We would be inclined to judge this work today as coming under the heading of the ethology rather than comparative psychology, chiefly because it is utterly lacking in experimental content and method. Carrying on a long tradition—and one to which Darwin himself was wed—Romanes presents *anecdotal* information on the mental life of lower organisms. In his Preface to this work, he is careful to observe the limitations of the "anecdotal method," and insists that he will only include those observations made by highly regarded naturalists whose objectivity is beyond question. He also anticipates those critics who will challenge the view that animals have any mind at all:

> Does the organism learn to make new adjustments, or to modify old ones, in accordance with the results of its own experience? If it does so, the fact cannot be due merely to reflex action . . . Of course to the sceptic this criterion may appear unsatisfactory, since it depends not on direct knowledge, but on inference. Here, however, it seems enough to point out . . . that it is the best criterion available; and further, that scepticism of this kind is logically bound to deny evidence of mind, not only in the case of lower animals, but also in that of the higher, and even in that of men other than the sceptic himself.[5]

What Romanes insisted is that all attributions of *mind* are inferential, since we can be sure only of our own minds and not those of others. The basis upon which we infer that Smith has fear is that Smith is doing the sorts of things we do when we are fearful. The same is the case when we attribute to Smith the properties of consciousness, motivation, memory, etcetera. Romanes was aware of the fact that some of these attributions are less firm than others, and all of them become weaker as we observe organisms lower and lower in the evolutionary series:

> [A]s the dawn of unconsciousness or the rise of the mind-element is gradual and undefined, both in the animal kingdom and in the growing child, it is but necessary that in the early morning, as it were, of consciousness any distinction between the mental and the non-mental should be obscure . . .[6]

The thesis developed by Romanes is that consciousness is the ultimate stage of mental evolution, as the primitive reflexes are the first stage. Instincts occupy a middle ground. Although a celebrated biologist in his own right, he resists the temptation to defend this thesis with facts drawn from physiology. He notes that, as functions move from the level of reflex to that of instinct, and from instinct to the genuinely rational, "the nervous processes engaged are throughout the same in kind, and differ only in . . . their complexity." [7] Thus he opposes biological reductionism and urges that the thesis be evaluated at the level of observable behavior.

What we discover in Romanes' work is a transition from introspective psychology to behaviorism, with Romanes himself reflecting the assets and the deficiencies of each. What his argument amounts to is this: (1) I begin by consulting the facts of my own consciousness, recognizing the factors that led me to certain feelings, actions, and thoughts. (2) I then take these introspectively yielded data and match them up with the behavior of lower organisms. I make careful distinctions between genuinely adaptive and learned behavior and the more rudimentary instincts and reflexes. (3) When I observe animals doing the sorts of things which I do as a result of my psychological processes, I infer that they too possess similar processes. Thus, Romanes concludes that ants practice slavery, that the termite queen summons an audience, and that the trap-door spider learns how to prevent illegal entries! What is *behavioristic* in this approach is the commitment to accept only observable behavior as the evidence of psychological functions. What is *mentalistic,* of course, is the commitment to translate this evidence into the language of consciousness. Together, these aspects of Romanes' psychology are illustrative of anthropomorphism, a very serious charge by modern lights. What this *ism* is said to be guilty of is unwarranted inference and the installation of the human mind as the standard by which to explain the behavior of nonhuman animals. Thus we find one of Romanes' contemporaries, the famous botanist William Lauder Lindsay, devoting two very large volumes to the phenomena of mental health and mental disease in animals, and urging the formation of mental hospitals for them. [8]

The antidote for Romanes' version of anthropomorphism was soon provided by his admiring critic, C. Lloyd Morgan in his *Introduction to Comparative Psychology* (1894). [9] It was in his work that Morgan developed his famous canon:

> In no case may we interpret an action as the outcome of the exercise
> of a higher psychic faculty, if it can be interpreted as the outcome of
> the exercise of one which stands lower in the psychological scale. (53)

In this same work, however, Morgan takes for granted that "It is with . . . states of consciousness that psychology has to deal" (25), and we therefore may place him within the mentalistic context of nineteenth-century psychology. Unlike

Romanes, however, he was more impressed by the potential usefulness of physiology to the study of psychological processes and by the need to rid scientific psychology of the consequences of its introspective language. We see both of these concerns in the following passages:

> [I]f we accept evolution as the true basis of explanation alike in biology and in psychology, we are justified in inferring that . . . concurrent with the community of nervous mechanism and its physiological functioning, there is a community of psychical nature and psychological functioning. (84)

> One of the greatest difficulties against which the student of zoological psychology has to contend is, that the language in which he needs must describe and endeavor to explain the mental processes of animals embodies the results of a vast amount of analytical thought. He has to employ phrases which imply analysis, to describe experiences which involve no analysis. (87)

In the first of these passages, Morgan offers anatomy and physiology as the grounds on which inferences may be plausible. That is, we only infer a psychological similarity between two animals when, in addition to the observable behavior, there is a similarity in their neurological apparatus. In the second passage, he further refines his cautions. The scientist who observes ants or bees or dogs brings to these observations a language rich in analytical content. To him, two colliding masses of ants are "armies," because only armies of human beings behave the same way. And a similar observation of human armies would lead the observer to conclude that the campaign was directed by a certain "strategy" and toward a certain "goal"; that actions of a given type were "heroic" and others "cowardly." Yet, these terms, "which imply analysis," originate in the observer and may be wrongly applied to events in which no such analysis is present. Indeed, even in human affairs many occurrences can be explained without any reference to mind or consciousness. This is not to say that human beings lack either, but that a scientific psychology may confine itself to the observed behavior and to the neurophysiological processes by which it comes about.

So far, the picture that emerges from this review of Morgan's comparative psychology is one of an utterly modern psychologist, behavioristically inclined, and confident that the theory of evolution and the science of physiology will give psychology all it needs to attain scientific status. But toward the end of his text he offers this:

> To the question, Is mental development in all its phases entirely, or even mainly, dependent on natural selection through elimination? I reply with an emphatic *no* . . . I see no evidence to show that com-

manding intellect, mathematical or scientific ability, artistic genius or lofty moral ideas, are attributable solely to natural selection. (355–356)

In this passage there is that element of conservatism found among a number of otherwise loyal Darwinists. Alfred Russell Wallace, for example, the co-founder of the theory of evolution, was also persuaded that an evolutionary account of ethics and of abstract artistic and mathematical creations simply fails.[10] It is only when we come to such radical evolutionists as Haeckel (1834–1919) that all such reservations are submerged in the triumph of monistic materialism. Thus, when Haeckel turns to those who would exempt art or ethics or sublime thought from the reach of evolutionary theory, he urges them to surrender their superstitions. As for the alleged exceptions:

> They merely cease to pose as truths in the realm of pure science. As imaginative creations, they retain a certain value in the world of poetry. . . . Just as we derive artistic and ethical inspiration from the legends of antiquity . . . so we will continue to do in regard to the stories of the Christian mythology.[11]

But this was merely polemical Darwinism, and the builders of modern psychology wisely avoided such material. Instead, they took from the theory of evolution what was most serviceable, and from Romanes and Morgan what was most in keeping with the tone and objectives of twentieth-century psychology. The first clear result of this was the Functionalist school discussed briefly in the next chapter. One of its major architects was James Rowland Angell (1869–1949), who would look back at the heat of the past and observe in the cool and confident language of the modern psychologist,

> Our whole tendency now-a-days is to recognize and frankly admit, that inasmuch as we must infer the psychic operations of animals wholly in terms of their behavior, we are under peculiar obligation to interpret their activities in the most conservative possible way.[12]

Angell had been William James' student at Harvard and John B. Watson's teacher at Chicago. Even in this brief passage, we can see the old mentalism and introspectionism receding, and the new objective behaviorism on the horizon.

Defining Psychology: Natural Science, Mental Science, Social Science?

The roots of modern neuropsychology are most firmly planted not in the purely speculative materialism of the eighteenth century, but in the clinical and experimental discoveries of the nineteenth. The contributions of Gall, Bell, Magendie, and others have already been cited (Chapter 10) in this connection. But it was only later in the nineteenth century that genuinely modern integrations of clinical and experimental findings were achieved. It is appropriate here to compare the old and the new views of mind and of mental illness within the general context of medicine and biology. A very useful basis for comparison is found in the history of law as it pertains to the "insanity" defense.

During the peaks of their ancient civilizations, both Greece and Rome wrestled with the question of legal responsibility and liability, and the laws of both made provision for diminished responsibility. Children, for example, were accorded special exemptions. But in dealing with adults, the presumption of personal responsibility was the rule, not the exception. In their developed form, the laws of Greece and Rome specifically outlawed the *vendetta* and the so-called "law of revenge" (*lex talionis*), and did so on the grounds that only the actor can be judged as responsible for the action in question. But what, then, of insane persons? Again, both Greek and Roman law acknowledged conditions of insanity and both granted a degree of relief to criminals judged to be insane. But the criterion of insanity was rather sharply set. In Rome, for example, the defendant had to be found to be *fanaticus* and *non compos mentis*. The former term is the equivalent of "wild," the latter, "no power of mind" or "no controlling mental power." In other words, the defendant had to qualify as something less than a human being.

This ancient "wild beast" standard was retained in Western law until the very beginning of the nineteenth century.[13] However, in the landmark British case of *Hadfield* (1800), this standard was directly and successfully challenged. Counsel for the defense argued that Hadfield, who was charged with treason (for attempting to take the life of George III), was laboring under a delusion, and that this delusion was very likely the result of brain injuries sustained in war.[14] Hadfield's acquittal illustrates the general willingness to accept neurological opinions as grounds for exoneration. We see, then, that as early as 1800 the "brain theory" of mind was sufficiently compelling to be decisive in a celebrated legal case. The point here is not that by 1800 even the man in the street had come to adopt a materialistic psychology, but that he was willing to concede that certain peculiar states of mind could be induced by pathological states of brain. Thus, the concept of free will was not abandoned, but was somewhat casually absorbed into the larger naturalistic framework. Hadfield, the reasoning went, did not act freely because his will was clouded by a delusion, and the delusion was conditioned by a diseased brain. Note that no such evidence was actually presented in this case; it was merely inferred. And note, too, that in the vast majority of similar cases, right up to our own day, the allegation of "brain disease" rarely is corroborated by clinical neurology in cases involving criminal offenses.

Within the more or less official councils of thought, however, there was (and is) a lingering debate on the extent to which mental life might be understood in purely neurological terms and by purely neurological methods. Throughout the nineteenth century, eminent spokesmen for both sides appeared and served up a wide assortment of arguments. Any brief attempt to summarize these runs the risk of libel, but we can at least strive to capture the essence of the competing views. On one account—let us call it the "mentalistic" account—the subject matter of psychology is consciousness and the best (if not the only) method by which it may be studied is the psychological method of self-examination. We call this approach "introspectionism," but it is important to recognize the subtler meaning of the term. As exercised by the earlier philosophical psychologists (for example Locke, Berkeley, and Hume), the introspective method was confined to an examination of one's own ideas and experiences on the assumption that all healthy persons had ideas and experiences in much the same way and according to the same basic principles. But with the advent of research on sensation, that is, actual experiments on perceptual processes, the introspective method was, as it were, externalized. Thus, the introspective method criticized by Wundt was the method of private or personal introspection of the philosophers. The method of *experimental* introspection, however, is nothing but the psychophysical methods developed by Fechner and praised by the same Wundt. Fundamentally, these methods proceed from the conviction that only a person having an experience can report it. Therefore, the proper study of such experiences (sensations, percepts, cognitions) necessarily depends upon objective observation and measurement of the self-examinations of the subject.

John Stuart Mill dubbed this the "psychological method" and he defended it against the utterly biologized psychology that Comte's *Positive Philosophy* demanded. Indeed, one of the reasons Mill had become disaffected with Comtean positivism—which he had stoutly supported earlier—was his conclusion that the positivist approach to a science of the mind was naive and constricted. The tension between Comte and Mill on this point was not a result of differing views regarding experimental science, for on this they were in complete accord. Instead, it arose out of differing views on the nature of explanation. Mill's defense of the "psychological method" was in keeping with his general and developed views on inductive science, his resistance to metaphysical speculation, and his disavowel of deductive methods when applied to matters of fact.

Let us now examine the other view, which, for the present purposes, will be called "reductive materialism." Its defenders begin with the proposition that all mental states, events, and processes originate in the states, events, and processes of the body and, more specifically, of the brain. To develop a true science of the mind, therefore, requires an understanding of the laws governing the organization and activities of the brain. For this, nothing more is needed than a careful examination of relevant clinical patients and a systematic program of research into the functions of the brains of the advanced species.

One of the most influential texts to arise out of this perspective (and to do so much to make it "official") was Henry Maudsley's *Physiology and Pathology of the Mind* (1867).[15] Maudsley is now known for many contributions. He was largely responsible for the creation of out-patient facilities in mental hospitals and for his powerful defense of the "medical model" of mental illness.[16] In the book of 1867, Maudsley paused to examine Mill's position:

> Mr. J. S. Mill has made a powerful defence of the so-called Psychological Method. In his criticism of Comte in the *Westminister Review* for April 1865, and in his "examination of Sir William Hamilton's Philosophy", he has said all that can be said in favour of the Psychological Method, and has done what could be done to disparage the Physiological Method. . . . [T]he admirers of Mr. Mill cannot but experience regret to see him serving with so much zeal on what seems so desperately forlorn a hope. Physiology seems never to have been a favourite study with Mr. Mill. . . . The wonder is, however, that he who has done so much to expound the system of Comte, and to strengthen and complete it, should on this question take leave of it entirely, and follow and laud a method of research which is so directly opposed to the method of positive science.[17]

In calling this perspective "reductive materialism," care must be taken not to confuse it with a new version of atomism or epicureanism or sensationism. The "reductive" feature refers to the requirement that the phenomena of mental life be reduced to the laws of neural function. Moreover, stress is laid to the dynamic nature of brain-function—its evolutionary character. Maudsley, for example, was as opposed to Mill's associationistic psychology as to his "psychological method":

> Infinite mischief and confusion have been caused by the habit of speaking of ideas as if they were the mechanical stamps of impressions on the memory, instead of as, what they truly are, organic evolutions in respondence to definite stimuli; our mental life is not a copy but an idealization of nature, in accordance with fundamental laws.[18]

Nor are we any longer confronting the odd "vibratiuncles" of a Hartley or the passive but "sentient statue" of a Condillac. The new neuropsychology is prepared to assimilate the entire range of psychological determinants as these come to affect the brain:

> When we are told that a man has become deranged from anxiety or grief, we have learned very little if we rest content with that. How does it happen that another man, subjected to an exactly similar cause

of grief, does not go mad? It is certain that the entire causes cannot be the same where the effects are so different; and what we want to have laid bare is the conspiracy of conditions, internal and external, by which a mental shock, inoperative in one case, has had such serious consequences in another. A complete biographical account of the individual, not neglecting the consideration of his hereditary antecedents, would alone suffice to set forth distinctly the causation of his insanity.[19]

In contrasting the positions of Mill and Maudsley, we catch glimpses of an issue that pervaded psychology throughout the nineteenth century, whether that psychology was taking place in the laboratory, in the lecture hall, in the clinic, or in the psychiatrist's private office. Together, naturalism, Darwinism, and the idea of progress had the effect of challenging that traditional confidence of the arm-chair psychologists. No one any longer believed that the full set of psychological laws could be discovered by a Lockean or Humean form of speculation. Nor was there the same willingness to dismiss "madmen" on the grounds that a science need not waste time on accidents and eccentricities. Each individual life was now judged to be a *life in progress,* an *evolved* life that could only be understood and explained by charting its unique experiences, its unique heredity, and its dynamic states of consciousness.

But even in this, the nineteenth century was caught up in one of those titanic contradictions for which the Victorian age is ungenerously remembered. On the other hand, there was that evolutionary naturalism concerned with the history and the destiny of the species as a whole. On the other hand, there was that libertarian ethic that nearly spent itself in striving to secure the freedom and defend the dignity of every individual. The ways invented by that century to eliminate the contradiction were elaborate, confused, confusing, and are very much with us now. As suggested in the previous chapter, Wundt's approach was to insist on two distinct sciences: one addressed to the individual mind (its "manifold of consciousness") and relying primarily on the methods of psychophysics, the other addressed to social aggregates and relying on the methods of historical and anthropological analysis.[20] The former science would be quite naturally tied to other natural sciences, and especially to biology. Wundt makes it quite clear in his *Lectures on Human and Animal Psychology,* however, that neither radical materialism nor radical idealism will have a place in this science.[21] The focus remains ever on the facts of *consciousness,* and not on the alleged neural or spiritual causes.

If we could see every wheel in the physical mechanism whose working the mental processes are accompanying, we should still find no more than a chain of movements showing no trace whatsoever of their significance for mind. . . . [A]ll that is valuable in our mental life still falls to the physical side.[22]

But Wundt's psychology of the individual mind is still "physiological" in the sense discussed in the previous chapter. It is a psychology of laws and principles which can be unearthed through the enlarged methods of psychophysics.

But on the matter of the second science, the "folk psychology," Wundt provides a most revealing side of the nineteenth century's struggle with determinism. The question at issue has to do with social man, the man of action, and not the idealized man examined in the perception laboratory. Now, it is all too common in discussions of the history of psychology to find Wundt described as a "voluntarist," and to make no connection between this and the balance of his psychological inquiries. We begin to detect the connection, however, when we recognize the implications contained in the above quotation and in those provided in Chapter 10. Wundt's commitment was to the psychology of *consciousness,* a science of mind as mind. When he insists in the above passage that "all that is valuable in our mental life still falls to the psychical side" even after we have studied brain mechanisms exhaustively, he is simply stating his essential defense of the psychological method. Thus, what is rejected is not so much the thesis of radical materialism but the correlated claim that the methods of the radical materialist are even germane to a scientific psychology. Again, let us be cognizant of Wundt's genuine respect for and thorough knowledge of the methods and findings produced by the neural sciences of his day. What he would not be led to by them, however, was the all too complacent belief that the problems of psychology would be (or could be) solved by them.

What was the source of his resistance? Quite simply, it was the facts of consciousness, itself, which established that significant human actions proceed from volition, and that volition was not explainable in terms of neural events.

To appreciate this analysis, it is necessary to understand what Wundt meant by "volition." If all he had meant was "desire" or "motivation," we would simply counter his claim with the well-established facts that connect brain activity with motivational states. But for Wundt, a motive is a uniquely psychological entity, not to be confused with biological drives or emotional states. A motive is a *reason for acting:*

> When we have taken account of every one of the external reasons that go to determine actions, we still find the will undetermined. We must therefore term these external conditions not causes, but *motives,* of volition. And between a cause and a motive there is a very great difference. A cause necessarily produces its effect: not so a motive. . . . [S]ince all the immediate causes of voluntary action proceed from personality, we must look for the origin of volition in the inmost nature of personality—in *character.* Character is the *sole immediate cause* of voluntary actions.[23]

Again, Wundt was not an anti-determinist, but he was an anti-mechanist. When he refers to "character," he refers to the complex creation of biological organization, cultural influences, hereditary predispositions, and that matrix of beliefs, opinions, attitudes, and feelings that give a person a unique identity. To understand this person, one must invoke the methods of historical science, not physical science. For to understand the person's *character* is akin to understanding an entire culture, and the determinants of the consciousness of that culture.

Many of these points can be found in the *Principles of Psychology* (1890) and other writings of William James (1842–1910), perhaps the most important figure in the history of American psychology. In James' books and essays, as in Wundt's, we discover that a scientific psychology begins with the facts of consciousness and proceeds to the laws of their organization. We discover also a general dismissal of the view that such facts and laws are (or even are in principle) reducible to physical or biological processes. And, finally, we discover a special place given to the *fact* of human will and the peculiar autonomy it displays in a variety of settings, not the least of which is the religious.

In the abridged (1892) version of his two-volume classic, James begins his presentation of psychology with allegiance to the biological perspective, and offers a materialism as a plausible hypothesis: "The immediate condition of a state of consciousness is an activity of some sort in the cerebral hemispheres." [24] He takes this hypothesis to be soundly supported by many medical and experimental findings and assumes therefore, "without scruple . . . that the uniform correlation of brain-states with mind-states is a law of nature." [25] His subsequent examination of the sensory and motor functions only tends to support this "working hypothesis." But then, when discussion turns to the matter of consciousness itself, James' loyalty to the hypothesis sustains first embarrassment and then abandonment:

> When psychology is treated as a natural science . . . "states of mind" are taken for granted, as data immediately given in experience; and the working hypothesis . . . is the mere empirical law that to the entire state of the brain at any moment one unique state of mind always "corresponds." This does very well till we begin to be metaphysical and ask ourselves just what we mean by such a word as "corresponds." This notion appears dark in the extreme. . . . [T]he difficulty with the problem of "correspondence" is not only that of solving it, it is that of even stating it in elementary terms. . . . We must know which sort of mental fact and which sort of cerebral fact are, so to speak, in immediate juxtaposition. We must find the minimal mental fact whose being reposes directly on the brain-fact. . . . Our own formula has escaped the . . . assumption of psychic atoms by *taking the entire thought* (even of a complex object) *as the minimum with which it deals on the mental side,* and the entire brain as the minimum on the physical side. But the "entire brain" is not a physical fact at

all! . . . Thus the real in physics seems to "correspond" to the unreal in physics, and *vice versa;* and our perplexity is extreme.[26]

James' friend, the distinguished philosopher C. S. Peirce (about whom more is said in the next chapter), had put the case more directly:

> The materialistic doctrine seems to me quite as repugnant to scientific logic as to common sense; since it requires us to suppose that a certain kind of mechanism will feel, which would be a hypothesis absolutely irreducible to reason—an ultimate, inexplicable regularity; while the only possible justification of any theory is that it should make things clear and reasonable.[27]

James, Wundt, Peirce, and their disciples shared an impatience with overly confident materialism, and an eagerness to rid psychology of the heavy metaphysical baggage that must be carried by those who insist that important questions are now settled once and for all. But these same attributes may be assigned to those like Maudsley who found in neurophysiology that very simplicity of description and economy of explanation for which philosophical psychologists had been searching for centuries.

The principle disputants in the free will-determinism controversy were drawn primarily from the quarters of "voluntarism" and "materialism," but the nineteenth century provided room for other perspectives as well. There were some observers who were convinced that the actual facts of human life were not properly comprehended or fully addressed by any of the prevailing doctrines; that introspection, confined to "states" of consciousness, was unable to discover the *laws* of mind; that Darwinism confused analogies with identities; and that materialism was hopelessly inadequate in the face of the complexity of social life. George Henry Lewes (1817–1878) is illustrative of the nineteenth-century thinkers who expected psychology to be a "positive" science, but who rejected the tidy *isms* promoted by their contemporaries. In his *Problems of Life and Mind: The Study of Psychology* (1874–1879),[28] Lewes argued more in the tradition of J. S. Mill and against those he judged to be uncritical in their acceptance of Darwinism, materialism, and introspectionism:

> The conditions of existence of mental phenomena are not only biological but also sociological studies. A serious investigation of these will serve to remove most if not all of the difficulties which make men cling to the spiritualist hypothesis, because they are profoundly impressed with the inadequacy of the materialist hypothesis. (Ch. IV, Sec. 61)

> We must quit Introspection for Observation. We must study the mind's operations in its expressions, as we study electrical operations

in their effects. We must vary our observations of the actions of men and animals, by experiment, filling up the gaps of observations by hypothesis. (Ch. VI, Sec. 75)

For Mr. Darwin's purpose it was needful that he should emphasise the position that "there is no fundamental difference between man and the higher animals in their mental faculties." . . . For our purpose it is needful to point out that, while there is no fundamental difference in the *functions* of the two, there is a manifest and fundamental difference in the evolved *faculties*. . . . When Comte affirms that there is nothing in Humanity, the germs of which are absent from Animality, the assertion requires qualification. Animals may be said to have the germs of our moral and intellectual life in somewhat the same sense that serpents have the rudiments of our limbs. (Ch. VIII, Sec. 101)

Note in these passages that Lewes is not so much denying determinism as he is rejecting mechanical versions of it and pointing to the wider social context within which determinative forces operate. Note, too, the insistence on the need for experimentation and the importance of objective observations of human and animal activity (behavior). Lewes does not oppose the study of animal psychology. He cautions, however, against overly "Darwinized" interpretations of such findings as may be produced by this study. In the end, the question of determinism is either meaningless or it admits of empirical test. In any case, the proper subject of psychology is the actual conduct of real persons in socially significant settings, their doings, their ideas, and the pressures operating on them.

As in the area of comparative psychology, so too in the nineteenth-century treatment of consciousness and mental life was psychology moving in the direction of the practical, the antimetaphysical, the antitheoretical. In some respects, the writings of Wundt, Lewes, James, Peirce, and the others diminished the general enthusiasm for a reductionistic physiological psychology. But the greater effect was to create a place for both psychology as a *natural* science and as a *mental* and *social* science. Thus, specialists did not suddenly appear, but certain methods and problems, which by their very nature came to be a collection of specialties, were gradually adopted.

The Darwinian perspective, too, tended to question the role of reductive materialism in psychology. With its emphasis upon dynamics, it opposed the traditional mechanistic theories of life and replaced them with a usefully general theory of functional adaptation. And with its emphasis upon progress through strife, it put the "life history" approach to psychological issues on a firm naturalistic foundation. More subtly, it imparted a pervasively pragmatic tone to psychological inquiry. It replaced the introspectionist's question, "What is this mental state?" with the functionalist's question, "What is it *for?*" And it tended to blur the age-old distinctions between instinct and rational choice, habit and purpose, accident and design. Again, Darwinism was rather the culmination than the origin of these positions.

Clinical Psychology and the Unconscious

The notion of unconscious sources of motivation is a very old one both in literature and philosophy. Even in the epics of Homer we discover dream-merchants sent by the gods to plant ideas and fears in the mind of the unsuspecting sleeper. So also is the more general notion that, underneath the veil of reason, our mental life actually proceeds according to the mandates of passion.

The history of theories of mental illness is a subject worthy of book-length treatment, but its essential character is revealed by two competing and conflicting assumptions about the very nature of the mind. One is, as we have seen in previous chapters, the naturalistic; the other is the spiritualistic. But in making this distinction, we are not always in a position to anticipate the corollaries that might be derived from these assumptions. Consider, for example, the "demonic possession" theories widely circulating from the late Middle Ages to the Elizabethan period, and the "lunar" theory of madness that replaced it. The argument for a demonic or spiritual possession of the mind has not always or often been based upon sheer superstition. Rather, it follows quite directly from a Cartesian-type theory of mind, a theory not inconsistent with Aristotle's account. Without carefully developing the argument, let us simply take the following propositions for granted: (1) The mind, in its highest level of activity, comprehends abstract principles, universals, and general concepts which are not given to or through the senses. (2) It is this rational faculty that allows rational beings to order their conduct in a purposeful manner and to bring their judgments into line with the realities of their lives and surroundings. (3) Since this rational faculty is in possession of knowledge that could not be gleaned from the senses, and since it pertains to that which is not material, the faculty itself cannot be a material faculty. (4) Since it is not a material faculty, it cannot be affected or influenced by anything that is purely material. Now, if all of this is accepted, it follows that the source of irrationality or genuine "insanity" must be immaterial, which is to say *spiritual.* That is, the only thing that can take possession of the rational faculty is that which is somehow like it and, as we have seen, the faculty in question is not like matter.

When we turn to the "lunar" theory, according to which states of madness are induced by the phases of the moon, we discover what was actually an early naturalistic theory of mental illness not unlike Mesmer's "magnetic" theory developed late in the eighteenth century. But note the grounds on which the rationalist (Aristotelian, Cartesian, etc.) would oppose such views. First, they entail the notion of "action at a distance" and offer no evidence of a medium through which the moon or the magnet might reach the mind. But more importantly, they require that purely physical forces influence that which has no physical features. To the rationalist, therefore, lunar and magnetic theories are simply witchcraft by another name, and are steeped in the "natural magic" of Renaissance heremeticism.

Mesmer (1734–1815), it must be recalled, always insisted that the magnetic trances he induced were explicable in physical terms, and that there was nothing magical or superstitious about them. His own medical training had been the best (he completed his work at Vienna under such notables as van Swieten) and, until "Mesmerism," he had been a highly regarded physician. But when his claims and theories were officially reviewed they were dismissed on a variety of grounds, including the "action at a distance" problem.[29] Even later, when such prominent medical men as John Elliotson (1791–1868) and James Esdaile (1808–1859) tried to convey to their British colleagues the great anaesthetic power of "mesmerism," they all but lost their standing in professional circles. Elliotson found and edited *The Zoist* (1843–1856) precisely because the scientific journals of his day would not publish studies reporting the salutary effects of mesmerism.

This discussion strives to underscore how completely the scientific establishment had come to subscribe to a mechanistic materialism, and how resistant this establishment was to anything that smacked of the old "metaphysics." In England there was a particular disinclination toward theories of any kind, let alone those advanced by the mesmerists. This same disinclination was fortified by the excessive claims of the new Romantic idealists—the direct descendants of Schelling, Goethe, Fichte, and Hegel—who were now seeking to replace empiricism and natural science with the Absolute Idea. On their construal, the mere physical facts of the universe hide the grander purpose behind it and fail to reflect the central fact of freedom, toward which the human spirit is unconsciously but inexorably propelled. The scientists of the time not only did not weigh these propositions very carefully, they developed a keen sense for detecting signs of the Absolute in any new idea presented for their judgment, and were quick to reject it should the signs be there.

Against this background, it becomes easier to understand two aspects of the psychoanalytic theory propounded by Sigmund Freud: first, the cool reception it received in the medical circles of the day; and, secondly, Freud's eagerness to translate, wherever possible, the psychoanalytic concepts into neurological ones. Before returning to this, however, it is instructive to examine the broad features of Freudian theory.

Freud was the product of that marvelously contradictory climate of German thought in which science was defined in the positivisitc, deterministic, and physicalistic language of Helmholtz, and in which philosophy was Hegelian.

Let us review the prevailing forces. Freud was born in 1856. He completed his doctorate (in neurology) in 1881. Wundt's laboratory had already been in existence and in highly productive existence for two years. Helmholtz, now sixty years old, was the senior statesman of German science. Evolutionary theory, which most of the leading antivitalist biologists adopted as fact, had already been in print for more than twenty years. Darwin himself was dead only six years. His most ardent and able disciples, Thomas Huxley (1825–1894) and Herbert Spencer, were both alive, and Spencer especially was promoting the psychology of "instincts." If we wonder how much of the new science was part of young Freud's education, we need only recall that his mentor in neurophysiology and neurology was Professor Ernst Brücke, himself a student of Müller's and one of Helmholtz's closest associates.

Philosophically the climate was overwhelmingly conditioned by Hegelianism. Psychology and phenomenology were two terms expressing the same subject: the subject of consciousness. The driving force in the universe was the free mind, a mind that had evolved from more primitive, nonreflecting, irrational substrates. Even Wundt, far from a Hegelian, had at least agreed that the subject matter of psychology was exhausted by the contents of consciousness. Wundt and Hegel also agreed on the central position of feeling in psychological life. If it is agreed that Freud had one of the great synthesizing minds of all time, we now have the list of major perspectives available for synthesis.

His energies as a senior graduate student were devoted chiefly to neurophysiological research under Brücke's direction. We know he hoped for a professional chair at the University of Vienna and that, as a Jew, his chances were negligible. The demands of marriage and family forced him into private practice in neurology, although, from the first, he never interrupted his research endeavors. Interestingly, he is cited in Wundt's *Principles* for a published article on aphasia that appeared in 1891.[30] A number of his patients suffered from "hysteria," which, at the time, was the diagnostic category invented to account for sensory and motor deficits occurring in the absence of detectable neuropathology. The French pathologist Jean Charcot (1825–1893) had revived and given respectability to the practice of hypnosis and was employing it in the treatment of certain hysterical cases as well as for anaesthetic purposes. Freud attended Charcot's lectures (1885–1886) at the University of Paris. He returned to Vienna, eager to apply the new technique, which had already caught the attention of Joseph Breuer, another of Brücke's former pupils. Freud would later write:

> Granted that it is a merit to have created psychoanalysis, it is not my merit. I was a student, busy with the passing of my last examinations, when another physician of Vienna, Dr. Joseph Breuer, made the first

application of . . . [hypnosis] . . . to the case of an hysterical girl (1880–1882).[31]

It was Breuer too who discovered, as one of his patients called it, the "talking cure" that later would receive the more impressive if less direct label "catharsis." While their medical relationship was not to last (though their friendship did), Freud and Breuer introduced the bare outlines of psychoanalytic theory in a series of jointly published papers beginning in 1895 and summarized in their *Studien über Hysterie*. Professor E. G. Boring gives a brief but searching insight into the theoretical bond between the two, Breuer being fourteen years the senior:

> Breuer held that a certain amount of the organism's energy goes into intracerebral excitation and that there is a tendency in the organism to hold this excitation at a constant level. Psychic activity increases the excitation, discharging the energy. . . . What Breuer and Freud had from it, however, was the conception of psychic events' depending upon energy which is provided by the organism and which requires discharge when the level is too high. Because Brücke had trained them into being uncompromising physicalists, they slipped easily over from the brain to the mind.[32]

Under hypnosis, the patient's hysterical symptoms could be transferred from one part of the body to another. The hysterically paralyzed hand could be made to move, but with the consequence of immobilizing the intentional use of the leg. Sight could be restored, but with deafness following in its wake. Since the patient showed no understanding of the causes of the problem and since hypnotic induction could relieve the symptoms, Freud and Breuer concluded that the mechanism of symptom-formation was *unconscious*. Contrary to two widely held views, the symptoms were not feigned in an attempt to receive pity, nor were they limited to women (even though the term "Hysteria" derives from the Greek for *uterus*).

The unpopularity of Freud's notions, differences of opinion between him and Breuer, and Breuer's desire to quit science for practice, all worked to drive the two apart. By 1900 their relationship was almost exclusively social. It was in 1900 that Freud's *Interpretation of Dreams* was published. In the following year, *The Psychopathology of Everyday Life* appeared. Thus, as the twentieth century began, he had a following. By 1920 there was a movement. By 1930 he was, and has since remained, the preeminent figure in modern theoretical psychology.

Central to Freud's system is the concept of "unconscious motivation. Herbart and Fechner had both speculated about unconscious processes. But Freud's use of the concept was among the first to be scientific. That is, in the sense in which the term was introduced in the first chapter, a "scientific explanation" was generated by Freud's treatment of the idea. It became a "covering law" from which one could deduce certain consequences. His recourse to the concept was based on the need to discover a force or an agency by which behavior might be controlled independently of the patient's will. When Breuer's patient was relieved by the cathartic method, it was clear to Freud that the power of particular memories to control behavior was reduced once the actual memories were revived. Accordingly (and here the metaphor of the machine becomes reality), he reasoned that painful thoughts are *repressed,* that they come to reside in the unconscious, and that the symptoms are the work done by the repressed elements—that is, psychic energy is conserved through symptom-formation. Only through a conscious reliving of the former trauma, an exhumation of it from the unconscious reliving of the former trauma, an exhumation of it from the unconscious recesses, can the symptoms finally be removed. Short of this, they can only be moved about the body, disguised, accepted. Freud's own words could not be more descriptive:

> *Hysterical patients suffer from reminiscenses.* Their symptoms are the remnants and the memory symbols of certain (traumatic) experiences.[33]

Freud's quick renunciation of hypnosis is to be understood not only in terms of his Helmholtzian bias—he described hypnosis as "fanciful" and "mystical"—but also as the result of his inability to bring a number of patients "under." He abandoned it completely in favor of the cathartic method and its complement, free association. Reasoning that neurotic symptoms are the result of incomplete or unsuccessful repression, Freud sought to discover the traumatic episode in the patient's life by a sort of subterfuge. This involved an analysis of so-called slips-of-the-tongue (*parapraxes*), automatic or free associations, and, most salient of all, dreams.

For Freud the determinist, there was no more reason to consider dreams and paraphraxes as uncaused than to consider normal speech or wakefulness uncaused. Thus, the "Freudian slip," no less than the disguised plots and fantasies of the dream world, under the penetrating lights of psychoanalysis, can be shown to have a direct bearing on the remnants of childhood traumas. These result from the very evolutionary progression that the human psyche undergoes on its way to an adult stage. Utterly consonant with Darwinism, Freudian theory rests on the notion of instinctual biological drives, that impel the individual to act in such a way as to survive. The principle governing or defining all these drives is that of *pleasure.* The exercise of this principle is sexual gratification which, in its most advanced expression, involves heterosexual relations for the express purpose of procreation. However, the individual arrives at this level only after successfully passing through more primitive stages of gratification at any one of which the individual might be arrested by trauma. The stages are identified in terms of the particular source of pleasure: oral, anal, genital, and phallic. The sexual "energy" devoted to this endless search for pleasure is the "libido," which operates in the service of that most primitive and survivalistic element of psychological beings, the "id." Since the instincts of the id are such as to lead to incestuous, murderous, and purely egoistic actions, no human society could survive its unrestrained expression. Every society, then, must "socialize" its young: develop in them a conscience, or "superego," which will direct the individual not away from gratification but toward socially acceptable means of gratification. The compromise between the impulses of the id and the constraints of the superego results in the self of whom we are aware, the "ego."

On the way to mature sexual motivation (this being the motive to procreate through heterosexual encounters), the child must select appropriate objects. The male child's natural proclivity is for his mother, who has been abidingly associated with gratification. However, to court one's mother is, simultaneously, to court a showdown with one's father, the effect of which is or may be castration. An Oedipal complex results, in which the boy can only succeed in removing the threat of castration by becoming his father's equal, or by shifting his quest to another object. Pathological conditions result from the failure to resolve the tensions implicit in the Oedipal stage.

Even from this mere sketch of the Freudian theory of personality, we are able to appreciate the synthetic features of the system. The human personality *evolves.* Its origins are animalistic, survivalistic, Bentham's pleasure principle has been raised to the level of a medical or neurological reality. The engine of psychological growth is energy, which behaves according to the same sorts of laws prevailing in the physical world. As with physical energy, psychic energy is conserved, directed, and partitioned, but never destroyed. With Fichte and Schelling, Freud saw the world as a set of polarities, with the forces at one pole opposing those at the other, an id struggling for supremacy over a superego, a "life force" (*eros*) in heedless conflict with a "death wish" (*thanatos*), natural instincts driving the organism toward the next stage in the face of societal taboos and rites that might otherwise frustrate instinctual expression— and with, through all of this, Freud constantly asserting that the ultimate topic of psychological concern is consciousness in that daringly Hegelian sense: the knowing, feeling, mind in search of an Absolute that might bring it peace and harmony.

A group was soon to form around Freud and to create the Freudian "school" of psychoanalysis. Membership in it was a fluctuating affair, primarily because of Freud's insistence on strict orthodoxy and because of the pettiness of his reactions to disagreement. On the whole, the departures from orthodoxy were largely matters of emphasis rather than fundamental disagreements over the central terms of the theory. At least at the outset, those who might otherwise be called "Freudians" were distinguishable in terms of the relative importance they gave to social as opposed to instinctive determinants, and to where in the hierarchy of instincts they placed the several contained in Freud's theory. Carl Gustav Jung (1875–1961), for example, was more self-consciously "Darwinian" in his conviction that the unconscious contains not only the repressed elements of an individual's personal experiences but also the *collective* unconscious elements characteristic of the entire race. Thus, every male has an idealized or "archetypical" unconscious picture or conception of the female (*anima*), and every female has the corresponding male archetype (*animus*) within her unconscious. Accordingly, and in contrast with Freud's notions, the unconscious cannot be understood solely or even primarily through an analysis of personal history, nor can neurosis be understood solely or even primarily as the consequence of sexual forces.[34] Alfred Adler (1870–1937) departed from Freudian orthodoxy also and, like Jung, founded his own "analytical" school of psychoanalysis. Central to Adler's position is the notion of a "will to power," and the dynamic interaction taking place between an ever-evolving self and the shifting fortunes held out by society. Thus, in Adler we discover a movement away from Freudian determinism and its implicit dependence on physiology and the natural sciences.[35] Both the Jungian and the Adlerian alternatives were more formal breaks with Freudian orthodoxy, but long before either there were psychoanalytic perspectives quite at variance with the theory for which Freud would become so famous.

To pick up the non-Viennese lineage of psychoanalysis, we turn to France and the productive career of Pierre Janet (1859–1947), who received his doctorate under Charcot, whom he succeeded as head of the Psychological Laboratory at the Salpêtriere in 1890. Jung himself had gone to Paris to study with Janet, whose international repute was already great by 1906 when he was invited to give a course of lectures at Harvard. In addition to his own original thoughts on the nature and causes of mental illness, the value of examining Janet's work from a historical perspective is that it sheds additional light on the problems Freud was to have with the scientific community. In this connection, Janet's most original work is also one of the few to have been translated into English—"The Mental State of Hystericals"[36]—and we will briefly consider it here.

Janet's chief interest was the diverse phenomena of hysteria to which Charcot had already devoted his energies and to which Freud would soon devote his. Hysterical symptoms ran the gamut from sleep-walking and morbid fear to "automatic writing" and profound losses of sensitivity and locomotion. The darkness that surrounded these disorders is suggested by the very term derived from the Greek word for *uterus*. Indeed, until Charcot had documented a number of cases of what he called "virile hysteria," the medical consensus stood behind the claim that only women were afflicted by it.

Given the nature of hysterical symptoms, sufferers generally sought the opinion and treatment of neurologists. But as Charcot and Janet demonstrated so convincingly, genuinely hysterical patients displayed no signs whatever of neuropathology. Moreover, although the symptoms limited neurological disturbances, they did not behave like such disturbances when carefully observed. Hysterical anaesthesia is illustrative. When anaesthesia is the result of neurological disease, all the reflexes associated with the now-lost sensations are also eliminated. But in hysterical anaesthesia, the sensitivity is gone while the reflexes remain generally unaffected. Moreover, hysterical anaesthesias are highly mobile and variable, changing not only in severity but in their location. The patient may have no sensitivity in the upper extremities on Thursday, and then none in the lower extremities on Friday—at which time the upper extremities display normal sensitivity.

The scientific question that arises in the face of such phenomena concerns first the issue of method and then that of explanation. Janet at this point was intimately familiar with the research of Wundt and his colleagues at Leipzig, and was convinced that their methods were unsuitable. The general aloofness of the Leipzig group toward complex medical problems—their "purist" stance on the nature of science—caused them, in Janet's view, to suffer a self-imposed naiveté. Even their notion of elementary sensations is largely stripped of psychological content, since their focus is on the verb (I *see*, I *feel*, I *hear*) to the exclusion of the subject (*I*).

> [T]he words "I, me" . . . are terms enormously complex. This is the idea of personality; that is to say, the reunion of presentations, the remembrance of all past impressions, the imagination of future phenomena. It is the notion of my body, of my capacities, of my name, of my social position, of the part I play in the world; it is an ensemble of moral, political, religious thoughts, etc.; it is a world of ideas. . . .[37]

Thus, Janet will not adopt the experimental methods of Wundt, for what concerns Janet is a malady of the *personality* as it reveals itself in perception or action or emotion; not a malady of perception per se. But if the methods of introspection and psychophysics are insufficient, must not those of the physiologist be used if the entire enterprise is to remain within a scientific context? This, of course, was the central issue, and what would come to be called the "psychoanalytic" method was designed to settle it. In Janet's hands, the method proceeds, as he says in his Introduction, from a "strict determinism" which ignores the study of "philosophic problems," and which begins with an objective description of the patient "in and by himself."[38] The overarching determinism accepted by Janet dictates that every mental phenomenon originates in the brain and that all mental pathology is of cerebral origin. There is no inconsistency between this thesis and the fact that the hysterical symptoms are not neurological. The latter term generally suggests some disease in the sensory or motor pathways or some lesion in a region of the brain. Janet's position is that hysteria does not come from conditions of this sort, but from the overall functional features of the brain as a whole. Specifically, he traces the symptoms to some "fixed idea" controlling the patient's mental life and so "narrowing the field of consciousness" as to render the patient inaccessible to external events. The paralyzed patient is unable to "represent" movement in the cerebral centers because of the uncoupling that has taken place between the personality and the real world.

In writing a brief Preface to the first (French) edition of Janet's text (1892), the great Charcot tells readers that his former student "wished to unite as completely as possible medical studies with philosophical studies," but in these comments Charcot was referring more to his own increasingly psychological perspective than Janet's. Throughout the text, Janet is critical of premature attempts at theory and of all philosophical approaches to the clinical problems with which he is dealing. Each patient must be treated uniquely and not according to some overly general theory of science of the ipsedixtis of the philosopher. When Janet requires the concept of the unconscious, he ties it to specific symptoms, to the observed behavior of the patient. We see this in his discussion of the fixed ideas:

> 1st. These ideas may develop completely during the attacks of hysteria and express themselves then by acts and words. 2nd. In dreams more or less agitated which take place during sleep and natural somnambulisms and which often happen unexpectedly; it is at such moments that fixed ideas are wholly confessed. 3rd. One of the best processes consists in causing the patient to enter artificially into a state similar to the preceding ones—namely an imposed somnambulism. . . . 4th [W]hen it is possible to employ it, the process, which consists of utilising automatic writing is more exact than all the others.[39]

But when Janet turns to the recent publication on hysteria by Freud and Breuer, he finds several deficiencies, even though the authors graciously acknowledge Janet's priority. His complaint with their article is they attempt to account for all hysterias with explanations that actually apply only to certain cases and that they underestimate the complexities of therapy:

> We formerly showed that . . . one could not treat the hysterical . . . before having reached those deep layers of thought within which the fixed idea was concealed. We are happy to see to-day MM. Breuer and Freud express the same idea. It is necessary, they say, to make this provocative event self-conscious; bring it forth to the full light. The [symptoms] disappear when the subject realises those fixed ideas. We

do not believe that the cure is so easy as that. . . . The treatment is unfortunately of a much more delicate nature.[40]

At the time Janet was writing these words, Freud had not yet developed his psychosexual theory of neurosis, but sufficiently kindred ideas were abundant enough for Janet to address the allegedly erotic element in hysteria:

> This erotic disposition exists, we think, in hysteria just as all possible fixed ideas do. We have collected, out of a hundred and twenty observations, four where it plays a role altogether predominant. There is nothing very strange in this. Amorous passions, sexual desires, must exist with these persons, most of them young, as they do with others. Hystericals hear people talk of love, see the evidence of it, read descriptions of it;—why should their mind, so ready to receive impressions, so docile to all influences, resist this one. . . . [P]atients speak very often of lover, husband, rape, pregnancy, etc. Now, you cannot put into your dilirium what is not in your mind. These things greatly preoccupy people of that age and sometimes of another age. It is quite natural that they should manifest them in their dreams.[41]

We discover in these passages that very tentativeness and common-sensical tone that would be so lacking in Freud's books and essays. Where Janet is willing to take the facts as they come, Freud would be nearly fretful in his attempt to fit the facts to his theories. To his contemporaries, there was something entirely unscientific about his very manner of discourse but, to posterity, it was precisely this manner that attracted attention, loyalty, and even devotion. Freud often thought of World War I as a kind of laboratory in which his central claims received macabre validation. In certain aspects, it was this war that gave a morbid cogency to Freudian theory among startled and saddened intellectuals. What is clear, however, is that the chilly reception his ideas suffered initially is not to be explained in terms of Victorian proprieties. Sexuality had been part of the clinical literature for decades by the time Freud made it a fixture in his writings. Anthropologists, at least since late in the eighteenth century, had reported on various primitive lusts and taboos, and had noted the centrality of sexual symbolism in many cultures. And the theory of unconscious motivation, with its ancient and even biblical roots, had already received refinements from Charcot and Janet before Freud had begun to assemble a theory. Freud's main problem—and here we must demystify that portrait of Gothic struggle painted by his friends and friendly biographers—was that his way of thinking was perceived as anachronistic. This not only accounts for the early disregard paid to his works, but to his own life-long attempt to confer scientific respectability on them. The scientific community had learned to live without the Absolute, without metaphysics, and without those cosmic integrations of an earlier age at just about the time Freud's psychoanalytic ruminations were published. Psychology particularly was earnestly seeking its place in science and so was especially embarrassed by a theory in which the ratio of facts to assumptions was so small.

If the earliest audiences for Freudian psychology were small and somewhat hostile, posterity more than compensated. But "clinical psychology" is more than and is different from "Freudian psychology." At the outset, it was medical psychology and, until the short-term triumph of psychoanalytic theory, it remained medical psychology. A distinction must be made, however, between this specialty and both the older and the more current forms of physiological psychology. As the term is used here, medical psychology addressed itself primarily to the phenomena of mental illness and to the application of experimental procedures by which these might be explained. Physiological psychology, both in Wundt's sense and in the current sense, has been chiefly concerned with those law-governed processes that characterize normally behaving, normally perceiving organisms.

A further distinction may be drawn along national lines, for in significant respects it was France that took the lead in medical psychology, Germany in physiological psychology. To give substance to this distinction, it should be recalled that Henry Maudsley was not an experimentalist or a "scientist" in the usual sense, but a practicing physician engaged in the sorts of things ordinarily associated with psychiatry. Thus, to the extent that clinical psychology identified itself with the natural sciences, it was an experimentally oriented discipline presented as something of an alternative to the introspective and psychophysical schools of Germany. It took the phenomena of the diseased mind not simply as conditions to be treated, but as suggestive of the principles by which mind as such could be understood. Disease, after all, is one of nature's experiments and the keen observer can learn as much from experiments of this sort (if not more) than can be gathered from any number of (unnatural, antiseptic, etc.) Leipzig studies. This was a general attitude in French psychology, but the psychologist who spoke most convincingly in its behalf was Alfred Binet (1857–1911).

Binet was born in Nice and received his degree from the Sorbonne under Professor Beaunis, who, upon retirement, turned his laboratory over to Binet. In the course of his abbreviated career, Binet co-founded France's most prestigious psychological journal, *L'Année Psychologique,* wrote authoritative texts on hypnotism, multiple-personality, and experimental psychology, and inaugurated those methods of intelligence-testing which came to define the field of mental tests and measurement. He was a close friend to Pierre Janet, a correspondent with William James, a regular author in international journals, and perhaps the person most responsible for keeping clinical psychology within the natural science framework at a time when theory-spinning threatened it with expulsion.

The Binet best known to psychologists is the co-founder of the Binet-Simon test of mental abilities, the test that would surface in America as the "Stanford-Binet." The test itself had an interesting history. It came into being after the Education Ministry called for a means by which retarded children could be singled out for special education in the Paris school system (1904). In not too many years, the test prepared by Binet and Simon were in use throughout the world and were even obligatory in certain states in America.[42]

In attending too closely to this phase of Binet's work in psychology, however, the contemporary psychologist often neglects not only Binet's larger contribution but the very basis upon which he came to develop tests of intelligence. Binet's interest was in the dynamics of mind, its development, disorders, and practical functions. Even in his clinical application of hypnosis, the dominating motive was to employ hypnosis as an *experimental* means by which to induce states of mind. In one of his earlier essays, "Mental Imagery" (1892), he discusses hypnosis as a method of "intellectual and moral dissection"[43] permitting the insertion of thoughts and ideas in the mind without having to rely on the subject's perceptual and sensory functions. It also permits the scientist to create cognitive states in the normal mind rather than waiting for the interesting clinical case. In his widely read text *On Double Consciousness,* he complains that while the universities continue to teach "an antiquated science, whose only method is that of introspection," French psychology is firmly tied to the natural sciences and has "left the investigations of psychophysics to the Germans."[44]

When we turn to his work on mental testing, we discover the same underlying considerations. Both in *The Intelligence of Imbeciles*[45] and *The Development of Intelligence in Children,*[46] Binet speaks of mental tests as forming an integral part of the "psychogenetic method." His primary interest in mental retardation arose from the judgment that the retarded individual—no matter the chronological age—presented a mind that was fixed at a given stage of development and was thus characteristic of that stage. In other words, what makes retardation pathological is not that its processes are unlike those found in normal persons, but that they are found only in much younger persons. Accordingly, careful tests of the retarded mind provide a key to our understanding of the normal evolution (genesis) of mind per se. Nonetheless, Bient recognized the immense variability displayed by this evolution and the need for a scientific psychology to address the facts of individual differences. In his essay "Mental Imagery," we find him praising Francis Galton for the statistical techniques that permit comparisons of individuals. Toward the conclusion of the essay, Binet observes:

> The whole present tendency of psychological research is to show, not that the mental operations of all person are of a similar nature, but that immense psychological differences exist between different individuals.[47]

Implicit in this passage is a criticism of the entire epoch of philosophical psychology and its residual, the introspective method. Experimental hypnosis, on the contrary, permitted inquiries into the differences between minds and allowed the investigator to create, repeat, and reverse mental states. Similarly, tests of mental functioning constituted a kind of indirect experiment in which mental age was the dependent variable and chronological age the independent variable. In *The Intelligence of Imbeciles,* Binet underscores the role of abstract thought in all genuinely mental operations and draws attention to the *functional* nature of this thought. Unlike the introspectionists, who labor so tirelessly to isolate the structural elements and atoms of experience, he states,

> we oppose [the] counterpart, that which gives action as the end of thought and which seeks the very essence of thought in a system of actions.[48]

Through the efforts of Binet in France and William James in America, the emerging science of psychology was given a broad middle ground between the psychophysical investigations of Germany and the increasingly narrow physiological investigations of the early neuropsychologists. This broad middle ground was occupied often by attention to psychopathology, but—at least with Binet and James—more often by research and theory devoted to normal cognitive functions. The practical and "pragmatic" tone of their writing was not at the expense of the mental, but it was at least indirectly supportive of a behavioristic (action-oriented) science.

The specialties incubated by those discussed in this section are no longer appropriately placed under the heading, "Clinical Psychology." What unites the otherwise diverse theories and practices of Freud, Janet, Binet, James, Jung, and Adler is first a recognition of mind as an entity with a past and a future, a dynamic entity whose principles of operation are not likely to be discovered simply through an analysis (introspective or psychophysical) of short-term sensations. Secondly, all of them shared in the conviction—but on different grounds—that genuinely psychological phenomena had to be dealt with at the level of psychology rather than at the (largely unknown) level of neurophysiology. Janet and Freud both subscribed to a "cerebrogenic" theory of hysteria, but both insisted that the disease must be described and treated at the psychological level. As a group, then, they resisted both introspective and biological reductionism. Finally, all of them argued in behalf of a *useful* scientific psychology; one able to explain mental phenomena as these occur in the real world and to actual persons.

By today's standards, William James would scarcely qualify as a clinical psychologist, even though he was medically trained and wrote often on clinical cases. Binet, too, was an experienced clinician whose essays and books often focus on aspects of psychopathology, but Binet would no longer qualify as a clinical psychologist. In putting them under this heading, therefore, is the present chapter guilty of an abuse of language? I think not. Instead, I would point to the even greater degree of specialization that has taken place in the past half century, the tendency to follow only one of the lines of inquiry instead of the many pursued by each of these men in his own professional life. For Binet, however, it would have made no more sense for a psychologist only to administer tests of intelligence than for a physician only to record temperatures. Mental tests were conceived as a means, not as the end, of scientific activity. So also with Janet, Freud, and Jung: The aim of psychoanalysis is not simply (if laudably) to cure a patient, but to come to understand the laws of mental life.

What we have seen in this chapter is the further refinement of general notions passed on late in the eighteenth and early in the nineteenth centuries. The refinement led to specialization both in the choice of problems and the choice of methods. Between 1870 and 1920, the distinct fields of personality, clinical psychology, comparative psychology, physiological psychology, and developmental psychology were all established. So, too, was that sociological point of view urged by Lewes, exercised by Wundt, and pervasive in the scholarship of the late Victorian period—the point of view that would soon become formalized as social psychology. We have, then, the bridge between the somewhat loose and hopeful "scientific psychology" and the more familiar fields and schools of contemporary psychology. What is missing from the record, however, is that further distillation that imparts the very special flavor of today's psychology. In the next chapter, therefore, we will examine three representative—dominant—orientations in modern psychology: behavioristic, Gestalt, and physiological. In Chapter 10, the nineteenth century was viewed through a wide-angle lens; in Chapter 11, more narrowly and in its latest stages. In the final chapter, we will be concerned with a briefer span, roughly 1920 to 1950, in which today's psychology was given its most immediate direction.

The Nineteenth Century's Invention

It was Alfred North Whitehead who credited the nineteenth century with having "invented the method of invention." It was the first period in history in which very nearly every important philosophical figure recognized the essential function of experimentation in the search for truth. It was also the first century in which the majority, the vast majority, of serious works in science were completely devoid of theological colorations.

The record of the century is particularly commendable in regard to psychology. When we examine the topics now filling the literature in professional psychology, we are hard pressed to find one that was not put forth—often in a form still to be improved upon—by those whose efforts we have examined in this chapter. Physiological psychology, little more than the product of polemicism in the eighteenth century, became a science in the hands of Flourens, Gall, Bell, Magendie, Helmholtz, and Wundt. Comparative psychology was invented by Spencer, Darwin, Romanes, and Morgan. The psychology of individual differences is the creation of Binet and Francis Galton, as are several of the statistical procedures needed for such studies. Cognitive and Gestalt psychologies are so intimately tied to phenomenology that only a purist could deny Hegel and the neo-Hegelians the title of "founders." "Freud," "Janet," "Jung," and "the unconscious" are near-synonyms. Our sense of what an experimental science is and ought to be is taken over, with only the slightest modifications, from J. S. Mill, and the general attitude toward the status of science remains largely the one advocated by Auguste Comte and his positivist disciples. Our fascination with hedonistic ethics, with the possibility of shaping the world through the processes of reward and punishment, is linearly traceable to Jeremy Bentham and the utilitarian movement. Even our vaunted "humanistic" psychologies, with their focus on "self-actualization," personal growth, and individual freedom, have never improved upon the original formulations by the German Romantics. Contemporary psychology then is largely a footnote to the nineteenth century.

Notes

1. Consult the entry "Rorarius" in Bayle's *Dictionary*. It can be found in *Historical and Critical Dictionary* (Selections), translated by Richard H. Popkin, Bobbs-Merrill, Indianapolis, 1965.

2. A six-volume English tradition of Gall's *On the Functions of the Brain and Each of Its Parts* was published in Boston in 1835. I have collapsed these, without abridgement, into Vols. XVI, XVII, and XVIII (series A) of *Significant Contributions to the History of Psychology*, edited by D. N. Robinson, University Publications of America, Washington, D.C., 1978. In my separate Prefaces to each of these volumes, I attempt to locate Gall's priority and influence in the areas of physiological psychology, comparative neuroanatomy, and fetal and neonatal anatomy.

3. Representative reviews are reprinted in "Darwinism," in *Significant Contributions to the History of Psychology*, Series D, Vol. IV.

4. Romanes, George, *Animal Intelligence*. The edition of 1883 is reprinted in *Significant Contributions to the History of Psychology*, Series A, Vol. VII.

5. Ibid., pp. 4–6.

6. Ibid., p. 13.

7. Ibid., pp. 12–13.

8. William L. Lindsay, *Mind in the Lower Animals in Health and Disease*, 2 vols., London, 1879. The two volumes are reprinted in *Significant Contributions to the History of Psychology*, Series D, Vols. VI, VII.

9. C. Lloyd, Morgan, *Introduction to Comparative Psychology*, London, 1894. This work is reprinted in *Significant Contributions to the History of Psychology*, Series D, Vol. II.

10. See especially the last chapter (XV) of his *Darwinism: An Exposition of the Theory of Natural Selection with Some of Its Applications,* Macmillan, London, 1889. In that chapter, Wallace argues that "It is impossible to trace any connection between (the musical faculty) and survival in the struggle for existence," and offers "Independent Proof that the Mathematical, Musical, and Artistic Faculties have not been Developed under the Law of Natural Selection" (pp. 468–469 of the edition of 1897).

11. This is taken from Ernst Haeckel's *Last Words on Evolution,* translated by Joseph McCabe, London, 1906. Reprinted in *Significant Contributions to the History of Psychology,* Series D, Vol. III, p. 111.

12. Angell's essay appeared as a journal article ("The Influence of Darwin on Psychology") in 1909, and is reprinted in *Significant Contributions to the History of Psychology,* Series D, Vol. IV. The quoted passage is on page 169.

13. For a brief review of this history, consult Ch. 2 of my *Psychology and Law: Can Justice Survive in the Social Sciences,* Oxford, New York, 1980.

14. This case of *Hadfield* and related British cases are given in Series F ("Insanity and Jurisprudence"), Vol. VI, of *Significant Contributions to the History of Psychology.*

15. Maudsley's text is reprinted in Series C, Vol. IV, of *Significant Contributions to the History of Psychology.*

16. For this defense, see his *Responsibility in Mental Disease* (1876) which, with Phillipe Pinel's *Treatise on Insanity* (London, 1806) is reprinted in Series C, Vol. III, of *Significant Contributions to the History of Psychology.*

17. Ibid., p. 37.

18. Ibid.,p. 187.

19. Ibid., p. 197.

20. The unabridged two-volume translation of Wundt's *Elements of Folk Psychology: Outlines of a Psychological History of the Development of Mankind,* London, 1916, is reprinted in Series A, Vol. XV, of *Significant Contributions to the History of Psychology.*

21. W. Wundt, *Lectures on Human and Animal Psychology,* translated by Janet Creighton, London, 1894. Reprinted in Series D, Vol. I, of *Significant Contributions to the History of Psychology.*

22. Ibid., p. 446.

23. Ibid., pp. 432–433.

24. W. James, *Text Book of Psychology,* Macmillan, New York, 1892, p. 5.

25. Ibid., p. 6.

26. Ibid., pp. 462–464, abridged.

27. C. S. Peirce, "The Architecture of Theories," *The Monist,* January 1891, pp. 161–176.

28. The separate work, *The Study of Psychology,* was published in London (1879) and is reprinted in Series A., Vol. VI, of *Significant Contributions to the History of Psychology.*

29. The assembled committee of reviewers included the aging Benjamin Franklin, from whom Mesmer had expected a more tolerant and informed hearing.

30. Wundt, *Lectures,* p. 308. Here, in a footnote, Wundt refers to Freud's "Zur Auffassung der Aphasien" (1891).

31. S. Freud, "The Origins and Development of Psychoanalysis," *American Journal of Psychology*, 21 (1910).

32. E. G. Boring, *A History of Experimental Psychology*, Appleton-Century-Crofts, New York, 1951, p. 709.

33. Freud, *Origins*.

34. The readiest source of these views is *Psychological Reflections*, selected and edited by J. Jacobi, New York, 1953.

35. Two representative works are *Inferiority and Its Psychical Compensation* (translated by S. E. Jellife, New York, 1917), and *Practice and Theory of Individual Psychology* (translated by P. Radin, New York, 1927). Adler's rejection of (Freudian) determinism is most clearly set forth in his *The Neurotic Constitution*, (1912).

36. P. Janet, *The Mental State of Hystericals*, translated by Caroline Corson, New York, 1901. This translation is reprinted in Series C, Vol. II, of *Significant Contributions to the History of Psychology*.

37. Ibid., p. 35.

38. Ibid., p. xiv.

39. Ibid., pp. 280–281.

40. Ibid., p. 412.

41. Ibid., p. 215.

42. It is one of the ironies of history that these tests, which sixty years ago were made mandatory in the interest of fairness and impartiality, are now under legislative and judicial attack.

43. The essay "Mental Imagery" was initially an article in *Fortnightly Review* (1892) and has been reprinted in Series B, Vol. IV, of *Significant Contributions to the History of Psychology*.

44. Binet's *Alterations of Personality* (English edition of 1896) and his *On Double Consciousness* (English edition of 1890) are given together in Series C, Vol. V, of *Significant Contributions to the History of Psychology*. The quotation is taken from p. 12 of the latter.

45. A. Binet, *The Intelligence of Imbeciles* (English translation, 1909). This has been reprinted in Series B, Vol. IV, of *Significant Contributions to the History of Psychology*.

46. A. Binet and T. Simon, *The Development of Intelligence in Children* (English translation, 1911). This is reprinted in Series B, Vol. IV, of *Significant Contributions to the History of Psychology*.

47. Ibid., p. 104.

48. Ibid., pp. 153–154.

12 Contemporary Formulations

Method as Metaphysic

To describe contemporary psychology with the abrasive term "footnote" is to tilt with the error of *scholastica successionis civitatium,* which we pledged to avoid in the very first chapter. No contemporary psychologist diffidently searches the annals of nineteenth-century scholarship in order to discover the problems or methods appropriate to psychology. The most casual inspection of the courses and the texts in psychology, at both the undergraduate and the professional levels, will prove beyond doubt, if not beyond concern, that very little of the debt to the nineteenth century is consciously acknowledged. The aspiring psychologist might be expected to know something about "Mill's methods" and that Wundt founded the first laboratory devoted exclusively to psychological research. It is also the expectation of a faculty that its students will be conversant with the general features of Darwinian biology and the sensory-physiological theories Helmholtz. But no one is asked any longer to pore over the works of Bain and Spencer, Fichte or Schelling, Kant or Hegel. Even William James is presented as a museum piece, and E. L. Thorndike as the author of a law and the prophet of a method that have both changed beyond recognition. No introductory course in psychology and certainly no concentration or "major" in psychological studies is considered complete or even respectable unless the great old names are periodically trotted out, dusted off, congratulated for having seen farther than most, and then gently returned to their crypts as psychology gets on with more pressing matters. Usually, as anyone willing to take the time to glance through the more popular general and historical texts will see, the list of great old names is expeditiously contracted. "The Greeks" of course, are never neglected. Perhaps a few lines will be devoted to the fact (except it isn't a fact) that not very much of intellectual consequence took place from the fall of Rome until the Renaissance except, perhaps, for St. Augustine

and Thomas Aquinas. The "modern era" is then announced, homage is paid to
Locke and Descartes and, with philosophy now out of the way, the study of
psychology can begin. It is just this approach to the foundations of the discipline that more or less guarantees to each generation of psychologists the privilege of rediscovering some of the more compelling ideas in the history of
thought. It also confers on psychology that state of perpetual youth which is
proclaimed by most of its leaders in all of its ages. Thus, Titchener could write
in his very widely read *Primer,* "Psychology is a very old science; we have a
complete treatise from the hand of Aristotle (384–322 B.C.). But the experimental method has only recently been adopted by psychologists." [1] From this,
one might gather that science in ancient days had adopted the experimental
method but that the science of psychology had not. The fact, of course, is that
experiments in most fields of science did not really begin to take precedence
over pure speculation until well into the eighteenth century and, at this time,
experiments in perception began with the rest. By all relevant standards, the
first psychology laboratory was installed late, but not by more than fifty or
seventy-five years. In fact, university laboratories like the modest one created
by Wundt were quite rare until the nineteenth century, and appeared in Germany relatively late.[2] The government of the French Revolution had executed
Lavoisier in 1793 and only allowed the French Academie des Sciences to re-
new its activities in 1795, when everyone was convinced that libertarian rhetoric would not solve France's problems. It was, however, not until the educational reforms introduced by Napoleon that French universities became centers
of excellence in experimental science. In Germany such endeavors began even
later, but when Fechner looked for data to support his science of psychophysics, he had to look no further than E. H. Weber's *Der Tastsinn und das Ge-
meingefühl* (1846), which was already printed in a popular handbook.[3] To repeat, psychology's adoption of an experimental outlook occurred at very nearly
the same time that that outlook was generally shared by the community of
sciences, and that time was the nineteenth century. It will not do, then, to insist
upon the youth of psychology even as an experimental endeavor. Psychology
is young in the sense of still conducting its affairs in the absence of a unifying
theory of the kind advanced by Copernicus, Galileo, or Newton. To the extent
that this is the case, we must be prepared to accept the possibility, though dis-
turbing, not simply that psychology is young as a science but that it is not yet
a science at all.

What defines the very character of science is not the mechanical application of one or another method, but a much larger narrative in which methods
are chosen because of their transparent relevance to a widely perceived problem. The methods adopted by Archimedes in ancient Greece, Newton in
seventeenth-century England, Darwin in the middle of the nineteenth century,
and Einstein early in the twentieth have very little in common at the descriptive

level. In a word, there is no single scientific method at all. There is, however, a quite systematic relationship between the identification of a problem of scientific consequence and the subsequent choice from among available methods of observation and measurement. What establishes this relationship is a theory rich in ontological or in explicative possibilities. A theory rich in ontological possibilities is one which, when found to be valid, clarifies and may even reduce the domain of really existing entities. We no longer believe that heated objects rise because, as a result of heating, they take on a substance called "levity." We no longer explain phenomena by referring to the properties of phlogiston.

The history of experimental psychology has not followed the pattern of the developed sciences and has deviated from the pattern by ever wider margins with the passage of time. Just what overarching theory is tested by contemporary experiments in psychology? What *problem* has been solved? What effect have psychological hypotheses or findings had on the size and nature of the ontological domain in which all genuinely psychological phenomena are to be found? What has been characteristic of experimental psychology is the adoption of a rather prosaic set of experimental "controls" and a repeated-measures paradigm. In a wide variety of settings, this method of procedure has yielded fairly stable functional relationships between dependent and independent variables under conditions generally so unlike the domain of interest as to render generalizations jejune. It is a credit to the psychologists of the nineteenth century that they valiantly undertook to apply such methods to psychological phenomena, for only by attempting to develop psychology in such a fashion could the limits of the method be assessed. It is less of a credit to the legions that have dutifully imitated these efforts for the better part of a century.

The error of *scholastica successionis civitatium* is committed when, in the absence of clear evidence, it is assumed that the reappearance of an idea at a later date in a different cultural context must be the result of its having been borrowed from an earlier one. We would fall into this trap were we to suggest, for example that a contemporary behavioral scientist, studying the effects of shock or food-reward on the acquisition of behavior, was a disciple of Jeremy Bentham's. Even if it could be shown that the early twentieth-century architects of behaviorism were directly inspired by utilitarian writings, it would not follow that those currently engaged in such work are disciples in any useful sense of the term. Rather, a quite different claim is made here. The claim is not that the nineteenth century provided contemporary psychology with an irresistible legacy but that contemporary psychology just *is* nineteenth-century psychology in its most global respects and that current departures from nineteenth-century perspectives have not been compelled by scientific data or theory. To the extent that contemporary psychologists reject the methods and the terminology of the immediate past they do so for reasons other than those that have produced analogous decisions on the part of physicists, biologists, and chemists. Contemporary physicists do not devote energies to the search for phlogiston because they have established that there is no such thing. They do not labor to create perpetual-motion machines because the conservation of energy legislates against them. The geneticist does not design experiments to test the inheritance of acquired characteristics because, within all but lethal ranges of environmental variation, the molecular biology of the gene is stable and therefore will not be altered by learning or practice. Shifting emphases in psychology are not based on the same considerations. In the first two decades of the present century, the most influential figures in American psychology were William James and E. B. Titchener. Titchener, in his *Primer,* offered a psychology not too far removed from that of his mentor, Wundt. Psychology was to be an experimental analysis of consciousness:

> [W]hen we are trying to understand the mental processes of a child or a dog or an insect as shown by conduct and action, the outward signs of mental processes, . . . we must always fall back upon experimental introspection . . . we cannot imagine processes in another mind that we do not find in our own. Experimental introspection is thus our one reliable method of knowing ourselves; it is the sole gateway to psychology.[4]

William James in much the same spirit, begins his *Text Book of Psychology* by defining the subject as the "description and explanation of states of consciousness."[5]

Now, it is unmistakable to anyone surveying the contemporary psychological scene that there is hardly a vestige of the program envisaged by Titchener and James. The "rules of introspection" presented by the former are applied in no laboratory, appear in no advanced treatment of the discipline, form no part of the modern psychologist's training. The same may be said of James' division of the discipline into Sensation, Cerebration, and the Tendency to Action.[6] But observe the difference between this shift in emphasis or complete abandonment of interest and the changes that have occurred in physics and biology. We *do* have minds, we *are* conscious, and we *can* reflect upon our private experiences because we *have* them. Unlike phlogiston or the inheritance of acquired characteristics, these phenomena exist and are the most common in human experience. The absence of orthodox Wundtians or Titchenerians or Jamesians, therefore, cannot be attributed to the disappearance of their subjects. Rather, it is to be understood as the result of the inability of the accepted method of psychological inquiry to address these subjects. The contemporary psychologist, if only insensibly, has made a *metaphysical* commitment to a method and has, per force, eliminated from the domain of significant issues those that cannot be embraced by that method.

The method itself is not simply some variant of the experimental method. Titchener and James both subscribed to that. The method referred to here is broader than a set of actions or procedures. It includes a way of thinking about problems and a way of talking about them. The method needs a label and the one most commonly applied to it is empirical, but custom can be observed only with reservation. The historic empiricists, Locke, Berkeley, Hume, Mill, and James, would almost certainly be misled by the term as it is applied to this metaphysical commitment. A psychology striving to rid itself of mental predicates is one they could scarcely fathom. We may even suspect that their reaction to the notion of psychology as a "behavioral" science would vary from incredulity to whimsy. "Empirical," if the contemporary usage is to be captured, must also suggest measurement, practicality, impersonality, ethical neutrality and (ironic) "antimetaphysical*ness.*" Contemporary journals, whether devoted to neuropsychology, clinical practice, animal learning, or family counseling, strive to reflect these "empirical" features.

To what extent is this pervasive aspect of contemporary psychology also a "footnote" to the nineteenth century? To answer this, we must once more pause to examine the impact of German idealism and the response to it.

The philosophy of the Enlightenment had cut deeply into what were once held to be the rational justifications for faith. In England and France, scholarship turned decidedly in a secular direction whence it has never retreated. The failure of the French Revolution to change the general conditions of life endured by the French and the excesses of that Revolution promoted political conservatism on both sides of the English Channel. Napoleon's rise to power and the wars resulting from it led to a variety of unconnected religious, social, and political movements. In England there was a definite tightening up of traditional class distinctions and a growing enmity toward those liberal philosophies that many blamed for the problems in the world. The objective and ultrarational philosophies of the British empiricists were judged to have taken people away from religion and to have driven a wedge between daily life and the transcendent by which daily life becomes meaningful. To some, the only solution to this crisis in faith was to be found in the philosophies of idealism then overtaking Germany. Schelling's "nature philosophy" found a British audience as sympathetic as any to be found in his own country. The most eloquent and forceful leaders of this British idealist movement were the "Lake poets" and, especially, Coleridge (1772–1834) and Wordsworth (1770–1850). More keenly than any, they sensed the weakening of historic values, and the creeping relativism in ethics and morality. Listen to Wordsworth calling up, from a time before Hume, the hero England lost:

> Milton! thou should'st be living at this hour;
> England hath need of thee; she is a fen
> Of stagnant waters . . .
>
> "London, 1802"

In his "Ode to Duty" and his "Character of the Happy Warrior," there is the same assertion of traditional ideals—a call to higher purposes. Then he crowns his pleading with the fifth stanza of his *Intimations:*

> Our birth is but a sleep and a forgetting;
> The Soul that rises with us, our life's Star,
>   Hath had elsewhere its setting,
>   And cometh from afar;
>   Not in entire forgetfulness,
>   And not in utter nakedness,
> But trailing clouds of glory do we come
>   From God, who is our home . . .

Coleridge, with an even larger following, pressed the message against the breast of countless thousands with his *Ancient Mariner,* whose speaker found it "sweeter than the marriage-feast / . . . To walk together to the kirk / With a goodly company." Here were the poets of nature, demanding a return to natural feelings including those of duty, morality, and love of the unseen but well-known God who is accessible only to what Byron in "The Prisoner of Chillon" would describe as the "Eternal Spirit of the chainless Mind."

If one is to comprehend the spirit of Victorian England, it is necessary to try to unite the ostensibly contradictory forces of romance and industry, naturalism and experimentalism, freedom and duty, Puritan moral simplicity and imperial social opulence, and sweatshops and philanthropy. Not only were these polarities constant, but each occurred on a truly mammoth scale. While coal miners choked to death for pennies a day, John Constable painted landscapes one could nearly walk through. As Charles Bell teased out the spinal nerves and as Gall and Spurzheim sought to transform morality into neurology, Tennyson depicted the twisted fate of all materialists in his *Lucretius,* in which the main character, convinced of his own insignificance, wracked by his "poor little life that toddles half an hour," kills himself.

There is a sense, and a defensible one at that, in which the Darwinian revolution was, itself, the melding of naturalist, romantic, and materialist tensions: man evolving from the slime of nature and rising, through instinctual duty to his own survival, to a position of temporary mastery over all that lives; man, "selected" by nature to play his present part. Not even Darwin, himself, whose works would come to serve as an authority against romantic idealism, was spared the influence of the Lake poets. He concludes *The Expression of the Emotions in Man and Animals* with the hypothesis that the very expression of an emotion (behaviorally) intensifies it and that this self-intensifying tendency has great adaptive value to the animal. Then, at the end of this long treatise, in which he has explored the minutiae of the facial muscles and bones of the jaw, Darwin finds support from a judge possessing "wonderful knowledge of the human mind."[7]

> Is it not monstrous that this player here,
> But in a fiction, in a dream of passion,
> Could force his soul so to his own conceit,
> That, from her working, all his visage wann'd;
> Tears in his eyes, distraction in's aspect,
> A broken voice, and his whole function suiting
> With forms to his conceit? And all for nothing!
>
> Hamlet, ii, 2

In the romantic poetry of Wordsworth and Coleridge, in the literary allusions of Darwin, we find the sign of the internal conflicts raging in Victorian England. For the actual substance, we must turn to the essayists of the middle and late nineteenth century—to Matthew Arnold, whose *Culture and Anarchy* sought to commit an industrial and ruthlessly economic system to "sweetness and light," good manners, taste, reason, and a sense of place,[8] and to J. S. Mill, whose *On Liberty* proclaimed the essential intellectual freedom of every individual and the right of that individual to be unconstrained by the state in all respects, short of the harm he might cause to others. And there was John Ruskin, whose style and penetration were briefly sampled in Chapter 6, who wrote of the history of art in such a way as to make architecture a lesson in morality— Ruskin, who reasserted the ideals of the Renaissance or, at least, what he perceived to be those ideals when he studied the buildings and paintings of the period. No single chapter, let alone several pages, can do justice to the restlessly agile and durable intelligence of the Victorians. Their period must be introduced, even this sparingly, in order for us to recognize the popular attitudes the scientific mind was fighting. In the most general respect, it was fighting Hegelianism, but, at the same time, it was finding in that same Hegelianism the most ardent of defense of freedom and progress. The romantic minds of this same period were fighting science and materialism, but, in these very movements, the ideals of liberty, safety, dignified work, and freedom from want were preserved as objectives. The character of contemporary psychology is to be found in the failure of these two nineteenth-century forces to find a means of reconciliation. Only a divorce would end the dispute. It was Helmholtz who summarized matters most candidly:

> It has been made of late a reproach against natural philosophy that it has struck out on a path of its own, and has separated itself more and more widely from the other sciences which are united by common philological and historical studies. The opposition has, in fact, been long apparent, and seems to me to have grown up mainly under the

influence of the Hegelian philosophy, or, at any rate, to have been brought out into more distinct relief by that philosophy. . . . The sole object of Kant's "Critical Philosophy" was to test the sources and the authority of our knowledge, and to fix a definite scope and standard for the researches of philosophy, as compared with other sciences. . . . [But Hegel's] "Philosophy of Identity" was bolder. It started with the hypothesis that not only spiritual phenomena, but even the actual world—nature, that is, and man—were the result of an act of thought on the part of a creative mind, similar, it was supposed, in kind to the human mind. . . . The philosophers accused the scientific men of narrowness; the scientific men retorted that the philosophers were crazy. And so it came about that men of science began to lay some stress on the banishment of all philosophic influences from their work; while some of them, including men of the greatest acuteness, went so far as to condemn philosophy altogether, not merely as useless, but as mischievous dreaming. Thus, it must be confessed, not only were the illegitimate pretensions of the Hegelian system to subordinate to itself all other studies rejected, but no regard was paid to the rightful claims of philosophy, that is, the criticism of the sources of cognition, and the definition of the functions of the intellect.[9]

All who entered psychology during the turmoil and all who have come since it was (temporarily) settled were to choose between some version of Hegelianism and the inductive science of Mill. Even the phenomenology of Brentano and Husserl,[10] so radically different from what Hegel had in mind, would be forged into a "descriptive psychology," more philosophy than psychology and never an "empirical" science. Thus, in Europe, where the idealist tradition was deepest, the psychologist could either become a Wundtian, a neo-Hegelian, or a physiologist in psychologist's clothing. In England and America the alternatives were further reduced. One was either a philosopher or an experimentalist. To fail to be the latter was to fail to be a psychologist. Note that the historical development was just that: historical and *not* scientific. No logical proof had been discovered by which it could be shown that a rationalistic psychology would fail. No experimental finding had made it clear that we lack a moral sense or a link with God or a love of beauty. No surgical procedure had established that the psychological dimensions of human life were readily reducible to neural mechanisms. Even "Mill's methods," now installed as the essential equipment of the new science, could claim neither the validity conferred by logic nor the reliability demanded by science, at least as these methods were applied in the psychology laboratory. Rather, what had taken place was the adoption of a metaphysical position not on the nature of *truth* but on the nature of *psychology*. The decision was made that psychology was no more than a certain kind of method, an "experimental" method, and that its subject matter would contain only those entries amenable to this method. Listen to another of the Leipzig graduates, Theodor Ziehen, introduce psychology in 1895:

> The psychology which I shall present to you is not that old psychology which sought to investigate psychical phenomena in a more or less speculative way. That psychology has long been abandoned by those whose method of thought is that of the natural sciences, and empirical psychology has justly taken its place.[11]

E. W. Scripture, another Leipzig Ph.D., writing from Yale in 1897 put it this way:

> The development of a science consists in the development of its means of extending and improving its method of observation. The great step that has lately been taken in psychology lies in the introduction of systematised observation, by means of experimental and clinical methods.[12]

Ziehen speaks of a "method of thought" endemic to the natural scientist, and Scripture announces the introduction of an improved method of observation. But what is this "method of thought" and how had the method of observation been improved? We may turn to Titchener for an answer:

> The rules for introspection are of two kinds: general and special. . . . Suppose, e.g., that you were trying to find out how small a difference you could distinguish in the smell of beeswax; that is how much greater the surface of the stimulus must be made if the sensation of smell is to become noticeably stronger. It would be a special rule that you should work only on dry days; for beeswax smells much stronger in wet than in fine weather. . . . The general rule of experimental introspection are as follows: (1) Be impartial . . . (2) Be attentive . . . (3) Be comfortable . . . (4) Be perfectly fresh.[13]

It is through these "methods" that Titchener hoped to discern the structure of consciousness: that is, to advance structuralism as that division of psychological science having the same role as anatomy in the biological sciences. This "science" didn't last and, indeed, couldn't last. The most that structuralism could ever have hoped to accomplish was the rediscovery of what every man, woman, and child know to be true during every waking hour of daily life. Having insisted upon a rupture with philosophical tradition and having affirmed the status of this new venture to be that of a "natural science," the founders of experimental psychology had to contract the domain of problems to . . . the smell of beeswax. Somehow, the laws of association would ultimately permit a coalescence of such findings into a complete description of the "elements of consciousness." The claim, for a while, was that the goal was reachable: that patient observation of "impartial, attentive, comfortable, and fresh" subjects would yield a natural science of the mind, that, indeed, the very observation *was* the natural science of the mind. But there was no "covering law," no theory worth the name, no independent set of measurements against which to validate the psychophysical methods. The data emerging from Wundtian studies—whether back at Leipzig or at Titchener's laboratory at Cornell—did not behave the way scientific data are supposed to. Even the most comfortable subject, try as he may, had trouble "introspecting" identically on separate occasions.

Notwithstanding the painfully apparent liabilities of the Wundtian-Titchenerian studies, the experimental method remained at the core of psychology, and still does. In a way, it is a method in search of a subject and, in the remainder of this chapter, we will review several of the possibilities unearthed along the way.

Behaviorism

We will call John B. Watson (1878–1958) the "father" of behaviorism, but only after acknowledging that fatherhood entails grandparents, at least one mate, and offspring. To this, we must add the fact that children are not to pay for the sins of the parents and that acquired characteristics are not inherited. And, since historical analysis involves a good deal more than genealogy, we leave the "paternalism" metaphor by remarking that the significant fact of behaviorism is not its authorship but its reception.

When Kurt Koffka (1886–1941) presented his *Principles of Gestalt Psychology* (1935) as, among other things, a rebuttal of behaviorism, he observed that Americans possessed a very high regard for science, "accurate and earthbound" science, which produced in them,

> an aversion, sometimes bordering on contempt, for metaphysics that tries to escape from the welter of mere facts into a loftier realm of ideas and ideals.[14]

He was, no doubt, reflecting on an American psychology that had begun to turn away from the problem of consciousness and toward the objective measurement of behavior. But in 1935 the trend was only a beginning. America, after all, was the country of William James and John Dewey, the country that Titchener had allowed to host structuralism. But it was not James or Dewey, and it was surely not Titchener, who called forth Koffka's stricture. Nor, we must observe, was it the mere fact that *behavior* was the subject of growing interest. And it certainly was not the very prominent place held by animal psychology in America, for Wundt never legislated against such an interest. In fact his *Lectures on Human and Animal Psychology* (1894) explicitly recommended it:

> The study of animal psychology may be approached from two different points of view. We may set out from the notion of a kind of comparative physiology of mind, a universal history of the development of mental life in the organic world. Or we may make human psychology the principal object of investigation. Then, the expressions of mental life in animals will be taken into account only so far as they throw light upon the evolution of consciousness in man. . . . Human psychology . . . may confine itself altogether to man, and generally has done so to far too great an extent. There are plenty of psychological text-books from which you would hardly gather that there was any other conscious life than the human.[15]

But American behaviorism was not just a commitment to study animal psychology, nor was it restricted to that aesthetic decision to examine behavior instead of something else. The behaviorism of John B. Watson was no less than the insistence that a scientific psychology must concern itself *only* with behavior and must abandon all interest in consciousness, mental states, introspection, unconscious processes, and other "ghosts." He announced the *ism* with unblemished lucidity in 1913:

> Psychology as the behaviorist views it is a purely experimental branch of natural science. Its theoretical goal is the prediction and control of behavior. Introspection forms no essential part of its methods, nor is the scientific value of its data dependent upon the readiness with which they lend themselves to interpretation in terms of consciousness. The behaviorist, in his efforts to get a unitary scheme of animal response, recognizes no dividing line between man and brute. The behavior of man, with all of its refinement and complexity, forms only a part of the behaviorist's total scheme of investigation.[16]

Structuralism, from Watson's point of view, is indefinite and self-serving in its methods, hopelessly out of control in its data, and committed to a mission that cannot succeed because it cannot ever end. There is no limit to the number of "experiences" one may have, especially since each may be attended by anywhere from "three to nine states of clearness of attention." Watson had, despite his search, never found a physician or lawyer or man of commerce who had even once found a use for the methods or findings of the structuralists. Clearly (to Watson), a venture so incapable of contributing to the practical affairs of life can have only a short future.

Having thus disposed of Titchener and the entire Wundtian tradition, Watson turned his attention—which was at least at the ninth state of clearness—to that *functionalist* school identified with James and Dewey. Watson frankly admits, and a good many psychologists have shared the admission, that he has never been able to understand just what functionalism is supposed to be, in contrast to Titchener's psychology. James, in his *Text Book of Psychology,* had complained of empirical associationism and of those who would describe a river in terms of "pailsful, spoonsful, quartpotsful, barrelsful, and other moulded forms of water."[17] He argued that the "stream of consciousness" cannot be arbitrarily fragmented to suit the needs of the experimental psychologist and that, therefore, any fractionation of consciousness into its structures could only lead to a distorted sense of what consciousness is. To discover the nature of consciousness, one must appreciate what it is *for,* in the Darwinian sense. One must, that is, discern its function, the part it plays in permitting man to adapt to the demands of the environment. This was James' position in 1882 before Titchener's structuralism even appeared, and when Titchener cites James in his *Primer,* it is invariably either in support of one of his own propositions or to borrow a *mot* from the master phrasemaker. John Dewey (1859–1952) also attacked elementalism, arguing that the notion of reflex "units" fails to appreciate the coordinated nature of successful (i.e., *functional*) behavior.[18] But neither Titchener nor Wundt may be said to have been naive on this point.[19] What disturbed Watson was not that the functionalists found something lacking in structuralism but that they had yet to recommend a plausible method of correcting it.

Watson, in a blizzard of criticism and revolutionary rhetoric, struck no compromise with any of his immediate predecessors. Whether the subject was infant care[20] or the nature of comparative psychology[21] or prescriptions for all psychological inquiry,[22] the message was the same: any branch of natural science must concern itself with the prediction of natural events; a science can study only that which can be observed; mental states and private experiences do not exist in the world of the publicly verifiable; and behavior alone is the object of a truly scientific study. Wundt and Titchener, therefore, went wrong from the outset by assuming that the adoption of an experimental point of view was sufficient to install an enterprise as science. They made a proper methodological decision but selected for their study a subject that never could have achieved the status of scientific subjects. At each key point, according to Watson, the Wundtian and Titchenerian psychologies were given the wrong emphasis: the emphasis upon man instead of the animal kingdom in general; the emphasis upon experience instead of action; the emphasis upon existential rather than evolutionary considerations; and the emphasis upon theoretical instead of practical considerations. Behaviorism, as defined by Watson and as understood ever since, is devoted to the redirection of each of these emphases.

It is difficult to assemble the various factors that led Watson to his manifesto. It is even more difficult to isolate the factors responsible for the early success of behaviorism and for the mounting attention it has received over the past half century.[23] He received his graduate education at Chicago during Dewey's tenure, but his subsequent attacks on the functionalism of the Chicago school suggest that Dewey's role was not a positive one. Dewey criticized the very "reflex arc" concept that Watson's later words would rely on heavily. The philosophical pragmatism of James and C. S. Peirce[24] was, by now, nearly an "official" American philosophy, and its insistence on the restriction of scientific terms to "observables" supported Watson's growing dissatisfaction with traditional formulations. A central tenet of Peirce's pragmatism is that the meaning of any concept, as applied to any *thing* in the natural world, can be no more than the *behavior* of that thing in a variety of clearly specified conditions. It was James who brought the ideas of Peirce, or at least the Jamesian reconstruction of these ideas, to the attention of large numbers of American philosophers and psychologists. Without belaboring the line of succession, we might simply observe the strong pragmatistic flavor of Watson's definitions and criteria of scientific discourse.

Far more important, however, than Dewey or James were the published works of E. L. Thorndike (1874–1949), who was a student of James' and whose *Animal Intelligence*[25] was a landmark in the history of so-called behavioral analysis. The work appeared in 1898 and described a series of experiments concerned with learning and memory in cats. Thorndike, with makeshift but serviceable equipment, plotted the speed with which animals escaped from a box in order to obtain food placed outside the box. He generated a series of "learning curves" which showed systematic improvement with increased practice. On the basis of these and related findings, Thorndike presented his famous "law of effect," according to which behavior is determined by its consequences. Behavior leading to "satisfying" states of affairs is more likely; behavior leading to unsatisfying states, less likely. Watson did not applaud Thorndike's choice of terms, finding the law of effect too mentalistic, but he did applaud the objective methods of measurement and the general demystification of the discipline.

Thorndike's "provisional" laws of learning, which he put forth as standing out "*clearly in every series of experiments on animal learning and in the entire history of the management of human affairs,*"[26] were of the sort we do not see any more in psychology:

> *The Law of Effect is that:* Of several responses made to the same situation, those which are accompanied or closely followed by satisfaction to the animal will, other things being equal, be more firmly connected with the situation, so that, when it recurs, they will be more likely to recur; those which are accompanied or closely followed by

discomfort to the animal will, other things being equal, have their connections with that situation weakened, so that, when it recurs, they will be less likely to recur. The greater the satisfaction or discomfort, the greater the strengthening or weakening of the bond.

The Law of Exercise is that: Any response to a situation will, other things being equal, be more strongly connected with the situation in proportion to the number of times it has been connected with that situation and to the average vigor and duration of the connections.[27]

There is little in either of these "laws" that could not be gleaned from Locke or Hume of Bentham or, for that matter, Aristotle. They are the classical laws of association with the addition of Darwinian and Benthamist principles. The difference, of course, is that the laws in Thorndike's case are supported by experimental findings. Still, it is not likely that anyone would have objected to either law even had experimental data not been presented. Practice makes perfect, as the aphorism goes, and we do tend to do the sorts of things we find satisfying. In their stated form, the laws are richly mentalistic and, as a result, pose no threat to those who believe that psychology is the science of mind. This was not Thorndike's position, but his two laws allow such a construction. Under the law of effect, he employs terms such as "the same situation" and "satisfaction"; under the law of exercise, "a situation." These are psychological terms and fit easily into the introspective tradition whether Thorndike wanted them there or not. It is this fact that invited Watson's dissent:

Most of the psychologists . . . believe habit formation is implanted by kind fairies. For example, Thorndike speaks of pleasure stamping in the successful movement and displeasure stamping out the unsuccessful movements.[28]

Instead of this, Watson elects to use the language of the reflex physiology, recently developed by Ivan Pavlov (1849–1936). English translations of Pavlov's works began to appear after Watson's behavioristic lectures of 1912, but by the time of the 1930 edition of Watson's *Behaviorism,* the complete Pavlovian system had been rendered into English.[29] We are not to make too much of Pavlov's effect upon Watson's thinking. The essentials of the latter's psychology were established before he learned of the Russian's theories and, even after mastering Pavlov's theories of conditioning, Watson could still observe:

Most of the psychologists talk, too, quite volubly about the formation of new pathways in the brain, as though there were a group of tiny servants of Vulcan there who run through the nervous system with hammer and chisel digging new trenches and deepening old ones. . . .

> Since the advent of the conditioned reflex hypothesis in psychology
> with all of the simplifications (and I am often fearful that it may be an
> over-simplification!) I have had my own [views].[30]

His own views involved the acceptance of the conditioned reflex as the "unit"
of behavior and the notion that all more complex forms of behavior were compounded of these units. Prudently he resisted the temptation to set tiny servants
to work in the brain. His explanation of psychological processes (i.e., the determinants of behavior) was to be of the descriptive variety, remaining relatively neutral on questions of physiological detail and relatively hostile on
questions of mental referents. Psychology is to determine the manner in which
elemental conditioned reflexes are built up to form complex habits. Our reactions to the world are to be understood in terms of these reflexes. Even our
most vaunted capacity, that of language, is but the product of conditioned reflexes involving the laryngeal musculature.[31] Through conditioning (and in
Humean fashion) *anything* can come to elicit a given response provided that it
has been presented in conjunction with an unconditioned stimulus. By association, the previously neutral stimulus becomes the *substitute* for the unconditioned stimulus.[32]

> The importance of stimulus substitution or stimulus conditioning can-
> not be overrated. . . . So far as we know now . . . we can take any
> stimulus calling out a standard reaction and substitute another stimu-
> lus for it.[33]

Since any desired behavior can be thus secured, Watson is prepared to reject
even the concept of instincts, noting that we have been credited with a large
number of them but that no recitation of the list tells us very much. He offers
the example of the boomerang that returns to the place from which it was
thrown. Its behavior is the result of its composition. Organisms also are composed of biological systems that are structured in such a way as to respond in
stereotypical fashion to particular environmental features.[34] However, we need
not invent a new instinct each time we observe a new pattern of behavior or a
new stimulus-response connection. In fact, only when we study the infant, devoid of a conditioning history, are we in a position to say anything about innate
dispositions. Watson, studying such infants, was persuaded that their only native, psychological apparatus are the primitive forms of the emotions of rage,
fear, and love.[35] He accepts these because, in the infant, he can find behavioral
manifestations of them. He rejects out of hand James' introspective method for
studying the emotions, and he chortles at the long list of instincts that William
McDougall (1871–1938) has made so popular.

Watson had begun his crusade as early as 1912 and had provided a seminal
article by 1913 and a textual guide by 1924. In the Introduction to the 1930

edition of his *Behaviorism,* he reflected on the evolution of the behavioristic perspective thus:

> [W]ithout behaviorism being overtly accepted, its influence has been profound during the eighteen years of its existence. To be convinced of this, one needs only to compare the contents of our journals title by title for 15 years before the advent of behaviorism and during the past 15 to 18 years. . . . Today, no university can escape the teaching of behaviorism . . . the younger generation of students demands at least some orientation in behaviorism.[36]

Watson, in this judgment, was correct. By 1930 the signs were unmistakable. Psychologists had divided into two camps—and "camps" is the apt description. In one were to be found the self-appointed scientists of the profession and, in the other, all the rest! The "scientists" were not yet *behavioral* scientists, but they were students of animal behavior, of conditioned reflexes, of brain-behavior relations. Introspection was a fading method. Freud's influence was becoming international, but psychoanalysis was still so mentalistic, so terminologically obtuse, that it was no threat to the new science of "real" psychology. There were, of course, still some problems. Watson's system was based on the notion that the full range of so-called purposive behavior could be accounted for in terms of the accretion of reflex connections. This seemed utterly implausible to anyone not ordained into the *ism.* Watsonian psychology also left out a topic that even the "scientists" were not willing to part with: *perception.* Watson's data, in the context of his ambitious system, were strikingly sparse and his proposed methods were daringly ambiguous. Promising to create "dentists" by Pavlovian conditioning, for example, left a good part of the psychological community incredulous, another part perplexed, and still another part in stitches. Quite simply, behaviorism could not have survived in the form bequeathed by Watson. Revisionism was in the air and it found its most articulate expression in B. F. Skinner's *The Behavior of Organisms* (1938).[37] Since its publication, the text has influenced American experimental psychology as much as any single work in the history of the discipline. Professor Skinner (1904–1989) has revised the initial formulations of his behaviorism and has extended the system to embrace issues of a broad, social character.[38] Notwithstanding the alterations, behaviorism in Skinner's treatments has retained the following features.

Psychology is a natural science whose subject matter is restricted to the observable behavior of organisms. The aim of the science is the prediction and control of behavior. It does not strive to complement the biological sciences, nor does the validity of its principles depend upon findings in biology or neurophysiology. "Neurology cannot prove (behavioral) laws wrong if they are valid at the level of behavior. Not only are the laws of behavior independent of neurological support, they actually impose certain limiting conditions upon any science which undertakes to study the internal economy of the organism." [39] As a descriptive science concerned with (Humean) regularities between environmental antecedents and behavioral consequents, psychology need not strive toward theoretical systematicization. Its law is the law of effect with mentalism removed, that is, the law of effect expressed in purely operational terms. That which increases the probability of the behavior that precedes it is, by operational definition, a positive reinforcer. That which reduces the probability of the behavior that precedes it is a negative reinforcer. The behavior of interest to psychology is that which operates on the environment and thereby affects the survival of the organism. Pavlovian reflexes, while conditionable, tend to involve sub-systems of the organisms. They are, of course, intimately associated with the balance of the organism's adaptive capacities but do not directly and immediately result in those adjustments of the environment that *operant* behavior produces. Where the creation of a dentist by Pavlovian conditioning appears implausible, the very appearance of a dentist is prima facie evidence of the success of operant conditioning. Consciousness, free will, intention, and the like, for purposes of a scientific analysis of the determinants of behavior, need not be considered. The terms, themselves, are merely "verbal operants" invented by a society that finds mentalistic interpretations reinforcing.

The foregoing position does not occur in an intellectual vacuum. Nor does its reception, whether positive or hostile. The groundwork for "Skinnerian" psychology had been laid years, even centuries, earlier, perhaps as early as Ockham's rejection of universals. More proximately, it may be said to have begun with Hume and to have reached its first great plateau in the functional biology of Darwin, the utilitarianism of Mill, and the pragmatism of James. For James, knowledge is utility. At the time James was discovering C. S. Peirce, Ernst Mach (1838–1916) was promoting a similar movement in Vienna, a movement away from metaphysics (especially Kant's) and toward practicality. Peirce had already written his little classic, *How to Make Our Ideas Clear* (1900), offering such antimetaphysical notions as, *"To say a body is heavy means simply to say that it will fall."* Einstein's challenge to complacent Newtonians even had physicists fearful of the metaphysical residue of "natural philosophy." Then, too, there was World War I, with its heightened nationalism and jingoism. Scholarly detachment could not survive that war which so intensified the long rift between Continental and English-speaking philosophy, psychology, and ethics. Recall Koffka's observations on the American devotion to "earthbound science" and aversion to "ideas and ideals." We are not to assess the scientific status of behaviorism either in terms of its origins or in terms of the social factors that may have led to its popularity. But to be assessed as a science, it had to come into being, and we would be remiss not to take note of that early twentieth-century atmosphere that was especially conducive to the behavioristic way of thinking. Watson's books and articles have the same anti-Hegelian ring (without ever mentioning Hegel) as the articles and books appearing at the same time in the philosophy literature of the English-speaking world.[40] Scientism had replaced idealism in intellectual circles, and psychology, forced again to choose between Nature and Spirit, now had the universal support of the sciences in choosing the former.

Modern Behaviorism: B. F. Skinner

Watson's dream or hope was not to be realized through his own writings, less through the questionable findings and haphazard methods punctuating his polemical tracts. But his groundwork was followed by a devastating world war in which a villainous enemy has raised hereditarianism to the level of a political ideology. In the allied countries, the years during and after World War II were not hospitable to genetic or instinctual theories of psychology. Apart from the technical or scientific merits of the case, by the 1940s there was a far more receptive audience for the environmentalistic psychology that Watson had earlier defended. The man and the moment were right for each other, as B. F. Skinner's books and articles introduced a far more sophisticated and promising version of behavioristic psychology.

Many have been attracted to behaviorism, even from the time of Watson, in part because it is free of the difficulties suffered by various forms of psychological materialism. Behaviorism does not regard itself as under any obligation to speak of brains or minds. As early as 1938, in his very first book (*The Behavior of Organisms*), B. F. Skinner announced his psychology's independence in regard to neurology, and those who subscribe to the so-called Skinnerian version of behaviorism have remained neutral or indifferent on the question of the role of brain-talk in psychology. There is no contradicting the fact that the independent status of psychology as a historically unique discipline is better protected by the canons of behaviorism than by any alternative formulation. It has surrendered (without regrets) the mind to philosophy, the body to biology, and personality to the clinicians. In certain respects, behaviorism in Skinner's hands liberated itself from the goals and methods of nineteenth-century science more successfully than has any other branch of the discipline. Skinner and his disciples declared themselves against attempts to invent grandiose theories, to provide satisfying explanations, to unite psychology with all other sciences, to unearth "mechanisms," to penetrate the "mind." In place of these historic and resoundingly failing missions, the behaviorists proposed to provide a descriptive science of behavior permitting the prediction and control of the actions of organisms.

The maxim that directs behavioristic inquiry is one or another form of the "law of effect," and this law is as close as they come to out and out theorizing. The law is no longer expressed in that richly psychological language used by Thorndike at the end of the nineteenth century. Stripped of all mentalistic features, the current versions of the law declare only that certain stimuli ("reinforcers") alter the probability of the response that produces them. With the unadorned ethic of the utilitarian, behaviorists ask to be judged by what they can do, not by what they say or fail to say about the mind or mental life. The Watsonian legacy is most pronounced on this point. Watson could only be pleased to note, too, how behaviorists have introduced the methods of operant conditioning into classrooms, psychiatric clinics, vocational schools, industrial training programs, counseling chambers, and even playgrounds. Techniques of behavior-modification are as at home in penal institutions as they are in the animal laboratory, or wherever there is a captive audience. Their application is recommended with the same confidence whether the "organism" is a white rat, a convicted felon, an autistic child, a performing seal, a schizophrenic patient, or a difficult student.

Critics of Skinnerian behaviorism have not been in short supply. Their attacks have been broad, often Delphic, seldom dispassionate. Humanists have condemned it as a "dehumanizing" psychology, theologians as a godless one, ethicists as amoral. Philosophers in the tradition of the British empiricists have found it quite compelling, those in the Hegelian tradition, quite absurd. Political theorists in the line beginning with Bentham have judged it to be quite sound, those in the Kantian succession, blatantly fascistic. Not even Freud attracted so diverse an assembly of friends and enemies. Far from chastening the advocates of behaviorism, all this attention and condemnation have called forth bold resolve, most recently evidenced in Skinner's *Beyond Freedom and Dignity,* which was intended as a scientific justification of his utopian scheme.[41] The most persistent complaint against behaviorism is that it fails to recognize the rational, volitional, and intentional elements of human behavior. That is, in its devotion to unearthing the environmental causes of behavior, it ignores the private reasons and motives of the behavior person. The accounts of behavior it offers are therefore limited to the "whens" and "wheres," while remaining stunningly silent on the matter of "why." In refusing to confront Smith's reason for doing X, the behaviorist neglects what some consider to be the most important psychological determinant of all: motivation. The behavioristic rejoinder is deceptively simple. What, it asks, is added to a description of the publicly observable relations between responses and reinforcers when the concept of "motivation" is supplied? What can the term "motive" mean beyond the efficacy of a stimulus to control the behavior that obtains or avoids it? In summarizing the archetypical behavioristic account of explanation itself, we might reach a fuller understanding of the totality of the behavioristic system. The following is a synthesis of many behavioristic works addressed to this question of *reasons*-explanations versus *causes*-explanations.

Neither praise nor blame is entailed by a causal account of behavior. When we attribute Smith's walking, talking, and working to the effects of reinforcing stimuli, we remove the behavior from a judgmental domain and locate it in the domain of natural phenomena. However, when we impute reasons to Smith, we earn the right to judge Smith's intentions as something above and beyond the mere actions and above and beyond the mere consequences of these actions. We can say, for example, that Smith intended to kill Jones even though the bullet missed "its target." We can say, further, that Smith is a "bad sort," wicked, worthy of our wrath, and a candidate for punishment. Our own histories of reinforcement are such that we seek control over the behavior of others, and the right to judge is precisely this kind of control. We are able to elevate ourselves to nobility of purpose as long as we can reduce Smith to an inferior station. Thus, one of the causes of our search for "reasons" is no more than the ability to control others. Then, too there, is the component of social desirability in talk of "reasons." This is rooted in religious traditions in which priests and witch doctors are thought to possess unnatural powers and understandings. In modern society, the residual of this tradition takes the form of applauding those able to "see more deeply" and "comprehend more richly" the "true" reasons of Smith's behavior. By retaining these notions, we are in a position to reward ourselves with "perceptiveness" and "insight" and other marks of distinction. Not content to deal with the Smith before us, we report that we have found the INNER SELF of Smith, who is really responsible for what Smith does. Not everyone, of course, has such powers of intuition. Those of us who, unlike the common run of humankind, can see further, unearthing the quasi-private, brooding, and mysterious motives of the person-behind-the-person, now have a special status. It is in the nature of a special status that those who have it are in more complete control of the distribution of "available resources," that is, reinforcers.

Causal explanations are however, at least in principle, available to everyone. A society restricting the study of the psychological person to the publicly observable and verifiable relationship between environmental events and behavioral outcomes is one in which each person has the same ability as does anyone else to judge and to decide. Just as the search for and the discovery of "reasons" place the observer in a position of privilege, belief in reasons as opposed to causes gives the actor a special power. Reasons, being private and, therefore, "one's own," put the actor in control, whereas causes put nature in control. Industrial societies especially place a great premium on competition. Secondary reinforcers are distributed in such a way as to increase the behavior of the laboring community. Titles and special distinctions are able to exert as great a control over behavior (including verbal behavior) as, in more primitive societies, food and shelter do. Accordingly, modern citizens are controlled in such a way as to emit those behaviors that seem to remove them from the direct control of others. It becomes increasingly important for them to refer to their "freedom" and "dignity," since such imputations carry special status. Such persons are not merely animals working for food but are now self-motivated, self-actualizing, and rational creatures; all this is proved by our asserting *reasons* for actions over and against natural causes. Those who criticize causal accounts by inserting "reasons" into Smith's "mind" are merely putting certain words into Smith's vocabulary, and these words get into his vocabulary in precisely the same way that bar-presses enter the behavioral vocabulary of the laboratory rat. The purely descriptive, causal account, on the other hand, allows a specification of the environmental sources of behavior control. This specification is, in principle, complete, predictively powerful, and pragmatically successful. It is not improved by the inclusion of inner selves, motives, reasons, spirits, wants, beliefs, or even neurons. To account for behavior is to offer descriptions of contingencies in the observable world, not references to unobservable, intangible, goals, dispositions, or intentions. Goals, if referred to at all, are understood as objects in the environment, not "in" the person, and they can be specified only after the fact. They may be said to have been reached, not sought. Until society recognizes the essential truth of the foregoing, it will continue to praise and blame, to create heroes and villains, to be wracked by centuries of war, and to enjoy only accidental peace. Worse, it will look to geneticists and neurosurgeons to create "good stock" or "healthy brains," without realizing that all social problems are never more than behavioral problems and that most of these can be solved by methods we now possess.

> An experimental analysis shifts the determination of behavior from autonomous man to the environment—an environment responsible both for the evolution of the species and for the repertoire acquired by each member. . . . Is man then "abolished"? Certainly not as a species or as an individual achiever. It is autonomous inner man who is abolished, and that is a step forward.[42]

Examining this thesis for the first time, the reader suspects that something has been left out, but can find no fatal flaw. On further examination, one begins to suspect, however, that the apparent invulnerability of the thesis may be the result of its failure to say anything! We approach this possibility with caution. We begin by recognizing behaviorism's Darwinian foundations. It shares with all systematic psychologies since Wundt (and these include Freudian formulations as well) the conviction that human psychological processes have evolved, are present in lower forms of life, and are the outcome of transactions between the organism and its environment. In a nonmentalistic way, it subscribes to one or another form of "pleasure principle" that is judged to be responsible for the validity of the law of effect. What survives of the early Watsonian formulation is the emphasis on practical application and objectivity, anti-mentalism, and indifference to philosophy. What survives of Pavlovian theory are the principles of generalization, reinforcement, and extinction. Unlike most earlier empiricist psychologies, it has only the loosest ties to materialism, and these are of a conceptual rather than an operational or a programmatic nature. Each of these features of behaviorism exposes it to a particular set of objections, some of which are more telling than others.

When a psychological system or thesis is presented as one of Darwinian complexion, it is important to realize that this is not a demonstration of its validity but a call for clarification. We require clarification because that which is "Darwinian" is not reducible to a small set of descriptions or propositions. Evolutionary theory is, itself, in the process of evolving, and is far from an adequate scientific theory. Indeed, as we get further away from the research and writings of molecular biologists, we find that evolutionary principles are thrown about in a way that can only be described as "loose talk." There is no "thing" in the world that is "selection," nor have we explained a phenotypic outcome by casually referring to "natural selection." The term "environment," which psychologists tend to use as if it were a voltage or a weight, stands for a continuously changing and extraordinarily large and complex assortment of conditions inevitably described in terms borrowed from the practices and prejudices of the observing culture. For the organism capable of locomotion, the "environment" is never really permanent. For the organisms with memory, it is never really gone. Moreover, once behaviorists commit themselves to any version of evolutionary theory, they are required to accept the fact and the implications of genetic heterogeneity by which there can be no specification of selection pressures without a simultaneous specification of the hereditary nuances of the organisms under study. Ultraviolet radiation, which is a central feature of the environment of the honey bee, does not illuminate the human world. Thus the effective environment is not what is loosely called "the environment," and the behaviorists who promise to bring behavior under the control of the "environment" by altering the latter may find that they have a new set of conditions to worry about each time the genotype of a behaving organism differs from the one they have just studied. What makes this fact important is the traditional indifference of behavioristic psychologists to genetic considerations. Indeed, it is frankly ironic that a system of psychology described as "evolutionary" should have such a spotless record of aloofness toward hereditary factors. The common reply to those who question this neglect of genetics is that the law of effect applies to all organisms, regardless of hereditary peculiarities, since survival is impossible for organisms unable to secure rewards and avoid aversive stimulation. But, like "Darwinian," the term "law of effect" does not refer unambiguously to a set of propositions or demonstrated facts. As formulated by Thorndike, it is so rich in mentalistic terms and so tautologous in its logical structure that no modern behaviorist could possibly subscribe to it.[43] That we and other animals tend to do the things that give us pleasure is a fact that will embarrass no philosophical or psychological position on human and animal nature. But to say, that is, that "X" is a reinforcer whenever it alters the probability of responses that precede it is to open up the domain of reinforcers so thoroughly that it would be impossible or at least impractical to locate the effective stimulus in an environment space now filled with a veritable infinity of "reinforcers."

One of Skinner's many important findings is that pertaining to the effects of schedules of reinforcement. It has been found that behavior brought under the control of reinforcers that have been applied irregularly during the acquisition of a response is extremely resistant to extinction. The ability of random reinforcement to result in virtually unextinguishable responding is one of the more striking demonstrations in all of psychology. Yet there is no formulation of the law of effect that permits one to predict such an effect, nor is the effect logically deducible from the law that is asserted as covering it. Thus, even if we accept the modified, neurophysiological version of the law of effect—that is, the version that would allow scientific status—we are unable to use the law in framing explanations of behavioral events of interest. We are inclined to conclude that the behavioristic thesis either lacks a covering law or that the covering law does not cover. While this deficiency may have little bearing upon the practical utility of the behavioristic thesis, it ensures that that thesis will never be able to provide a scientific explanation of the phenomena it purports to embrace. To repair this defect Skinner showed something of a surprising flirtation with biology.[44] In his later writings he attempted to present behavior as something akin to "digestion," but the net effect of the analogy is to move his psychological system back into the nineteenth century, when the metaphor of the machine was the reigning theoretical device.

In all, behaviorism from Watson to Skinner offered more than it delivered, and called upon psychologists to abandon too much of what inspired the development of the subject from the time of the ancients. What was left out were just those cognitive, problem-solving powers, those feats of creativity and self-expression that are marks of the mental. The problem has always been one of preserving the subject-matter while still developing *scientific* methods and theories. One alternative to the behavioristic approach, designed to compensate for its apparent deficiencies, has been Gestalt psychology, a forerunner of today's *neurocognitive* perspective.

"Gestalt" Psychology (the Continental Reply)

The names associated with the founding of Gestalt psychology are those of Max Wertheimer (1880–1943), Kurt Koffka (1886–1941), and Wolfgang Köhler (1897–1967). The three worked together at the Psychological Institute of Frankfurt for several years, beginning in 1909. All psychologists are familiar

with Wertheimer's "discovery" of the "*phi* phenomenon": the apparent continuous movement produced by two different stimuli, partially separated, and
illuminated successively at brief intervals. Since strolioscopes were available
as children's toys by 1910 and since the earliest motion pictures had been
filmed some twenty-years before, we must be careful about the sense in which
we use the term "discovery." It was not *apparent movement* that was discovered by the Frankfurt group but a new approach to psychology, an approach
based on such perceptual phenomena as *phi*.

We can locate the spirit of Gestalt psychology as well as its philosophical
orientation by reading Ivan Pavlov's *Criticism of Köhler's Idealistic Concepts*
(1935).[45] It is not that there is anything especially "idealistic" in the Gestalt
movement. None of the major treatises by the Gestalt psychologists contains
an acknowledgment of Hegel's or Berkeley's influence. No Gestalt psychologist has ever denied the existence of matter or has suggested that the fundamental stuff of the universe is mental. But physiology by 1935, like psychology, had declared its opposition to metaphysical pronouncements. Pavlov is
just one of the more famous spokesmen of that materialistic and associationistic school that judged any departure from the orthodox interpretation to be
"idealism." If the Gestalt psychologists are to be united, in a loose way, with
the German idealist tradition of Kant, Hegel, and the neo-Hegelians, it must be
through the notion of mental categories by which sense-data become organized
percepts. At the risk of simplification, we might draw a parallel between Thorndike's experiments and those of Wertheimer, Koffka, and Köhler. It is this:
Thorndike borrowed the philosophical principles of association, combined
them with the utilitarian emphasis on the "pleasure principle," and created an
experimental environment in which these principles could be demonstrated.
The Gestaltists accepted the Kantian-Hegelian principle of the pure categories
of the understanding, brought them to bear on studies of visual perception, and
thereby provided a laboratory demonstration of the role of the mind in organiz-
ing and transforming the raw facts of experience. The *phi* phenomenon, then,
was just a means of displaying the essential premise of Gestalt psychology:
perception is the result of an interaction between the physical characteristics of
stimulation and the mental laws governing the experiences of the observer. The
setting for demonstrating *phi* is a dark room in which one can present two
illuminated stripes placed several inches apart. An interval can be found between the illumination of one stripe and that of its neighbor such that the observer reports, not two stripes, but the movement of one stripe from left to
right, or right to left, depending on the order of illumination. The important
point is this: nothing in the experimental arrangement, except the phenomenon
itself, would lead to the prediction of apparent movement. That is, there is no
physical feature of the environment that permits the prediction of the effect. A
purely stimulus-bound description of the laboratory setting will be devoid of

any allusion to motion. The motion is created by the observer. It is one's perception of motion, not one's response to motion that is the object of study. In short, it is the study of a mental as opposed to sensory or behavior state.

Far from being an idealistic psychology, Gestalt psychology has always accentuated the role of brain dynamics in accounting for the large number and variety of Gestalt phenomena. Köhler insisted that the relationship between perception and neurophysiology was "isomorphic," meaning that the structural features of the percept were matched by the structural features of the brain's functional organization:

> [W]e are inclined to assume that when the self feels in one way or another referred to an object there actually is a field of force in the brain, which extends from the processes corresponding to the self to those corresponding to the object. The principle of isomorphism demands that in a given case the organization of experience and the underlying physiological facts have the same structure.[46]

The Gestalt critique of behaviorism is uncompromising.[47] The behavioristic reliance upon physiological reflexes is judged to be not only simplistic but at variance with even the physiological facts. While presuming to model itself after physics, behaviorism fails to take from physics its most celebrated discoveries, those that allow a comprehension of dynamic processes. Even the very modern forms of behaviorism that labor so diligently to avoid commitment to any physiological theory or body of knowledge are still, by Köhler's lights, wed to the old associationistic principles of learning and their hedonistic corollaries. This is why behaviorism must always be embarrassed by the abilities of animals performing in settings that have not been trivialized by the demands of behaviorism. On this latter point, Köhler presents the results of his studies of chimpanzees that he conducted from 1913–1917 at the Anthropoid Station in Tenerife.[48] The experiments say more about the Gestalt view of psychology than many of the words written to describe the system. If the rat in a box is the image created by the term "behaviorism," then the chimp with two sticks in his hands is the symbol of the Gestalt laboratory.

It is axiomatic that all one will learn of the abilities of an animal is what the testing situation permits the animal to perform. Köhler's apes were called upon to solve problems. They would, for example, be confronted by a situation in which food was hung too high above them to be reached. Boxes were then strewn about the cage. The chimp soon solves the problem by stacking the boxes and climbing on them to reach the food. Then there is the most famous of the "insight" studies, in which the animal is given two sticks, neither long enough to reach food placed beyond the cage. After some random activity and a good deal of looking at the sticks, the chimp suddenly joins them together, creating a single stick of twice the length, and proceeds to haul in his reward. The animal will also break off branches to achieve the same end. So, too, will they imitate the behavior of a human engaged in such tasks. And from such findings, Köhler concludes that the processes

> *occur in chimpanzees, exactly as in man* . . . it is these "impressions," which are not at all "something that has been read into" the chimpanzees, but which belong to the elementary phenomenology of their behavior. . . . If *this* is an anthropomorphism, so then is the sentence: "Chimpanzees have the same tooth formula as man." [49]

Later formulations of the Gestalt position, particularly those by E. C. Tolman (1886–1961),[50] have retained the original emphasis upon cognitive as opposed to stimulus-response aspects of the psychology of learning. Tolman's *Cognitive Maps in Rats and Man* (1948)[51] is a summary and interpretation of a variety of Gestalt-like experiments on maze-learning and problem-solving in which the experimental animals seem to behave in terms of a "mental image" or "map" of the experimental situation rather than in terms of the purely associationistic demands of the situation. Here and elsewhere, Tolman distinguishes between *performance,* which is under the control of rewards and punishments, and *learning,* which occurs whenever a complex organism has perceptual commerce with the immediate environment. Rats permitted to run freely in a maze come to solve the maze more quickly on subsequent occasions when food-reward is introduced than do animals without the original "irrelevant" experience. This so-called "latent learning" is assumed to violate the law of effect which requires reinforcement if learning is to occur. Similarly, an animal receiving reward by responding to, say, a circle five inches in diameter and not rewarded for responses to one that is two-and-a-half inches will subsequently choose one of ten inches over one of five inches. That is, after originally learning the choice of "5" vs. "2 1/2," the animal, given the new choices "10" vs. "5," does not choose the "5" (with which all previous rewards were associated) but, instead, chooses the larger. This, according to the Gestaltist, requires us to assume that what was originally learned was a *relationship* and not merely a physical value. This is taken as an instance of "transposition," in which the relational properties are abstracted from the stimulus elements. A more common instance of transposition occurs in music, where the listener will recognize a melody played in different keys despite the fact that, from key to key, the actual notes are of completely different frequencies.

From the Gestalt point of view, latent learning and transpositional learning introduce fatal flaws into the traditional behavioristic perspective. The former is judged to be telling evidence against the law of effect and the latter against associationism. While as opposed to introspective psychology as any behaviorist, the Gestalt psychologist finds nothing in behaviorism that warrants its claims to superiority. It is one thing to assert that the Wundtian-Titchenerian psychology is "subjective," but quite another to deny the relevance of immediate experience to the study of man—and all other complex organisms. It is one thing to assert that the conditioned reflex comes about by virtue of the formation of reflex associations among cortical neurons but quite another to suggest that the brain is capable of *only* such "connections." It is one thing to demonstrate how practice and reward affect performance, but a very different matter to submit that practice and reward are the *only* determinants of learning. In the light of these reservations, the Gestalt theorist offers his own findings and hypotheses:

1. Organisms do not merely respond to their environments; they have *transactions* with the environment.
2. The "environment" is not just the physical objects proximate to the animal but the outcome of an interaction between the organism's perceptual *fields* and these physical objects. The effective stimulus, therefore, must always be specified from the organism's point of view.
3. The relationship between experience and action on the one hand and brain physiology on the other is *isomorphic*. Accordingly, to the extent that the experience or the performances of the organism are not of a reflex nature, the underlying brain physiology also must not be of a reflex nature. Reflex organization within the brain is but one of the many forms of organization available to so complex a system.
4. Perception, as with all other fundamentally biological processes, is governed by laws or principles of organization. The laws of perception are such that the organism will impose upon the physical environment a certain form (*Gestalt*) or organizational quality (*Gestalt-Qualität*). It is by virtue of this filtering and transforming of stimulus elements that the organism is able to deal with the demands of the environment in an economical and orderly way. In the following drawing we see two collections of six lines but perceptually, we "see" something other than six lines on the right. We see three *groups* of two lines. *Grouping* is but one example of the imposition of *Gestalt-Qualität* upon the elements of sensation.

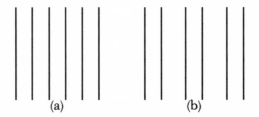

(a)   (b)

It is by virtue of the same principles of perceptual organization that we perceive a *constancy* of experience in the face of changing stimuli. Dinner plates are seen as round no matter what the angle of regard. Their circularity is apparent even though, in any plane other than the normal, they project ellipses on the retina. Shape-constancy violates the predictions of geometric optics because the latter does not take into account the *perceptual* principle of constancy. For the behaviorist or introspectionist to dismiss such effects as "illusory" is tantamount to their rejection of the bulk of human experience from the domain of psychological inquiry.

5. The naive *phenomenalism* of J. S. Mill, Titchener, and others must be abandoned. There is not a simple relationship between the world of matter and the world of experience. We do not "see" objective stimuli, we transform them. If we can say, with Mill, that matter is the permanent possibility of sensation, then we must also say that sensation is the permanent possibility of perception. For Köhler, "sensory organization constitutes a characteristic achievement of the nervous system. This emphasis has become necessary because some authors seem to think that, according to Gestalt Psychology, '*Gestalten,*' i.e., segregated entities, exist outside the organism and simply extend or project themselves into the nervous system. This view, it must be realized, is entirely wrong." [52]

Like behaviorism, Gestalt psychology has splintered into a generous assortment of derivative enterprises. Not every contemporary "cognitive" psychologist expresses allegiance to Köhler's formulations, just as many "behavioral scientists" are quick to deny kinship with Watson—and even with Skinner. Clark Hull (1884–1952), for example, offered a behaviorism rich in mathematical notation and closely tied to physiology and evolutionary biology, but neither of these features characterizes the workaday activities of the current "operant" psychologist. In the Gestalt tradition, many investigators are carefully examining the nuances of perceptual organization, the rules of information-processing, and so forth, without worrying over psychophysical isomorphism or *Gestalt-qualität*. Around the world and with ever increasing comparability, laboratories are engaged in essentially descriptive work designed to establish the extent to which environmental modifications lead to alterations in the measurable features of behavior. Theoretical tensions have, at least for the moment, subsided and this must be appreciated as something of a victory for that modern behavioristic injunction against theorizing, physiologizing, and mathematizing.[53] The essentials of the Gestalt system still survive and flourish in the work of such European psychologists as Jean Piaget, whose theories of cognitive and moral development take recourse to neo-Hegelian concepts of evolutionary stages and to Gestalt concepts of cognitive organization and neuroperceptual isomorphism. Psycholinguistics, too, received its impetus from the cognitive elements of the Gestalt psychologists, but, like Piagetian psychology, it is considered by many behavioral psychologists to have little to do with psychology, proper. To these psychologists, the idea that language-structure is somehow "given" sounds very much like the instinct theories of the 1920s and 1930s.

Physiological Psychology

Although it is true that behaviorism derived some inspiration from the writings of Ivan Pavlov, it is important to recognize that Pavlov hardly qualifies as a "behaviorist" in the sense in which that label has come to be applied. Pavlov was a physiologist by training and by commitment. His ambition was to articulate the laws of the central nervous system functions which accounted for the so-called psychological dimensions of life. His studies in Germany brought him in contact with the Helmholtzian school—principally with one of Helmholtz's colleagues, Carl Ludwig—but he was a declared physicalist before he ever left Russia.

In his speech accepting the Nobel Prize for pioneering research in gastric physiology, Pavlov (1909) introduced the concept of the conditioned reflex to the general scientific community. His important papers on reflex conditioning were not translated into English for several years, but the Nobel address then, as now, received wide attention. That the rest of his life would be devoted to the newly discovered phenomenon can be anticipated from his closing remarks on that occasion:

> In point of fact, only one thing in life is of actual interest for us—our psychical experience. But its mechanism has been and still remains wrapped in mystery. All human resources—art, religion, literature, philosophy and historical science—have combined to throw light on this darkness. Man has at his disposal yet another powerful resource—natural science with its strictly objective methods. . . . The facts and considerations which I have placed before you are one of the numerous attempts to employ—in studying the mechanism of the highest vital manifestations in the dog, the representative of the animal kingdom which is man's best friend—a *consistent,* purely scientific method of thinking.[54]

Pavlovian psychology, some trivial aspects which have recently been rediscovered and raised to the level of a "science" with the name "biofeedback," forms very little of contemporary behavioral psychology outside Russia. As a method of conditioning autonomic activity, it has had a great impact on the field of psychosomatic medicine. But the actual conditioning procedures themselves rarely appear in contemporary research except as part of some larger issue. It is generally agreed that autonomic conditioning takes place, that it does so in the manner first reported by Pavlov, and that it needs no more confirmation than do Galileo's findings. Even in his 1909 speech, Pavlov was quick to note that "it has long been known that the sight of tasty food makes the mouth of a hungry man water."[55] Thus, even from the outset, the conditioning *effect* could hardly be called revolutionary. It is this fact that makes it difficult for the nonpsychologist to understand, as he reads modern textbooks, how psychology was "revolutionized" by Pavlov. The point, of course, is that it was not his studies of the conditioned reflex that made Pavlov a figure to be contended with; it was the *theory* advanced on the basis of these studies. While Watson never had a sophisticated appreciation of the Pavlovian system, his behaviorism, as reviewed above, provides the essentials. We will not examine the details. In the broadest terms, the Pavlovian theory requires that all so-called psychic functions are reducible to reflex mechanisms within the brain. By the frequent association of a neutral stimulus with one having unconditional biological significance, the former comes to have the power of eliciting responses originally produced by the latter only. Stimuli thus associated are now conditional (or conditioned) stimuli. If they are presented repeatedly without the application of the unconditioned stimulus, they will lose their power of elicitation; that is, "extinction" will occur. Not only does the specific, conditioned stimulus acquire the power of the unconditioned stimulus, but those stimuli physically similar to the conditioned stimulus acquire this power by "generalization." Thus, if a tone of 1000 Hz is paired with the delivery of powdered food to the mouth, the 1000 Hz tone will acquire the power to elicit salivation. So, too, will tones of 900 Hz, 1100 Hz, etc. The magnitude of the conditioned response will diminish in proportion to the difference between the initial conditioned stimulus and the test-stimulus. Stimulus generalization is explained in terms of the "irradiation" of cortical responses to stimuli. A given stimulus (e.g., the 1000 Hz tone) establishes a region of maximum activity within the cortex, and this activity radiates, decrementally, to adjacent regions of the cortex. Accordingly, a reflex-association is established which is strongest between "food" and 1000 Hz and weaker between "food" and a tone different from 1000 Hz. Combined with conditioned excitation of the cerebral cortex, there may be conditioned *inhibition,* such as that produced by the application of a second stimulus on all trials when no reward is to be administered. Responses to the pair "S+" and "S°" (where "+" refers to the application of reward and "°" to its absence) will be weaker than responses to "S+" alone.

These principles—conditioning, extinction, radiation, excitation, and inhibition—are the central elements of Pavlov's biological psychology. Conceptually, it is scarcely different from the reflex-association of Hartley or, for that matter, Descartes, with the dualism removed. Procedurally, however, it is radically different from any of its philosophical ancestors because of its reliance upon laboratory investigation and quantitative measurement. Since the case for this psychology is stated against the background of data, criticism was to take an experimental form and would come chiefly from Karl Lashley (1890–1958).

Lashley and Watson were colleagues for a time at Johns Hopkins and even coauthored a paper on the environmental determinants of homing behavior in birds (1915).[56] Lashley's training was in anatomy, but his life was devoted to physiological psychology. He was neither a "mentalist" nor an "idealist," but he was quick to discern the deficiencies of that biological associationism fostered by Pavlov and, in less mature fashion, by his American disciples. If we were to summarize his role in twentieth-century developments in physiological psychology, we might say that he bore the same relationship to the Pavlovians that Flourens bore to the phrenologists. We are tempted to call him a Gestalt psychologist—a label that would not have offended him in the least—but his work possessed a rigor and systematic quality seldom found in the orthodox Gestalt tradition. Moreover, he did not merely speculate about the nervous system, he examined it directly. As with our treatments of other productive experimenters, we are not in a position to analyze the details of their research. Lashley's most important statements have been collected by his former students and may be consulted.[57] For historical purposes, a very brief review is sufficient.

Like Flourens, Lashley was sufficiently familiar with the clinical findings in neurology and neurosurgery to know that no simple relationship existed between a particular locus within the brain and complex psychological processes such as perception, learning, and memory. Like Köhler, he subscribed to a form of isomorphism regarding this relationship. His own studies of the effects of surgical destruction of brain regions in animals trained to perform various discriminations proved that complex abilities survive the removal or maceration of even extensive amounts of cortical tissue. But other studies also revealed very substantial effects when restricted regions of the brain were disturbed or removed. Lashley framed two broad principles to account for as much of the data as any general statement was likely to embrace: the principle of "mass action" and that of "equipotentiality." By "mass action" Lashley meant to convey the idea that when it comes to complex psychological processes, the brain functions as a whole and is to be understood as a whole. The principle of equipotentiality was invented to account for the otherwise perplexing fact that surgically produced deficits disappear in time. The deficits indicate that particular regions of the brain do serve specific functions, but postoperative recovery indicates that other regions of the brain are able to assume these functions when the primary area has been destroyed or removed.

If Pavlov's studies were careful, Lashley's were clever and even dramatic. A cat is equipped with an eye patch covering the right eye and is called upon to learn a visual discrimination: for example, to jump off a platform toward a circle but not toward a triangle. Once the animal has learned this to a criterion level of performance, the patch is removed from the right eye and placed over the left. On the very first test-trial, the cat performs as well as it did on the last training trial. Since the optic nerve involved in the original learning is not involved in the test of transfer, we must be cautious about the sorts of neural "associations" we are to posit in attempting to account for learning. Clearly it cannot be the impulses in the optic nerve that have become associated with specific motor discharges. With respect to the latter, Lashley had another finding to offer. If the motor roots of the spinal cord are compressed, the limb on the treated side becomes paralyzed, and impulses from the cord to the peripheral muscles are blocked. In time, the limb recovers. Animals so treated can be trained to emit appropriate responses with the normal limb during the period in which the treated limb is immobile. Now the trained animals undergo compression of the motor root on the normal side such that the limb that performed the task no longer can. Shortly after this, the initially treated limb has recovered. How does the animal perform? The answer is, *perfectly*. The limb that was not and could not have been involved in the actual learning is used as effectively as the now immobile limb that had been the "trained" limb. Again, if the associationist theory requires that specific sensory and specific motor elements be combined for learning to occur, the theory is wrong.

With respect to generalization and the "irradiation hypothesis" designed to account for it, Lashley presented data from transposition studies to show that the effective stimulus is not necessarily the one most closely represented cortically. Not only that, but in some sensory systems, the sensory outcome is not topographically represented at all. The experience of loudness, for example, is not based upon the anatomical juxtaposition of progressive "loudness" centers.

Perhaps Lashley's most significant findings were in the areas of learning and memory. He demonstrated in a variety of different experimental settings that the animal's ability to acquire a complex behavioral repertoire and to reproduce it after a long retention interval was not systematically related to specific loci within the cerebral cortex. He remarked whimsically that, after searching for the "engram" of memory for many years, he was forced to conclude that learning was simply not possible! Behind the wry comment was the caveat that the brain is not one of La Mettrie's clocks, nor is man Condillac's sentient statue. More recent findings in physiological psychology have shown that Lashley may have been too pessimistic in his position on localization of function and that he was certainly misled by his focus on the cerebral cortex to the exclusion of subcortical mechanisms. These findings have not, however, led to a rejection of his central message.

Defining Psychology

The assets and liabilities of the various schools and systems of Psychology concern us only to the extent that they tell us something about intellectual history, including the history now being written by our own age. The essentials of the behavioristic thesis can be found in the scholarship of all previous ages. Its mechanistic flavor is unadulterated Hobbism. Its only "law" is to be found in the writings of Epicurus, Marcus Aurelius, Ockham, Gassendi, La Mettrie, Locke, Hume, Bentham, Mill, and numerous other distinguished commentators. Conditioning techniques have been developed to a level never attempted before, but this fact is one the historian must try to explain and not merely report. The techniques have been developed because many psychologists in the present century have labored to develop them and because modern society has supported their development. Endeavors which, in the seventeenth or eighteenth centuries, might have engaged only the animal trainer or those with domestic pets now find expression in university laboratories the world over. Manufacturers have made available devices by which shocks can be transmitted over great distances in order to modify the behavior of children in playgrounds, pets in yards, and patients in psychiatric wards. These devices are new, but the idea is prehistoric. We must ask why the modern world is so eager to control behavior so thoroughly and so precisely. Clearly, cultures have a way of installing "truths." The scientific rhetoric of the twentieth century would claim for science the authority once possessed by the gods and revealed in their works and words. Today, to speak "with authority" is to borrow (if not possess) the discursive features of scientific expression. Psychologists have been caught up in this at least since the middle of the previous century, though at the outset there was at least an awareness of the limitations.

If "scientific" psychologies of various stripes have a common ground, it is that upon which rejections of "autonomous man" have always been based. The neurocognitive psychologies would replace this entity with a congeries of processes, no one of which is under personal, conscious control. Behaviorism long ago targeted "autonomous man" for early retirement, taking the term itself to be laden with mysticism. The notion of "free will" in a determined universe is widely thought to violate every canon of parsimony, scientific unity, objectivity, and even the thermodynamics laws! Kant's categorical imperative seems painfully whimsical in a world that has gone to war twice in less than a century, a world now getting used to the idea of our determined, material, and temporary biologies which, like the planet hosting them, are accidents of a mindless creation. Modern people look back upon the romantic visions of Wordsworth, the simple faith of the medieval peasant, and the thrilling experiment begun in Athens the way they look back upon their own childhoods. But every stage of personal development, just as every period in intellectual history, finds the participants believing that they have uncovered truths unknown to all earlier times. Copernicus is the father of Ptolemy, Einstein, a more mature Newton. Our modern commitment to the methods and perspectives of science is no stronger than was the medieval commitment to the syllogism or the Renaissance confidence in natural magic. Each age finds an idea that seems to reveal something that is common to the major events that occupy that age. Science both leads and follows the ages in which it finds a home. The psychologists of antiquity looked at their age and discovered law, illusion, cultural variety, courage, deception, and fate. Their psychologies made room for all of these. Writers in the Middle Ages believed in a personal God of salvation and justice, author of immutable, rational, and inescapable laws. Medieval psychology quite naturally found the divine in the human mind, whose comprehension of rational certainties and universals redeemed the theology that accounted for its origin. And so on, down to our own time, the world we create comes, first, to serve as the metaphor of ourselves and, soon, as the reality of which we are but models or copies. Failing to find purpose and design in our own world, we question the existence of purpose or design in ourselves. Looking everywhere and discovering only matter in motion, we begin to see the same in the mirror. It is one of the quiet triumphs of the human mind that it can exhaust itself in attempting to refute its existence. With total abandon, as it were, we set out to establish the fiction of our autonomy, thereby establishing it still further.

It is not uncommon for the scientifically committed psychologist to express disbelief when scholars entertaining other perspectives insist that a scientific account of human action and human experience is not any nearer than it was in pre-Socratic times. There is, mixed with this disbelief, a conviction often stated that those who do not subscribe to the scientific accounts are clutching at the straws of metaphysical dualism and religious mysticism. It seems that this reaction to the scholar's reservation is prompted by the fear that such a reservation must be based on a denial of the validity of science itself. To believe, for example, that studies of the behavior of nonhuman organisms is not likely to illuminate the factors responsible for human conduct is, we are to understand, a rejection of Darwinian theory. To question whether neurophysiological findings bear directly on the question of psychic causation is to question determinism. There is a danger in this perspective, and the danger is greater to science than it is to those perspectives that are assumed to be hostile to the scientific agenda. The danger is that of stagnation, for, if there is a sure road to intellectual atrophy, it is paved with the complacent certainty that one's critics are deluded.

Consistently, the defenders of a scientific psychology have been ready to dismiss as "idealism" nearly the entire range of competing claims. Equally indiscriminately they have declared physics to be the ultimate arbiter, evolutionary notions to be unimpeachable, and practical success to be the final criterion of validity. But this cannot succeed. Fundamental debates about the possibility of a scientific understanding of things are not settled by appeals to science, for these are only begging the question. If it were the case (which it *isn't*) that mental events capable of bringing about behavior violated the second law of thermodynamics, then *so be it*—just in case there are such events and just in case they do, in fact, have behavioral consequences. A fact of experience cannot be undone by a theory that legislates against it. As for evolutionary theory, it is primarily a narrative account, a kind of natural history of life, and we have every reason to include ourselves in it. But there is nothing in the theory that sets limits on the phenotypes that might arise from evolutionary pressures, nor is there any requirement (quite the contrary!) that these phenotypes will be the same throughout phylogeny. Uniformity is not to be expected, and, when exceptions arise, they are not to be dismissed but explained. It is the theory that is on trial, not the facts.

To consider evolutionary theory a moment longer: Note that psychologists have tended (wrongly) to believe that the theory requires interspecies comparisons to be made along a quantitative continuum. But to accept this is to be trapped by the appealing but incorrect assumption that structural analogies entail functional equivalence. No one denies that such homologous structures as fins, wings, and limbs function equivalently in allowing movement; however, the scientific principles that must be invoked in rendering a complete description of the determinants of swimming, flying, and walking are not identical. Locomotion is achieved at every level of biological organization in the animal world and in portions of the plant kingdom as well. To note that very substantial qualitative differences occur across species, then, is not to deny Darwinian principles but to assert them. The Darwinian account can, for example, require all successful species to adapt their behaviors to the demands of their environments. It does not require that this behavioral adaptation be achieved in exactly the same way or even in accordance with the same principles. Adaptation for one species may involve changes in the color of the coat or skin; for another, seasonal migration; for another, hibernation. All of these are behavioral adaptations and all can be comprehended by the Darwinian system. Nevertheless, except for the fact that these are adaptive behaviors, there is hardly a qualitative similarity to be found among coloration, migration, and hibernation. It is one thing to insist that behavior must conform to the demands of the environment, but quite another to say that the conformity must be displayed in invariant fashion.

We realize, therefore, that to raise objections regarding the comparability of species in the matter of behavioral control is not the same as questioning the validity of evolutionary principles. This is not to say, however, that one is not allowed to raise such a question. If, for example, it could be shown that a particular aspect of human life is observed nowhere else in nature in any form even approximating its human expression, we would have to conclude either that selection pressures of an unknown variety were responsible or that the Darwinian theory is simply not equipped to address the fact. Some have speculated that human moral systems and language are of this sort; others, that consciousness is. This is not the place for an evaluation of such claims, although it is worth pointing out that the manner in which the child comes to display language has proved to be utterly resistant to behavioristic analyses. It is also worth noting that whether a creature is to be spared pain and suffering should be based on the capacity for such and not on some other capacity.

Skinner once complained that "no modern biologist or physicist would turn to Aristotle for help . . . but the dialogues of Plato are still assigned to students and cited as if they threw light on human behavior." [58] Elsewhere he has proclaimed that "The methods of science no longer need verbal defense; one cannot throw a moon around the earth with dialectic." [59] Such passages are illustrative of that contemporary psychological attitude that animates the discipline with the insouciance of late nineteenth and early twentieth-century positivism. Were this *ism* adopted, however, its first casualty would be psychology itself. At least for the time being, the only relationship between the laws of physics and the principles of psychological life is a metaphorical one. Of course we obey the laws of physics: when we jump from a height we fall; when we row a boat, we work against damping forces; when we strut across the moon, we weigh less than we do on earth. The validity of the laws of physics, however, does not bear on the question of the validity of the behavioristic thesis, which, in any of its forms, fails to resemble any physical law. It is an undeniable law of physics that the mass and density of the human body are such, and that the viscosity of the earth's atmosphere is such, that no one will ever be able to escape the pull of gravity by beating his arms up and down. We accept this law and proceed to invent the airplane. Our reason is not recruited in attempts to violate the very laws of nature that this reason discovered. Rather, it embraces these laws in fashioning methods and instruments that permit us to adapt more successfully to the demands of the environment.

Darwin's theory, modern genetics, neurophysiology, biochemistry, and the experimental analysis of behavior are of obvious relevance to any discipline that has accepted human psychology as its subject. Along the way, students of this discipline may have good reason to adopt as working hypotheses any number of metaphors borrowed from science and from the arts. What history teaches, however, is that progress is retarded when scholars fail to distinguish between the metaphor and the fact it is intended to represent. Condillac's sentient statue, La Mettrie's clocks, and the telephone switchboards of the early part of this century have really had no part either in the evolution of neurophysiological science or in refining our self-understanding. Of course, modern psychologists may very well ignore these lessons of history and proceed to deal with the subject in any way they see fit. They are, after all, autonomous.

Simplifications die hard. The logical behaviorism of the 1930s is no exception. Increasingly, however, the number of serious philosophers willing to contend that all psychological predicates finally reduce to statements about performance or "dispositions to behave" shrinks.[60] And the mind-body problem, which has refused to surrender to the techniques and devices of the laboratory, has been no more retiring in the face of philosophical positivism, logical or otherwise.

The past speaks to the modern world. It tells us not that science has failed but that it has limitations; not that psychology is pointless or immature but that, as with all intellectual ventures, it is incomplete and evolving. The history of civilization displays confusion and evokes regrets, but were it not for civilization there would be no standard of clarity and no basis for regret. The history of science, too, is rife with failure and banality, but were it not for this very history, we would not be in a position to make the judgment. Psychology's history is but a profile of the larger history of civilization. As a creation of human intelligence, it teaches by error as well as success. The systems of psychology we create, even when steeped in blunders and contradiction, are our creations and thus teach us something about ourselves. Behaviorism and psychoanalytic theory, neuropsychology and phrenology, utilitarianism and pragmatism—all reveal those qualities of mind, objects of hope, collections of values and perceptions unique to our species, one of myriad species. The present work began as an attempt to locate psychology in the history of ideas. It concludes with the obvious recognition that psychology is the history of ideas.

Notes

1. E. B. Titchener, *A Primer of Psychology,* Macmillan, New York, 1914, p. 32.

2. A brief but excellent discussion of this point is found in W. C. Dampier's *A History of Science* (Cambridge University Press, 1966), pp. 288–290.

3. Fechner, in his *Elemente der Psychophysik,* cites R. Wagner's *Handwörterbuch der Physiologie,* Vol. III, Sec. II, pp. 481–588, as the source containing Weber's findings and law (Gustav Fechner, *Elements of Psychophysics,* translated by Helmut Adler, Holt, New York, 1966, p. 15).

4. Titchener, *Primer of Psychology,* p. 32.

5. William James, *A Text Book of Psychology,* Macmillan, New York, 1892, p. 1.

6. Ibid., pp. 6–8.

7. Charles Darwin, *The Expression of the Emotions in Man and Animals* Appleton-Century-Crofts, New York, 1896, p. 366.

8. Matthew Arnold's *Culture and Anarchy* was first published in 1869 and can be found in any number of Arnold anthologies. Particularly good is the discussion by

J. Dover Wilson, *Arnold's "Culture and Anarchy"* (Cambridge University Press, 1932).

9. This quotation is taken from W. C. Dampier, *A History of Science* (Cambridge University Press), pp. 291–292. Helmholtz rarely addressed himself to the philosophy of science. His closest associates (e.g., DuBois-Reymond, Brücke, Ludwig) knew him to be opposed to vitalism in all its forms. Even his paper on the conservation of energy was written from a self-consciously antivitalistic perspective. However, he was the least polemical of the nineteenth-century physicalists, letting his outstanding contributions in science speak for themselves.

10. The phenomenological systems advanced by Brentano and by Husserl bear little relationship to what Hegel had called phenomenology. Recall that Hegel used the term to refer to that science concerned with the "manifold of consciousness." For Hegel, phenomenology was the science generated by the philosophy of the Absolute.

Franz Brentano (1838–1917) was a greater teacher than he was either a philosopher or a psychologist. His students included Husserl, Meinong, Carl Stumpf, and Christian Ehrenfels. Stumpf taught both Köhler and Koffka and was a very productive psychologist in his own right. Ehrenfels was the first to speak of *Gestaltqualität*. It was Brentano's *Psychologie vom Empirischen Standpunkt* (1874) that promised a "descriptive psychology" able to discover universal laws of the perceiving mind. Edmund Husserl (1859–1938) studied with Brentano. His phenomenological theories underwent several revisions over the years but were never far removed from the insistence that a scientific or an experimental psychology could only arrive at the contingent (intentional) and fallible aspects of perception. He emphasized the role of reflection in experience—as opposed to the mere description of experience—as the means by which a mental science might be cultivated. Husserl's influence is found most in that aspect of phenomenological psychology that rejects experimental inquiry in favor of intuitive and logical analyses. For a discussion of Husserl's phenomenology, see Herbert Spiegelberg's *The Phenomenological movement,* Vol. I (Mouton, The Hague, 1960), pp. 73–167.

11. Theodor Ziehen, *Introduction to Physiological Psychology,* Macmillan, New York, 1895, p. 1.

12. E. W. Scripture, *The New Psychology,* Scribner, New York, 1910, p. 2.

13. Titchener, *Primer of Psychology,* pp. 33–34.

14. Kurt Koffka, *Principles of Gestalt Psychology,* Harcourt Brace, Jovanovich, New York, 1935, p. 18.

15. Wilhelm Wundt, *Lectures on Human and Animal Psychology,* translated from the second German edition by J. E. Creighton and E. B. Titchener, Macmillan, New York, 1907, pp. 340–341.

16. John B. Watson, "Psychology as the Behaviorist Views It," *Psychological Review* 20 (1913): 158–177.

17. James, *A Text Book of Psychology,* p. 165.

18. John Dewey, "The Reflex Arc Concept in Psychology," *Psychological Review* 3 (1896): 357–370.

19. The pragmatic bent of the Americans was keenly observed by Alexis de Tocqueville. In his classic, *Democracy in America,* he notes that, "in America the purely practical side of science is cultivated admirably, and trouble is taken about the theoretical side immediately necessary to application. . . . But hardly anyone in the United States

devotes himself to the essentially theoretical and abstract side of human knowledge" (from the edition of 1848, Volume II, Part I, Chapter 10). Charles Sanders Peirce (1839–1914), briefly cited in the preceding chapter, was the founder of pragmatism and exerted a great influence on William James. He was a logician and mathematician, not a psychologist, and not especially pleased by the renditions of his system given by James. His principal epistemological concern was with meaning, and over the years he evolved a theory of meaning that finally became the pragmatic theory of meaning. It may be summarized in a variety of ways, but the summaries tend to trivialize the concept. The meaning of a term, on the pragmatic account, cannot go beyond the actual objective features of that which the term is used to denote. Any distinction, therefore, between the actual behavior of an object or event and the definition given for the term used to represent that object or event is, literally, nonsense.

20. John B. Watson, *Psychological Care of Infant and Child,* Norton, New York, 1928.

21. John B. Watson, *Behavior: An Introduction to Comparative Psychology,* Holt, Rinehart and Winston, New York, 1914.

22. See also his *Psychology from the Standpoint of a Behaviorist* (Lippincott, Philadelphia, 1919) and J. B Watson and W. McDougall, *The Battle of Behaviorism* (Norton, New York, 1929).

23. Titchener here distinguishes between reflex movements, instincts, and more complex integrated movements. He argues that even the reflex movements are far from simple and are not the earliest phylogenetic units of action at all (E. B. Titchener, *A Primer of Psychology,* pp. 171–182.)

24. In his chapter "Reflex Functions" (Ch. VI), in the *Principles of Physiological Psychology,* Wundt is careful to distinguish between the reflex actions and intentional or willful action. He cautions (pp. 240–251) against extending the concept of the reflex so far that it fails to explain anything. The edition of the *Principles* referred to is that translated by Titchener and published by Macmillan, New York, 1904.

25. In 1911, E. L. Thorndike published a collection of his experiments under the title "Animal Intelligence: Experimental Studies." This seminal collection has been reprinted as a facsimile edition by the Hafner Publishing Co., Darien, Conn., 1970.

26. E. L. Thorndike, *Animal Intelligence,* p. 244.

27. Ibid.

28. John B. Watson, *Behaviorism,* University of Chicago Press, 1924, p. 206.

29. Ivan Pavlov, *Conditioned Reflexes: An Investigation of the Physiological Activity of the Cerebral Cortex,* translated and edited by G. V. Anrep, Oxford University Press, London, 1927.

30. Watson, *Behaviorism,* p. 206.

31. Ibid., p. 225.

32. Ibid., pp. 22–39.

33. Ibid., p. 24.

34. Ibid., pp. 111–113.

35. Ibid., Ch. VII.

26. Ibid., Introduction, p. vii.

37. B. F. Skinner, *The Behavior of Organisms,* Appleton-Century-Crofts, New York, 1938.

38. Professor Skinner has published widely as a social philosopher, a behavioral technologist, a philosopher of science, a novelist, and a laboratory investigator. These varied talents are loosely combined in *Science and Human Behavior* (Macmillan, New York, 1956). His prescriptions and proscriptions for a behavioral science appear in "Are Theories of Learning Necessary?" (*Psychological Review* 57 [1950]: 193–216), and in "The Science of Learning and the Art of Teaching" (*Harvard Educational Review* [Spring 1954]: 86–97). His utopian vision is shared in *Walden II* (Macmillan, New York, 1948) and defended in *Beyond Freedom and Dignity* (New York: Knopf, 1971).

39. Skinner, *The Behavior of Organisms*, p. 432.

40. If the British philosophical reaction to Hegelianism is to be dated, we may cite G. E. Moore's *Principia Ethica* as the cogent starting point. In this work, Hegel's distinctions between "wholes" and the "sums of their parts" are rendered ambiguous and even meaningless (pp. 30ff.). Darwinian notions are incorporated into ethics and, following Thomas Reid, "natural language" and "common sense" are restored to talk about principles of human conduct (G. E. Moore, *Principia Ethica*, Cambridge University Press, 1903.)

41. Skinner, *Beyond Freedom and Dignity*.

42. Ibid., p. 205.

43. Thorndike, *Animal Intelligence*, Ch. XI.

44. B. F. Skinner, "The Steep and Thorny Way to a Science of Behavior," *American Psychologist* 30 (1975): 42–49.

45. An extensive collection of Pavlov's lectures and papers has been translated and placed in a volume titled *Experimental Psychology and Other Essays* (Philosophical Library, New York, 1957). The quotation is taken from p. 599 of this work, translated from a lecture given in 1935.

46. Wolfgang Köhler, *Gestalt Psychology*, Liveright, New York, 1947, p. 177.

47. Ibid., pp. 7–41.

48. Wolfgang Köhler, *The Mentality of Apes*, translated by Ella Winter, Vintage Books, New York, 1959. The first English edition was published by Routledge and Kegan Paul, London, 1925.

49. Ibid., p. 93.

50. Tolman always resisted the appellation "Gestalt psychologist," though he accepted that of "cryptophenomenologist." To discuss Tolman in connection with "later formulations of the Gestalt position" is simply to acknowledge the striking agreement between Tolman's major findings and conclusions on the one hand and the Gestalt position on learning and perception on the other. To the extent that Tolman was a "behaviorist," so indeed was Köhler. Both may be so described if all one means by the label is that both placed a high premium on the need for objective assessments of behavior. But used this way, the label ceases to be conceptually informing or historically notable.

51. E. C. Tolman, "Cognitive Maps in Rats and Man," *Psychological Review* 55 (1948): 189–208.

52. Köhler, *Gestalt Psychology*, p. 94.

53. See Skinner's "The Science of Learning and the Art of Teaching."

54. Pavlov, *Experimental Psychology*, p. 103.

55. Ibid., p. 141.

56. J. B. Watson and K. S. Lashley, *Homing and Related Activities of Birds,* Carnegie Institution, Department of Marine Biology, New York, 1915, Vol. VII.

57. Karl S. Lashley, *The Neuropsychology of Lashley: Selected Papers of K. S. Lashley,* edited by F. A. Beach, McGraw-Hill, New York, 1960.

58. Skinner, *Beyond Freedom and Dignity,* p. 3.

59. Skinner, "The Science of Learning and the Art of Teaching," p. 97.

60. See Professor Hillary Putnam's *Brains and Behavior* (in *Analytical Philosophy,* edited by R. J. Butler, Barnes and Noble, New York, 1965) for an excellent discussion of this.

Index of Names
Index of Subjects

Index of Names

Index of Subjects

Agent intellect, 105
Alienation (Marx), 288
Analytic propositions, 172, 220–21
Animal intelligence (Thorndike), 342
Analytic school, 319. *See also* Psychoanalytic theory
Animal spirits (Galen), 86, 238, 248
Anthropomorphism, 302
Apatheia, Stoic conception, 62
Archetypes (Jungian), 319
Asceticism, 71–72
Associationism: (Aristotle), 50–51, 110, 189; Berkeley, 186; Epicureanism and, 99; (Hartley), 249–50; (Hume), 177–78; (James), 341; (Locke), 161; (J. S. Mill), 265; (Plato), 27; and reflexes, 250; (Reid), 244, 343–44; (Wundt), 279–80
Atomism, 61, 69, 203
Averroism, 101–2, 143

Behaviorism, 267, 339–52
Bell-Magendie law, 274–75

Categorical imperative, 167, 225, 362
Categories: (Aristotle), 52, 74, 110, 128, 152; Kantian, 223–24, 283, 285
Category mistake, 195–96
Catharsis, 77, 316
Causation, (Aristotle), 53, 56–57; (Bacon), 158; (Hume), 177, 219–20; (Kant), 220–23; (J. S. Mill), 265–66; reasons and causes, 284–287, 348–350; (Reid), 183
Chivalry, 111–14, 118, 132, 145
Christianity: and Aristotelianism, 96–98;

early period, 68–87; Platonism, 74–78; Scholastic psychology, 98–106
Clinical psychology, 313–26
Cognition: (Aristotle), 52–53; Gestalt conceptions, 352–58; and Kant's categories, 221–24; (Piaget), 296; Scholasticism, 104–5. *See also* Knowledge
"Common sense" school (Reid), 180–87
Comparative psychology, 299–304, 308
Consciousness, 75, 190, 218; (Marx), 287–88, 301–2, 309, 325
Consequentialism: utilitarianism and, 244

D'Alembert's Dream, 262
Darwinism, 270–73, 299–304
Death: Homeric attitude toward, 16; medieval attitudes toward, 93–95
Deists, 208
Didaskaleion, 73–74
Dualism: Patristics and, 72, 115, 168; (Cartesian), 238–40

Egoism, 245
Emotion: (Aristotle), 55–58; (Augustine), 83; (Berkeley), 168; (Darwin), 271–72; (Galen), 85–86; humoral theory, 40, 86, 88; Platonic theory, 36–37; (Spinoza), 209–10; Stoic attitudes toward, 62–63
Empiricism, 129, 149–93; definition of, 152; (F. Bacon), 153–60; (Berkeley), 167–172; (Hume), 172–80; (Leonardo da Vinci), 139–41; (Locke), 160–67; (J. S. Mill), 263–70; (Ockham), 110; and utilitarianism, 187–88

379